Dynamic Modeling and Econometrics in Economics and Finance

Volume 23

More information about this series at http://www.springer.com/series/5859

Bettina Bökemeier • Alfred Greiner
Editors

Inequality and Finance in Macrodynamics

Springer

Editors
Bettina Bökemeier
Department of Business Administration
and Economics
Bielefeld University
Bielefeld, Germany

Alfred Greiner
Department of Business Administration
and Economics
Bielefeld University
Bielefeld, Germany

ISSN 1566-0419 ISSN 2363-8370 (electronic)
Dynamic Modeling and Econometrics in Economics and Finance
ISBN 978-3-319-54689-6 ISBN 978-3-319-54690-2 (eBook)
DOI 10.1007/978-3-319-54690-2

Library of Congress Control Number: 2017939652

Printed on acid-free paper

This Springer imprint is published by Springer Nature
The registered company is Springer International Publishing AG
The registered company address is: Gewerbestrasse 11, 6330 Cham, Switzerland

Preface

Inequality and finance are very topical issues that are not only debated in the scientific community but that concern politics and the society as well. Moreover, it is rather a broad topic related to social sciences, economics, health, education and many others. Nevertheless, for a profound scientific analysis of certain facets of inequality, it is essential to focus on partial views.

This book collects a variety of papers dealing with inequality and finance and addresses several aspects of dynamic macroeconomics and economic policy. The idea came up during the international workshop 'Macrodynamics and Inequality 2016' held at Bielefeld University, Germany, on March 22 and 23, 2016. The workshop was part of the seed funds project 'Inequality in Germany and the United States: Trends, Policies and Macroeconomic Implications', funded by the Deutsche Forschungsgemeinschaft (DFG). The project supported the collaboration of researchers from the Department of Business Administration and Economics at Bielefeld University, the Department of Management and Engineering at the University of Applied Sciences in Karlsruhe and the Graduate Faculty of the New School for Social Research in New York. The participants of the workshop are specialists in various fields of economics and are affiliated with different universities, research institutes and international organizations such as the International Monetary Fund (IMF) and the International Labour Organization (ILO).

Most of the contributions to this volume come from workshop participants. In addition, upon invitation we were able to win several additional scientists, all experts in their fields, to provide some of their current research on inequality and finance in the context of macrodynamic frameworks. This allows to cover a wide range of applications. Further, it combines theoretical and empirical approaches written by established researchers as well as by young scholars.

The book consists of a short introduction and of ten scientific papers. Except for the introduction, all contributions are presented in alphabetical order with respect to the first author's last name.

António Afonso and Mina Kazemi's paper assesses the public sector's expenditure efficiency by utilizing composite indicators and a non-parametric approach. Their study covers 20 OECD countries over 5 years on the macro level and also for core public sector areas. The results show that public sector spending is most efficient in Switzerland. Regarding core public responsibilities, health and education indicate to be more efficient.

Paulo Brito's contribution studies public debt, fiscal rules and growth dynamics in a theoretical setting. He analyses the dynamics of an endogenous growth model with productive public spending and public debt. The government sticks to a fiscal rule where the primary surplus is a function of the deviations of the actual public debt from a target value. It is demonstrated that the fiscal rule gives rise to impasse singularities implying the existence of over-determinate balanced growth paths and it constrains the basins of attraction of determinate balanced growth paths.

Davide Furceri, Jun Ge and Prakash Loungani empirically study the relationship between global financial integration and a rise in income inequality in low-income countries with a data set of 29 economies from 1970 to 2010. They find that capital account liberalization periods are followed by persistent increases in income inequality in low-income countries which is expressed in a 3% short-term and a 6% medium-term increase, respectively.

Alfred Greiner's paper focuses on the role of public debt and economic growth. The analysis is based on a basic endogenous growth model allowing for labor market imperfections and shows under which conditions public debt is neutral as regards the allocation of resources.

Elmar Hillebrand studies an endogenous growth model with technical change that is driven by R&D investments. A special focus is set on financial intermediation, which finances these endeavours. The paper demonstrates that the risk effect, reflected in interest rates, influences technical change: it enhances innovations in those sectors where the risk of failure is lower.

Atsumasa Kondo's paper assesses the part of the tax system for sustainability of public debt. From a theoretical model, he derives the critical level which is relevant for the balanced growth path and studies consequences if the ratio exceeds this level. Moreover, the findings show how these tax rates can be presented as a function of the initial debt ratio.

Wolfgang Kuhle's contribution analyses the effect of the demographic transition on different interest rates. The study employs an overlapping generations model to show that the change from high to low fertility lowers both rates, however, with a stronger effect on the risky rate.

Unurjargal Nyambuu's paper studies several aspects of energy resources, trade and finance for different types of countries. The analysis is done in a growth setting and discusses the implications of human capital investment with respect to inequality.

Christian R. Proaño and Benjamin Lojak address the relation between the fiscal policy of an economy and the financial markets for members of a monetary union with regard to factors that determine sovereign risk. With a model set-up and evaluation based on simulations, they study macroeconomic consequences of different perception of those determinants. Their findings show, for instance, that a too strict austerity policy of a country may affect its economic activity if the markets have a different view on the central targets.

Willi Semmler and Damien Parker study the rise of wealth disparity with the help of a formal model and apply it to empirical data. In a stochastic dynamic model with heterogeneous households, they analyse the drivers of the differences in net financial wealth and distinguish the effects of the returns on assets, of saving rates and of the borrowing capacity. The empirical results relate to the US economy and reveal that net wealth shares are shifting over time.

The volume aims at researchers and practitioners in universities and research institutes dealing with problems of this kind. Further, graduate students can benefit from the contributions presented in this book for their own research.

We thank all the authors for their contributions and the referees for their reports and comments helping to improve the individual papers and enhancing the quality of this book.

Bielefeld, Germany Bettina Bökemeier
Bielefeld, Germany Alfred Greiner
February 2017

Contents

Contributors

António Afonso ISEG School of Economics and Management, University of Lisbon, Lisbon, Portugal

Bettina Bökemeier Department of Business Administration and Economics, Bielefeld University, Bielefeld, Germany

Paulo Brito ISEG and UECE, University of Lisbon, Lisbon, Portugal

Davide Furceri International Monetary Fund, Washington DC, USA

Jun Ge International Monetary Fund, Washington, DC, USA

Alfred Greiner Department of Business Administration and Economics, Bielefeld University, Bielefeld, Germany

Elmar Hillebrand EEFA Research Institute, Muenster, Germany

Mina Kazemi ISEG School of Economics and Management, University of Lisbon, Lisbon, Portugal

Atsumasa Kondo Faculty of Economics, Shiga University, Shiga, Japan

Wolfgang Kuhle Max Planck Institute for Research on Collective Goods, Bonn, Germany

Benjamin Lojak Department of Social Sciences and Economics, Otto-Friedrich-University, Bamberg, Germany

Prakash Loungani International Monetary Fund, Washington, DC, USA

Unurjargal Nyambuu Department of Social Science, The New York City College of Technology, The City University of New York, New York, USA

Damien Parker Department of Economics, New School for Social Research, New York, USA

Christian R. Proaño Department of Social Sciences and Economics, Otto-Friedrich-University, Bamberg, Germany

Willi Semmler Department of Economics, New School for Social Research, New York, USA

Inequality in Germany and the US: An Introductory Note

Bettina Bökemeier

Abstract This introduction addresses various aspects of income and wealth inequality and their developments in Germany and the United States. In both countries the distribution of income and wealth has become more unequal during the past decades. We observe some changes in the figures especially for Germany indicating a more unequal situation at present. This holds for both, the income and the wealth inequality measures. Even though redistribution instruments reduce the Gini coefficient for the disposable income compared to the market outcome, there still is an almost steady increasing trend since the middle of the 1980s. Regarding wealth inequality, the ratio of mean to median net wealth reveals values above the OECD average for Germany and the United States. Both, income and wealth inequality are, however, more pronounced in the US.

1 Introduction

Inequality is related to many aspects and touches various disciplines, such as social sciences, economics, health and education. It may refer to a variety of dimensions. Even if the focus is set only on economic aspects, still there are many facets such as inequality of income, of wealth or wages—all parts are just as important as others. These issues not only involve discussions among scientists but also regard politics and the society as a whole. In economics the discussion has mainly started in the 1950s amongst others with *Kuznets*' seminal work Kuznets (1953, 1955) on the inverse u-shaped pattern between ongoing economic development (growth) and income inequality. As economic growth proceeds income inequality begins to rise and after a certain point the pattern reverses and inequality declines.

Income inequality analysis concentrates on the flow figure income, which may come from different sources such as wages or profits. It recently gained a lot of attention starting with the seminal book by Piketty (2014) and other publications

B. Bökemeier
Department of Business Administration and Economics, Bielefeld University, 33501 Bielefeld, Germany
e-mail: bboekemeier@wiwi.uni-bielefeld.de

B. Bökemeier, A. Greiner (eds.), *Inequality and Finance in Macrodynamics*, Dynamic Modeling and Econometrics in Economics and Finance 23,
DOI 10.1007/978-3-319-54690-2_1

(Stiglitz 2012; Fratzscher 2016, i.e.). Looking at the top and bottom of the income hierarchy, many studies find a rise in income inequality. Wealth inequality focuses on the differences in the stock of material assets. Many publications also show an increase in wealth inequality, which may even be larger than income inequality,[1] and wealth inequality indicates to be more concentrated, cf. Murtin and d'Ercole (2015). Again, these findings hold for the top end of the very rich as well as the bottom. Also, recent studies show that the impact of the financial crisis had different effects on wealth inequality across countries. Inequality in wages may for instance concern differences in the pay-gap or its evolution over time. These aspects are not explicitly considered here, see for instance Piketty and Saez (2003) for a discussion. They show that both, income and wages of the top 10% of the distribution reveal a u-shaped pattern over the twentieth century in the USA. In a more recent study for several countries Atkinson et al. (2011) still find that labor income (wages and salaries) accounts for a considerable part of the recent increase of income in the top end in several countries. However, they attenuate it a little in connection with the study by Wolff and Zacharias (2009): both, labor and capital income matter for the recent development. The focus on Germany and the United States presents a rather distinct choice of countries. The US are not only the largest economy world wide, they also represent the country of opportunities and fortune and refer to a public social security system with less pronounced services as compared to Germany. Germany, on the other hand, is the largest European economy and provides a marked social security system as well as redistributive measures. Thus, a US-German comparison may indicate several differences among the underlying public sector concepts. Plus, both are among the countries in which inequalities have increased considerably during the past years, cf. OECD (2011a). Moreover, both countries were affected by the financial crisis starting in 2008/ 2009. They implemented fiscal policy measures and stimulus packages to encounter the downswing and support the economy. Certainly, the crisis did not come along without impacts on the social situation and interfered with distribution and inequality aspects.

The remainder of this introductory note is structured as follows: the next section presents some empirical information. First, Sect. 2.1 describes some data and indicators of income inequality. Then, Sect. 2.2 presents some figures of wealth inequality. Finally, Sect. 3 summarizes the central aspects and concludes.

2 Some Data

The data presented in this section focus on Germany and the US, as it shows that these two usually belong to the group of countries for which inequality is rather pronounced. Moreover, as mentioned above, they established different social

[1]There is, however, a close pattern among wealth and income.

security systems and provide different public social services. So, the focus can be justified and a comparison may reveal central differences.

2.1 Income Inequality

As mentioned above, income inequality is one of the major aspects in the inequality debate. Many studies have analyzed the problem over the long run or for the top end, see for instance Piketty and Saez (2003) or Atkinson et al. (2011).

Instead of looking explicitly at the top of the distribution, here we use a usual global empirical measure for income inequality, the Gini coefficient. It is a summary measure of inequality which can take values between zero (complete equality) and one (total inequality). Table 1 and Fig. 1 show the current values in 5 year intervals

Table 1 Gini coefficient, market income

	1985	1990	1995	2000	2005	2010	2013
Germany	0.44	0.43	0.46	0.47	–	0.49	0.51
USA	–	–	0.48	0.48	0.49	0.50	0.51

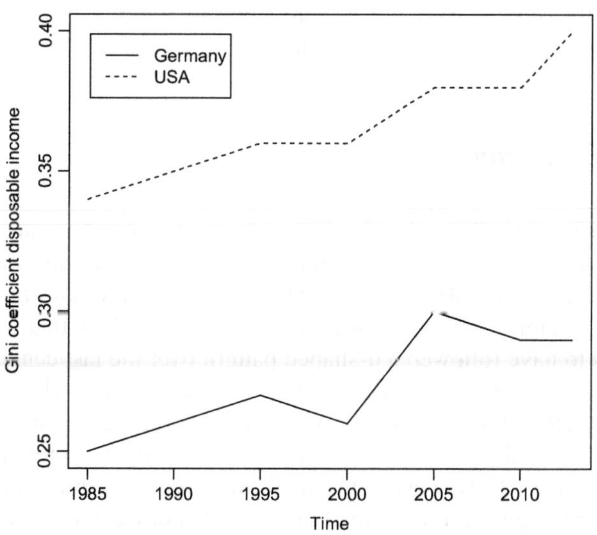

Fig. 1 Gini coefficient disposable income

from the mid 1980s to the most recent available data in 2013.[2] First, Table 1 presents the Gini coefficient for the market incomes. They show the 'raw' distribution of income, which results from the market process.

In 2013, market income inequality as measured by the Gini index was identical in Germany and the United States. In both countries the coefficient amounted to 0.51. However, comparing the developments since the middle of the 1990s reveals that the coefficient rose by 3 percentage points in the US whereas for Germany it rose by 5 percentage points. For Germany the Gini coefficient has risen by about 16% over the last 30 years. Indicating a much faster increase of inequality in Germany than the USA. Moreover, comparing these values with other industrialized countries shows that for instance in 2013 in Sweden the Gini coefficient before taxes and transfers was 0.44 or in the Netherlands 0.43, cf. OECD (2016a).

Figure 1 shows the Gini coefficient calculations for the disposable income, that is after taxation and transfer payments, for the same time intervals as above. Certainly, the values are lower than in Table 1, however, it is remarkable that the German value in 2013 of the disposable income is by 22 percentage points lower than the market income while for the US it declines only by half of that (11 percentage points). This indicates a stronger redistribution policy in Germany than in the USA.

This matters for the whole economy in the macro context. Relating inequality aspects to the performance of the economy, Atkinson et al. (2011, Table 1, p. 8f.) show that in the US the top income share (upper 1% of the distribution) account for a large part of the economic growth.[3] There seems to be a rather important influence of top incomes on the economy as a whole.

2.2 *Wealth Inequality*

Similar to the income discussion in the previous chapter, there are a number of contributions studying the wealth distribution such as Piketty and Zucman (2014) or Wolff and Zacharias (2009) for instance. Looking at the wealth-income ratio the concentration dynamics and inheritance, long-run studies show that wealth inequality also seems to have followed a u-shaped pattern over the last century, cf. Piketty and Zucman (2014). This especially holds for private wealth, while public wealth seems to be rather stable or declining. Comparing wealth concentration in Europe and the USA reveals that for a long time wealth inequality was more pronounced in Europe[4] than in the US. However, starting in the 1960s this trend reversed and the top wealth distribution is more unequal in the USA today compared to Europe.

[2]There are many ways for calculation and sources for Gini coefficients, see for instance World Bank (2014). Here, the data has been taken from OECD (2016a) for comparison.

[3]About 60% for the 30 years from 1976 to 2007, cf. Atkinson et al. (2011, p8f.).

[4]Piketty and Zucman (2014, p.17) especially look at France, Britain an Sweden, but argue that the pattern is rather typical for Europe.

Table 2 Mean and median net wealth per household in 2010, national currency

	Germany	USA
Mean net wealth	182,532	452,900
Median net wealth	51,358	62,500
Ratio	3.55	7.25

Recent comparable data on wealth inequality is for instance provided by the OECD, see Murtin and d'Ercole (2015). However, common distributional instruments such as the Gini coefficient are not applicable due to the large number of households without wealth or even debt, cf. Murtin and d'Ercole (2015, p.4). Therefore, the plot below presents some data for median/ mean wealth in the USA and Germany and some calculations the with those data regarding the inequality of the distribution.

The mean and median net wealth per household show some current numbers on the available wealth in the two countries, the data is presented below.[5]

A common measure of wealth inequality is the ratio of the mean and median net wealth, which is presented in the last row of Table 2. It shows that mean net wealth is about seven times higher than median wealth in the US and about four times in Germany. These are quite high numbers which indicate relatively high inequalities in the distribution of net wealth especially in the US. For instance in France and the UK these ratios take values of less than two (For comparison: the OECD wealth distribution database average = 2.5, cf. Murtin and d'Ercole (2015).) This indicates a rather strong inequality for both countries, however, especially pronounced in the US.

3 Conclusion

This note has addressed some aspects of income and wealth inequality in Germany and the USA. It shows that inequality is an important issue for modern economics. This holds true for the USA and Germany, both large industrialized countries with a strongly concentrated distribution of income and wealth. They belong to those economies in which inequalities have increased considerably during the past years, cf. OECD (2011a). The Gini coefficients for income show a rather unequal distribution in both the USA and Germany, with an increasing trend in inequality since the middle of the 1980s for both measures, the income before and after taxation. It also shows, that redistributive instruments indicate to be stronger in Germany than in the USA. Concerning wealth inequality the ratio of mean to median net wealth reveals values above the OECD average for both countries, however more pronounced in the US.

[5] All wealth data has been taken from OECD (2016b). All values refer to the most recent available data (2010) and current prices of national currency (GER= €, USA= US$).

From the macroeconomic point of view, these rising trends in inequality should not be neglected. For example, in the past years several studies have analyzed the nexus between income inequality and overall economic performance. In fact, most recent analyses conclude that high and rising income inequalities may inhibit long-term economic growth, cf. Ostry et al. (2014) or Behringer et al. (2016) for an overview. Moreover, a more unequal distribution of income and wealth may be worrisome from a social point of view as it is supposedly undermining social cohesion, cf. OECD (2011b, pp.7).

Acknowledgements I thank Christina Anselmann, Alfred Greiner and Silvia Melzer for their comments.

References

Atkinson, A. B., Piketty, T., & Saez, E. (2011). Top incomes in the long run of history. *Journal of Economic Literature, 49*(1), 3–71.

Behringer, J., Theobald, T., & van Treeck, T. (2016). *Ungleichheit und makroökonomische Instabilität: Eine Bestandsaufnahme* (pp. 1–36). Berlin: Friedrich-Ebert-Stiftung, available online: http://library.fes.de/pdf-files/wiso/12690.pdf, last access: October 5, 2016.

Fratzscher, M. (2016). *Verteilungskampf: Warum Deutschland immer ungleicher wird*. München: Carl Hanser Verlag GmbH Co KG.

Kuznets, S. (1953). *Shares of upper income groups in income and savings* (Vol. 55). New York: National Bureau of Economic Research.

Kuznets, S. (1955). Economic growth and income inequality. *The American Economic Review, 45*(1), 1–28.

Murtin, F., & d'Ercole, M. (2015). *Household wealth inequality across OECD countries: New OECD evidence*. OECD Statistics Brief No. 21, June 2015.

OECD (2011a). *Divided we stand: Why inequality keeps rising*. Paris: OECD Publishing.

OECD (2011b). *Inequality: Why the struggle matters*. OECD Observer No. 287 Q4.

OECD (2016a). OECD.stat, Social protection and well-being. Income distribution and poverty. http://stats.oecd.org/index.aspx?queryid=66670#, last access: September 13, 2016.

OECD (2016b). OECD.stat, Social protection and well-being. Wealth distribution. https://stats.oecd.org/Index.aspx?DataSetCode=WEALTH, last access: September 8, 2016.

Ostry, J. D., Berg, A., & Tsangarides, C. G. (2014). *Redistribution, inequality, and growth* (pp. 1–36). IMF Staff Dicussion Note SDB/14/02, available online: http://www.imf.org/external/pubs/ft/sdn/2014/sdn1402.pdf, last access: October 11, 2016.

Piketty, T. (2014). *Capital in the 21st Century*. Cambridge: Harvard University Press.

Piketty, T., & Saez, E. (2003). Income inequality in the United States, 1913–1998. *Quarterly Journal of Economics, 118*(1), 1–39.

Piketty, T., & Zucman, G. (2014). *Wealth and inheritance in the long run*. CEPR Discussion Paper No. DP10072.

Stiglitz, J. E. (2012). *The price of inequality: How today's divided society endangers our future*. New York: WW Norton & Company.

Wolff, E. N., & Zacharias, A. (2009). Household wealth and the measurement of economic well-being in the United States. *The Journal of Economic Inequality, 7*(2), 83–115.

World Bank. (2014). *All the Ginis, 1950–2012 (updated in autumn 2014) by Branko L. Milanovic*. http://go.worldbank.org/9VCQW66LA0. Last access: 5 Sept 2016

Assessing Public Spending Efficiency in 20 OECD Countries

António Afonso and Mina Kazemi

Abstract We follow the framework of Afonso et al. (Public Choice, 123(3–4):321–47, 2005), to look at the public expenditure of 20 OECD countries for the period 2009–2013, from an efficiency perspective. We construct Public Sector Performance and Public Sector Efficiency indicators and use Data Envelopment Analysis. The results show that the only country that performed on the efficiency frontier is Switzerland, Canada, Japan, Luxembourg and the United States are also more efficient performers. The average input-oriented efficiency score is equal to 0.732. That is, on average countries could have reduced the level of public expenditure by 26.8% and still achieved the same level of public performance. The average output-oriented efficiency score is 0.769 denoting that on average the sample countries could have increased their performance by 23.1% by employing the same level of public expenditure.

1 Introduction

Being the main element in the policy-making decisions, governments have a great responsibility to move the countries towards economic growth and to increase the social welfare. Confronting the constant budget constraints and employing the correct policies by governments is one of the crucial issues due to the pressures from globalization and ageing population on the countries budget on both expenditure and revenue sides (Deroose and Kastrop 2008). As a large share of the GDP is allocated to the public spending, improving the public spending efficiency is an important

A. Afonso (✉)
Department of Economics, ISEG/Lisbon—Universidade de Lisboa, R. Miguel Lupi 20, 1249-078 Lisbon, Portugal

UECE—Research Unit on Complexity and Economics, ISEG/Lisbon—Universidade de Lisboa, R. Miguel Lupi 20, 1249-078 Lisbon, Portugal
e-mail: aafonso@iseg.utl.pt

M. Kazemi
ISEG/ULisbon—University of Lisbon, R. Miguel Lupi 20, 1249-078 Lisbon, Portugal
e-mail: mminakazemi@gmail.com

B. Bökemeier, A. Greiner (eds.), *Inequality and Finance in Macrodynamics*, Dynamic Modeling and Econometrics in Economics and Finance 23,
DOI 10.1007/978-3-319-54690-2_2

issue that could help to ensure the sustainability of the public finances (Barrios and Schaechter 2008). Understanding how far the governments can increase their performance at the same spending levels simply by increasing their spending efficiency could help fiscal policy makers achieving sustained fiscal disciplines (Mandl et al. 2008).

This study is going to assess the public spending efficiency in 20 OECD countries during the period 2009–2013. The main reason of doing this work is to recognize how well and efficient these countries are performing from both input and output perspectives. First we constructed the composite indicators on Public Sector Performance (PSP) and computed the Public Sector Efficiency (PSE), and then we implemented a non-parametric approach called Data Envelopment Analysis (DEA) for six different models. The first two models are considering the efficiency of the government in a macro level and the other four models assess the efficiency of public expenditure in four different core areas of government performance: administration, education, health and infrastructure.

This work follows Afonso et al. (2005) with a slightly smaller country-sample due to the data availability, but with more recent data, and substituting FDH with the DEA approach. The reason that we preferred DEA to FDH is the possibility of assuming the convexity possibility. Hence, some of the DMUs that are efficient under FDH are not always efficient under DEA but the efficient DMUs under DEA are always efficient under FDH.

DEA results obtained from running model 1 and 2 show that Switzerland by applying the lowest amount of public expenditure could achieve the highest level of performance in this sample and it's the only country that is performing on the efficiency frontier with a significant distance from the other countries. The results of running the DEA for the other models suggest that governments of these countries are performing more efficiently in the health and education systems than in the administration and infrastructure functions.

Our results are highly in line with the results of the previous studies in this subject (e.g. St. Aubyn et al. (2009), Afonso et al. (2005), etc.) suggesting that the governments could get a higher level of performance by spending at the same level or that they could obtain the same level of performance by spending less. The average input-oriented efficiency score is equal to 0.732. That is, on average countries could have reduced the level of inputs by 26.8% and achieve the same outputs. The average output-oriented efficiency score is 0.769 denoting that on average the countries could have increased the level of their outputs by 23.1% by employing the same level of inputs.

Section 2 is a literature review. Section 3 introduces the methodology. Section 4 describes the results of the assessment and Sect. 5 concludes.

2 Literature

The literature on assessing the government spending efficiency has usually obtained the efficiency frontiers either by applying parametric or non-parametric approaches. Stochastic Frontier Analysis (SFA) is a popular parametric approach and Free Disposal Hull (FDH) and Data Envelopment Analysis (DEA) are the two non-parametric approaches that have been used by many researchers in order to obtain an efficiency frontier. It is worth mentioning that there haven't been too many studies in evaluating the public spending efficiency at an aggregate level.

Herrera and Pang (2005), applied FDH and DEA methodologies to compute the input and output efficiency scores of health and education public sectors of 140 countries for the period 1996 to 2002. Their results indicate that countries with higher spending levels obtained lower efficiency scores.

Afonso and St. Aubyn (2005), assessed the efficiency of the public spending for the education and health sectors across 17 and 24 OECD countries in 2000. They applied FDH and DEA approaches in order to compare the results of each method. For the education analysis they used hours per year in school and teachers per 100 students as inputs and PISA scores as output. For the health analysis they used the number of doctors, nurses and beds as inputs and infant survival and life expectancy as outputs. The results related to the comparison of these two techniques infer that some of the countries that were considered as efficient under FDH are no longer efficient according to the DEA results, and that countries could have obtained better results by applying the same level of inputs.

Afonso et al. (2005), computed the Efficiency scores for 23 OECD countries for 1990 and 2000 by constructing the PSP indicators and considering the PSP scores as an input measure and public expenditure as percentage of GDP as an output measure by applying the FDH methodology. The results of their studies show that small governments obtained better performance and efficiency scores compared to the larger ones. And larger governments could have obtained the same level of performance by decreasing the level of the public expenditure.

Sutherland et al. (2007), applied both non-parametric (DEA) and parametric (SFA) approaches to assess the public spending efficiency in primary and secondary education among OECD countries. The results of school-level efficiency estimated by them suggest a high correlation between the results of both approaches. Their results show that governments could gain higher efficiency scores by decreasing the expenditure levels and keeping the performance constant.

Afonso and Fernandes (2008), assessed the public spending efficiency of 278 Portuguese municipalities for the year 2001 by applying a non-parametric approach (DEA). They constructed a composite indicator of local government performance and considered it as the output measure and the level of per capita municipal spending as the input measure of the DEA. The results of the DEA implemented by them suggest that most of these municipalities could have achieved the same level of performance by decreasing the level of the public resources application.

St. Aubyn et al. (2009), applied a two stage semi-parametric (DEA and the Tobit regression) and a parametric approach (SFA) in order to evaluate the efficiency and effectiveness of public spending on tertiary education for 26 EU countries plus Japan and the US for two different periods (1998–2001 and 2002–2005). They conclude that to be considered as good performers countries do not necessarily need to increase their spending on higher education but need to spend efficiently.

Afonso et al. (2013), computed the Public Sector Efficiency (PSE) and conducted a DEA in order to assess the public expenditure efficiency for 23 Latin American and Caribbean countries for the period 2001–2010. The output measure suggested by them is the Public Sector Performance (PSP) scores computed by constructing the composite indicator of public sector performance. The input measure is the total public spending-to-GDP ratio. They conclude that the PSE scores have an inverse correlation with the size of the governments and also that these governments could achieve the same level of output with less government spending. Figure 1 summarizes the related literature.

3 Methodology and Data

This study's Database is compiled from various sources that are listed in Figs. 16 and 17 in Appendix. Figure 16 in Appendix lists several sub-indicators that are used for constructing the PSP indicators. These PSP indicators are then used as the output measure for the frontier analysis. Figure 17 in Appendix includes the data on various governments' expenditures area, which then could be used as the input measures for the efficiency analysis.

The methodology applied in this study includes three approaches. The first two sections explain how the PSP and PSE are constructed and the third section provides an intuitive approach to the Data Envelopment Analysis (DEA).

3.1 Public Sector Performance (PSE)

In order to compute the Public Sector Performance, we followed Afonso et al. (2005). They introduced the two main components of PSP, called opportunity indicators and the traditional Musgravian indicators.

The opportunity indicator that focuses on the role of the government in providing various and accessible opportunities for individuals in the market place contains four sub-indicators. These sub-indicators reflect the governments' performance in four areas, administration, education, health and infrastructure. The administration sub-indicator comprises the same indices as it had in Afonso et al. (2005), which consists of: corruption, burden of government regulation (red tape), judiciary independence and shadow economy. Besides that, we added another component called the property rights to the administration sub-indicator [following Scheubel (2015)] due to its'

Authors	Methodology	Country Coverage	Sample Period	Results
Herrera and Pang (2005)	FDH, DEA	140 countries	1996-2002	Applying a higher level of expenditures results in a lower efficiency scores
Afonso and St. Aubyn (2005)	FDH, DEA	OECD Countries	2000	Countries could obtained better results by applying the same amount on Inputs
Afonso et al.(2005)	FDH	23 OECD Countries	1990 and 2000	Smaller governments performed better than larger ones Larger governments could increase their performance by decreasing the usage of resources
Sutherland et al. (2007)	DEA	OECD Countries	2003	Governments could get a better efficiency scores by decreasing the spending and keeping the outputs constant
Afonso and Fernandes (2008)	DEA	278 Portuguese municipalit ies	2001	Most of the municipalities could achieved a higher level of output by applying the same level of input
St. Aubyn et al. (2009)	DEA, SFA	26 EU + Japan + US	1998-2001, 2002-2005	To be a better performer countries do not necessarily need to increase spending but spend efficiently
Afonso et al.(2013)	DEA	23 Latin American and Caribbean countries	2001-2010	Inverse correlation between the PSE scores and the size of the governments Government could achieved the same level of output by spending less

Fig. 1 Papers on the evaluation of the Public Spending Efficiency

important role in increasing the welfare and economic growth by providing a reliable environment for individuals and companies to invest. In order to measure the education sub-indicator, we used the secondary school enrolment rate, quality of educational System and PISA scores. For the health sub-indicator, we compiled data on the infant mortality rate and life expectancy. The infrastructure sub-indicator is obtained by using the measure "quality of overall infrastructure" as reported in the Global Competitiveness report. In order to focus on the structural changes we computed the 5-year (2009–2013) average of all the indices in constructing the opportunity indicators.

Total Public Sector Performance			
Opportunity indicators		**Standard ''Musgravian'' Indicators**	
Administrative	Corruption	**Distribution**	Gini index
	Red tape		
	Judicial independence	**Stability**	Coefficient of variation of GDP growth
	Property rights		
	Shadow economy		Standard deviation of Inflation
		Economic performance	GDP per capita (PPP)
Education	Secondary School Enrolment (gross %)		GDP growth
	PISA Scores		Unemployment
	Quality of educational system		
Health	Infant mortality		
	Life expectancy		
Public infrastructure	Infrastructure Quality		

Fig. 2 Total Public Sector Performance (PSP) indicator

The Musgravian Indicators consist of three sub-indicators: distribution, stability and economic performance. In order to measure the PSP of distribution sub-indicator, we used the 5-year average of the Gini Coefficient (2009–2013). For the stability sub-indicator, we used the coefficient of variation of 10-year (2004–2013) GDP growth and standard deviation of 10 years (2004–2013) inflation. For the economic performance sub-indicator, we used the 10-year average (2004–2013) of GDP per capita (PPP), GDP at constant prices and unemployment rate.

Figure 2 presents a list of the variables that we collected data on, in order to construct the PSP indicators. After having collected all data on all of the sub-indicators, we normalized all the measures by dividing the value of a specific country by the average of that measure for all the countries in the sample, in order to provide a convenient platform for comparing the results. The PSPs in each sub-indicator was then constructed by the aggregation of the measures related to each sub-indicator, after assigning equal weights to them.

In order to compute the total Public Sector Performance, we gave equal weights to each sub-indicator of opportunity and Musgravian indicators and aggregated them.

Assume there are p countries with n areas of performance, then we can determine the overall performance of the country i by:

$$PSP_i = \sum_{j=1}^{n} PSP_{ij}\ , i = 1, \ldots, p;\ \ with\ PSP_{ij} = f(I_k). \tag{1}$$

where $f(I_k)$ is a function of k observable socio-economic indicators I_k.

3.2 Public Sector Efficiency

In order to compute the Public Sector Efficiency, we take into account the costs that governments have in order to achieve a certain performance level. So, we now consider the Public Expenditure as the input and relate that expenditure to its' relevant PSP indicator. We consider the government consumption as the input in obtaining the administrative performance, government expenditure in education as the input for the education performance, health expenditure is related to the health indicator of performance and public investment is considered as the input for the infrastructure performance. For the distribution indicator we consider the expenditure on Transfers and subsidies as the cost affecting the income distribution. The stability and economic performance are related to the total expenditure. Then we weigh each area of government expenditure to its' relative output and compute the Public Sector Efficiency for each indicator and also the total PSE of each country as follows:

$$PSE_i = \sum_{j=1}^{n} \frac{PSP_{ij}}{EXP_{ij}},\ i = 1, \ldots, p \tag{2}$$

where EXP_{ij} denotes the government expenditure of the country i in the area j. Figure 18 in Appendix presents data on different categories of public expenditure (% of GDP) for the sample countries that are the computed 10-year average for the period 2004–2013.

3.3 Data Envelopment Analysis (DEA)

Data Envelopment Analysis (DEA) is an approach that assesses the relative performance and efficiency of a set of Decision-Making Units (DMUs) by using the linear programming methods in order to construct a production frontier. This method assumes the convexity of the production frontier. DEA's inceptions were first introduced by Farrell (1957) and the term DEA was used and became popular for the first time by Charnes et al. (1978).

DEA can be conducted for the input and output-oriented analysis by assuming that the technology is constant or variable return to scale (CRS or VRS). The constant return to scale DEA model doesn't consider the constraint of convexity and also under this assumption, the efficiency scores achieved from the both input- and output-oriented specifications are equal.

Suppose there are I Decision-Making Units (DMU), each DMU uses N inputs to produce M outputs. If X is the $N \times I$ input matrix and Y is the $M \times I$ output matrix for all the I DMUs, then x_i is an input column vector and y_i is an output column vector for the ith DMU. So for a given DMU the DEA model according to Banker et al. (1984) is as follow:

$$\begin{aligned} &Max_{\varnothing,\lambda}\varnothing \\ &Subject\ to - \varnothing y_i + Y\lambda \geq 0 \\ &x_i - X\lambda \geq 0 \\ &I1'\lambda = 1 \\ &\lambda \geq 0 \end{aligned} \tag{3}$$

where $\varnothing$ is a scalar and $\frac{1}{\varnothing}$ is the output-oriented efficiency score and satisfies $0 < \frac{1}{\varnothing} \leq 1$. According to Farrell (1957), if the efficiency score of a DMU is equal to 1, then the firm is performing on the efficiency frontier and considered as a technically efficient firm.

$\lambda(I \times 1)$ is a vector of constants that measures the weights for identifying the location of the inefficient firms. The constraint $I1'\lambda = 1$ is the convexity restriction imposed on the variable returns to scale DEA model.

Figure 3 plots an example of the CRS and VRS DEA frontiers for three different firms. As illustrated, firms A and B are located on the VRS efficiency frontiers so they are considered as efficient DMUs. Firm A is considered efficient under CRS and VRS but firm B is not performing efficiently under CRS. Firm C is considered

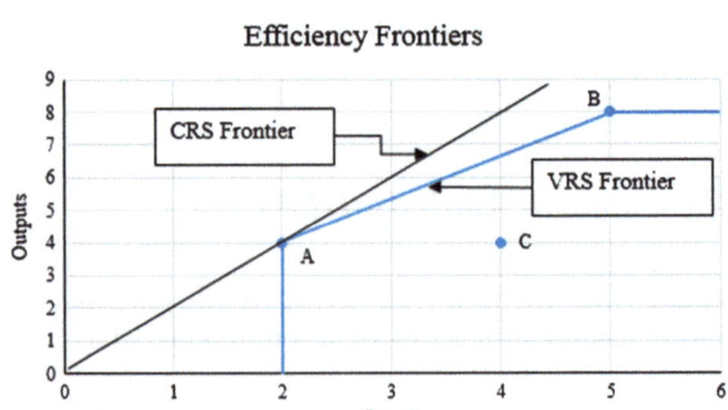

Fig. 3 Example of the DEA frontiers

inefficient because it could have achieved a higher level of outputs by employing a lower level of inputs (Coelli et al. 2005).

4 Empirical Analysis

The results are presented in three different sections. Section 4.1 presents the results from constructing and evaluating the PSP indicator and scores. Section 4.2 provides the PSE values and finally, Sect. 4.3 represents the efficiency scores and results of the conducted DEA models.

4.1 Public Sector Performance

As we explained in the methodology section, we constructed the composite indicator on the public sector performance by applying different variables for both Opportunity and Musgravian indicators. Figure 4 depicts the results of the PSP computations where countries with the PSP scores higher than 1 are considered as good performers. The PSP scores range from 0.56 to 1.30 suggesting that Switzerland is the best performer and Greece is the worst performer in the sample countries. The top four best performers are Switzerland, Luxembourg, Norway and Canada. The worse performers according to the results are Greece, Italy, Portugal and Spain.

Comparing the PSP results of each individual sub-indicator for different countries, we can observe that Switzerland and Luxembourg are the best performers in the administration area. Finland and the Netherlands are performing the best in education. In the provision of health almost all of the countries are performing well. Switzerland and Finland are the best performers in public infrastructure. We can also notice that in terms of income distribution, Norway and Finland are performing the best, in terms of stability Switzerland and Canada rank the best and Luxembourg has the best economic performance in the sample.

In order to check the robustness of the results and to check if different sub-indicators have different impacts on the final results of the PSP scores, we assigned a higher weight (2/3) to the Musgravian indicators and a lower weight (1/3) to the Opportunity indicators (instead of assigning equal weights to each indicator) by assuming that the Musgravian indicators have higher impacts on the overall performance of the public sector of a country.

The results of the robustness analysis are very similar to the PSP scores computed by assigning equal weights to each indicator. The countries that obtained a PSP score higher than average when assigning the equal weight to each indicator also achieved higher than average performance results by assigning different weights to Opportunity and Musgravian indicators. Similar results were also attained for the countries with a lower than average PSP scores.

Country	Opportunity Indicators					Musgravian Indicators				Total Public Sector Performance	
	Administration	Education	Health	Infrastructure	Opportunity PSP	Distribution	Stability	Economic Performance	Musgravian PSP	**Equal weights**	Different weights
Austria	1.11	0.97	1.00	1.09	1.04	1.03	1.27	1.24	1.18	**1.11**	1.13
Belgium	0.88	1.08	1.00	1.01	0.99	1.05	1.17	0.98	1.07	**1.03**	1.04
Canada	1.09	1.05	1.00	1.02	1.04	0.97	1.75	1.18	1.30	**1.17**	1.21
Denmark	1.07	1.06	0.99	1.04	1.04	1.03	0.84	0.88	0.92	**0.98**	0.96
Finland	1.16	1.11	1.00	1.11	1.09	1.06	0.69	0.90	0.88	**0.99**	0.95
France	0.95	0.98	1.00	1.10	1.01	0.99	1.23	0.85	1.02	**1.02**	1.02
Germany	1.02	1.01	1.00	1.07	1.02	1.01	1.11	0.96	1.03	**1.02**	1.03
Greece	0.61	0.85	1.00	0.78	0.81	0.95	0.01	-0.03	0.31	**0.56**	0.48
Ireland	1.04	1.08	1.00	0.84	0.99	1.00	0.63	1.06	0.90	**0.94**	0.93
Italy	0.63	0.88	1.01	0.74	0.81	0.97	0.46	0.45	0.63	**0.72**	0.69
Japan	1.09	0.98	1.01	1.04	1.03	0.95	1.00	0.98	0.98	**1.00**	0.99
Luxembourg	1.18	0.95	1.00	1.04	1.04	1.02	1.13	1.85	1.33	**1.19**	1.23
Netherlands	1.13	1.10	1.00	1.06	1.07	1.06	1.21	1.09	1.12	**1.09**	1.10
Norway	1.04	1.02	1.00	0.90	0.99	1.10	1.43	1.56	1.36	**1.18**	1.24
Portugal	0.77	0.94	0.99	1.05	0.94	0.94	0.29	0.37	0.53	**0.73**	0.67
Spain	0.76	0.95	1.00	1.01	0.93	0.95	0.70	0.66	0.77	**0.85**	0.82
Sweden	1.08	1.00	1.00	1.03	1.03	1.08	0.96	1.17	1.07	**1.05**	1.06
Switzerland	1.24	1.06	1.01	1.15	1.12	1.01	1.75	1.69	1.48	**1.30**	1.36
United Kingdom	1.08	0.99	1.00	0.94	1.00	0.97	1.09	0.97	1.01	**1.01**	1.01
United States	1.10	0.94	0.99	0.99	1.00	0.87	1.28	1.21	1.12	**1.06**	1.08
Average	1.00	1.00	1.00	1.00	1.00	1.00	1.00	1.00	1.00	**1.00**	1.00
Maximum	1.24	1.11	1.01	1.15	1.12	1.10	1.75	1.85	1.48	**1.30**	1.36
Minimum	0.61	0.85	0.99	0.74	0.81	0.87	0.01	-0.03	0.31	**0.56**	0.48

Fig. 4 Public Sector Performance (PSP) indicators, 2009–2013

Figure 5 depicts the results of the Comparison of our PSP results with the results obtained by Afonso et al. (2005) for 23 OECD countries for 2000. As we can see, Switzerland, Canada, Norway, United States, Germany, Belgium, France and the United Kingdom have improved their performance during these years.

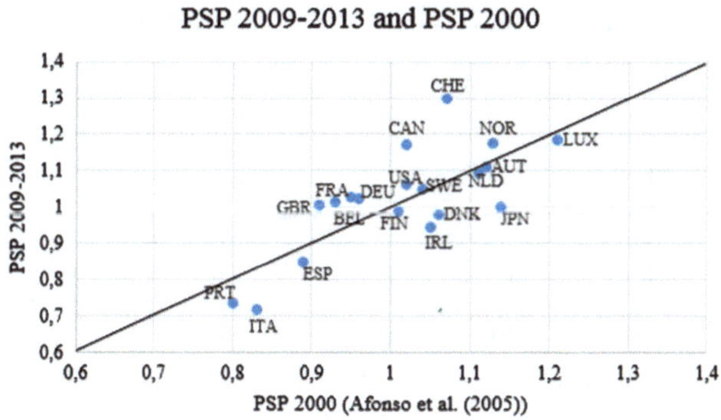

Fig. 5 Comparison of our PSP results with the results obtained by Afonso et al. (2005)

4.2 Public Sector Efficiency (PSE)

The following illustration shows the PSE scores that we computed by dividing the PSP scores of each country for different sub-indicators by the level of the relevant expenditure category. As we can see in Fig. 6, the PSE scores are ranging from 0.63 to 1.69. Switzerland is considered as the most efficient country among the 20 countries obtaining the PSE score of 1.69. On the other hand, Greece is considered as the least efficient country, obtaining a PSE score equal to 0.63. The other efficient countries followed by Switzerland are Luxembourg, Canada, Japan, Norway and Germany.

By considering the results of the computations of PSP and PSE at the same time, we can find that countries such as France and Sweden that are considered as good performers are not among the group of countries that are considered as efficient. Ireland on the other hand is not considered as a very good performer but performs relatively efficiently. Figure 7 illustrates these results by defining four quadrants in which these countries are situated.

Comparing the PSE results with the results obtained from the earlier work of Afonso et al. (2005) on the OECD countries, we observe that Switzerland, Luxembourg, Canada, Norway, Ireland, Austria, Germany, Belgium, Sweden and France have increased the level of their Public Sector Efficiency while the other countries obtained lower PSE scores (Fig. 8).

4.3 Data Envelopment Analysis (DEA)

We performed DEA for six different models assuming both constant and variable returns to scale. The summary of the results of these models is reported in Fig. 12. Model 1 assumes 1 input (the governments' normalized total spending) and 1 output (total PSP scores). The results obtained from analysing model 1 are illustrated in

Country	Opportunity Indicators					Musgravian Indicators				Total Public Sector Efficiency	
	Administration	Education	Health	Infrastructure	PSE Opportunity	Distribution	Stability	Economic Performance	PSE Musgravian	**Equal weights**	Different Weights
Austria	1.15	0.94	0.94	1.25	1.07	0.81	1.14	1.11	1.02	**1.05**	1.04
Belgium	0.77	0.94	0.94	1.56	1.05	0.89	1.03	0.87	0.93	**1.00**	0.97
Canada	1.06	1.11	1.02	1.13	1.08	1.36	2.02	1.36	1.58	**1.30**	1.41
Denmark	0.83	0.69	0.84	1.13	0.87	0.89	0.72	0.75	0.79	**0.84**	0.82
Finland	1.02	0.94	1.18	1.01	1.04	0.95	0.61	0.79	0.78	**0.93**	0.87
France	0.82	0.93	0.86	0.94	0.89	0.79	1.04	0.71	0.85	**0.87**	0.86
Germany	1.10	1.15	0.88	1.72	1.21	0.91	1.13	0.98	1.01	**1.12**	1.08
Greece	0.60	1.18	1.18	0.63	0.90	0.85	0.01	-0.03	0.28	**0.63**	0.49
Ireland	1.20	1.09	1.24	0.85	1.09	1.25	0.69	1.17	1.04	**1.07**	1.06
Italy	0.65	1.06	1.06	0.87	0.91	0.81	0.43	0.43	0.56	**0.76**	0.68
Japan	1.14	1.42	0.97	1.07	1.15	1.12	1.18	1.16	1.15	**1.15**	1.15
Luxembourg	1.45	1.41	1.20	0.87	1.23	0.98	1.23	2.02	1.41	**1.31**	1.35
Netherlands	0.92	1.10	0.84	0.93	0.95	1.40	1.23	1.11	1.25	**1.08**	1.15
Norway	1.04	0.79	0.97	0.79	0.90	1.18	1.53	1.67	1.46	**1.14**	1.27
Portugal	0.77	0.98	1.07	0.99	0.95	0.91	0.28	0.35	0.51	**0.76**	0.66
Spain	0.81	1.13	1.15	0.87	0.99	1.03	0.75	0.71	0.83	**0.92**	0.88
Sweden	0.87	0.81	0.94	0.82	0.86	1.09	0.86	1.04	1.00	**0.92**	0.95
Switzerland	2.31	1.09	1.09	1.33	1.46	1.20	2.44	2.37	2.00	**1.69**	1.82
United Kingdom	1.05	0.98	1.00	1.18	1.05	1.09	1.11	0.98	1.06	**1.06**	1.06
United States	1.40	0.94	0.94	0.89	1.04	1.01	1.51	1.42	1.31	**1.16**	1.22
Average	1.05	1.03	1.01	1.04	1.03	1.03	1.05	1.05	1.04	**1.04**	1.04
Maximum	2.31	1.42	1.24	1.72	1.46	1.40	2.44	2.37	2.00	**1.69**	1.82
Minimum	0.60	0.69	0.84	0.63	0.86	0.79	0.01	-0.03	0.28	**0.63**	0.49

Fig. 6 Public Sector Efficiency (PSE) indicators, 2009–2013

Fig. 9. According to these results, Switzerland is the only country that attains the efficiency score of 1, so it is considered to be the most efficient country of the sample in terms of the public expenditure. The least efficient country in the input-oriented analysis is France by attaining the efficiency score of 0.605 meaning that France

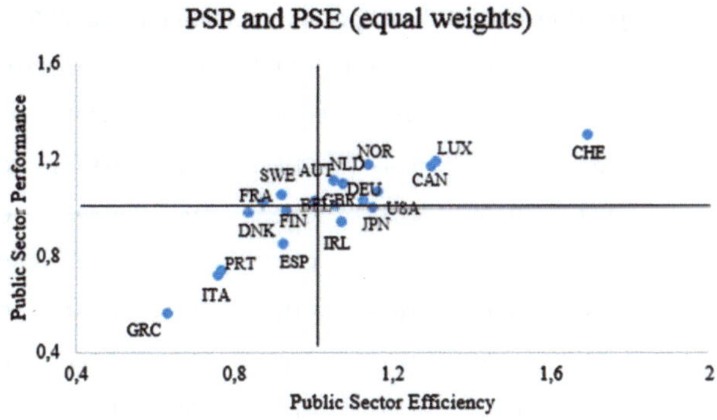

Fig. 7 Public Sector Performance and Public Sector Efficiency (2009–2013)

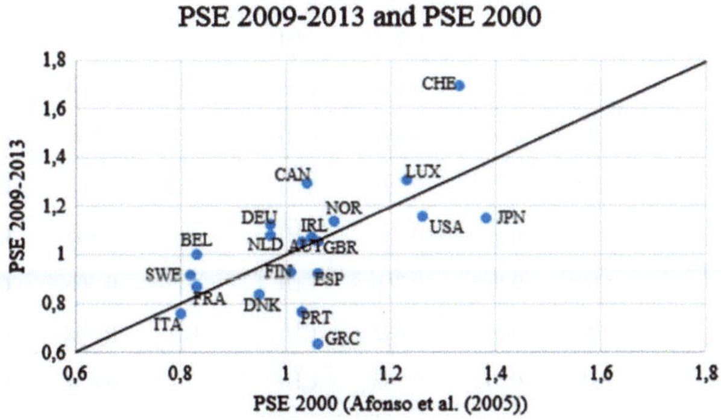

Fig. 8 Comparison of our PSE results with the results obtained by Afonso et al. (2005)

could have actually obtained the same level of outputs by reducing the amounts of inputs by 39.5%. Considering the results of the output-oriented analysis, Greece is attaining the efficiency score of 0.431, which leads the country to be the least efficient among the other countries. This indicates that Greece could have increased the outputs level by 56.9% and by consuming the same level of the inputs.

The average input-oriented efficiency score is equal to 0.732. That is, on average countries could have reduced the level of inputs by 26.8% and still achieve the same level of outputs. The average output-oriented efficiency score is 0.769 denoting that on average the sample countries could have increased the level of their outputs by 23.1% by employing the same level of inputs.

Figure 10 shows Model 1's variable returns to scale efficiency frontier. As we can observe Switzerland is the most efficient country and the only country that is performing on the efficiency frontier while the other countries are performing below this frontier.

Model 1 - 1 Input (Normalized Total Spending), 1 Output (Total PSP scores)									
COUNTRY		CRS	INPUT ORIENTED			OUTPUT ORIENTED		
			VRS	PEERS	RANK	VRS	PEERS	RANK
Austria	AUT	0.554	0.649	CHE	14	0,854	CHE	5
Belgium	BEL	0.505	0.637	CHE	16	0,792	CHE	9
Canada	CAN	0.745	0.828	CHE	4	0,9	CHE	4
Denmark	DNK	0.464	0.615	CHE	19	0,754	CHE	15
Finland	FIN	0.485	0.637	CHE	16	0,762	CHE	14
France	FRA	0.475	0.605	CHE	20	0,785	CHE	10
Germany	DEU	0.576	0.735	CHE	9	0,785	CHE	10
Greece	GRC	0.272	0.632	CHE	18	0,431	CHE	20
Ireland	IRL	0.572	0.791	CHE	5	0,723	CHE	16
Italy	ITA	0.376	0.679	CHE	13	0,554	CHE	19
Japan	JPN	0.652	0.847	CHE	2	0,769	CHE	13
Luxembourg	LUX	0.724	0.791	CHE	5	0,915	CHE	2
Netherlands	NLD	0.616	0.735	CHE	9	0,838	CHE	6
Norway	NOR	0.695	0.766	CHE	8	0,908	CHE	3
Portugal	PRT	0.389	0.692	CHE	12	0,562	CHE	18
Spain	ESP	0.512	0.783	CHE	7	0,654	CHE	17
Sweden	SWE	0.519	0.643	CHE	15	0,808	CHE	8
Switzerland	CHE	1	1	CHE	1	1	CHE	1
United Kingdom	GBR	0.565	0.727	CHE	11	0,777	CHE	12
United states	USA	0.691	0.847	CHE	2	0,815	CHE	7
Average		0.569	0.732			0.769		
Minimum		0.272	0.605			0.431		

Fig. 9 DEA results (Model 1), 2009–2013

Model 2 assumes 2 outputs, the Opportunity PSP scores and the other one is the Musgravian PSP scores and 1 input, the governments' normalized total spending. According to the results, Switzerland is the only efficient country and France (in the input-oriented analysis) and Greece (in the output-oriented analysis) are again obtaining the least efficiency score among all the countries. The results of this model are quite similar to the results we obtained from implementing DEA on Model 1 (Fig. 11). The production possibility frontier of this model is illustrated in Fig. 31 in

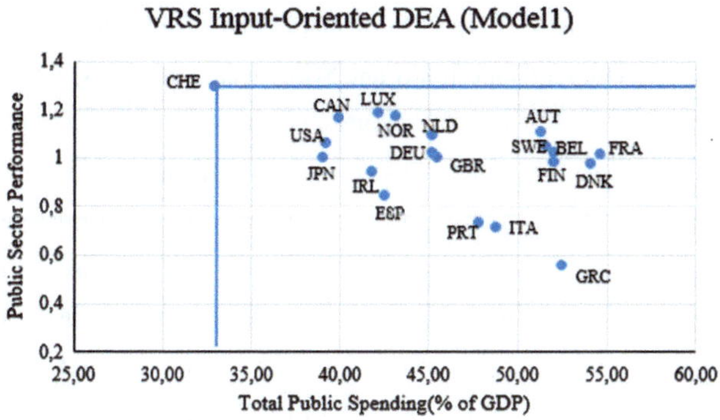

Fig. 10 Production Possibility Frontier (Model 1)

Appendix. Due to the existence of two outputs and one input we could only plot the production possibility frontier assuming that there exist constant returns to scale.

DEA was also conducted for the other four models. These models try to evaluate the efficiency of each country in different areas of governments' performance. Figure 12 shows the summary of the results of these evaluations. Results of the Model 3 which focuses on the administrative performance suggest that governments on average could have reduced the level of their consumption by 44% and still got the same level of administrative performance. The only country that had an efficient administration is Switzerland (Fig. 20).

Model 4 results suggest that the same education performance could have been achieved by lowering the level of expenditure on education. The results show that Finland, Japan, Luxembourg and the Netherlands are performing on the efficiency frontier (Fig. 21).

Model 5 considers the efficiency of the public health system. The results of the DEA implemented on this model show that there exist four countries on the frontier that are considered to be efficient. These countries are Ireland, Japan, Luxembourg and Switzerland. On average the sample countries could decreased the health expenditure by 16.1% and attained the same level of health performance or they could had increased their performance by 0.8% with the same level of health expenditure. This shows that these countries on average are performing most efficiently in the health sector when compare to the other sectors (Fig. 22).

The results of implementing DEA on Model 6 that considers the efficiency of public infrastructure shows that Germany and Switzerland are the most efficient countries in the sample in terms of public infrastructure, and on average all these governments could have reached to the same level of infrastructure outputs by decreasing the public investment by 32.7% (Fig. 23).

Model 2 - 1 Input (Normalized Total Spending), 2 Output (Opportunity and Musgravian PSP scores)								
COUNTRY		CRS	INPUT ORIENTED			OUTPUT ORIENTED		
			VRS	PEERS	RANK	VRS	PEERS	RANK
Austria	AUT	0.602	0.649	CHE	14	0,929	CHE	4
Belgium	BEL	0.563	0.637	CHE	16	0,884	CHE	15
Canada	CAN	0.768	0.828	CHE	4	0,929	CHE	4
Denmark	DNK	0.571	0.615	CHE	19	0,929	CHE	4
Finland	FIN	0.62	0.637	CHE	16	0,973	CHE	2
France	FRA	0.546	0.605	CHE	20	0,902	CHE	12
Germany	DEU	0.669	0.735	CHE	9	0,911	CHE	11
Greece	GRC	0.457	0.632	CHE	18	0,723	CHE	19
Ireland	IRL	0.699	0.791	CHE	5	0,884	CHE	15
Italy	ITA	0.491	0.679	CHE	13	0,723	CHE	19
Japan	JPN	0.779	0.847	CHE	2	0,92	CHE	8
Luxembourg	LUX	0.735	0.791	CHE	5	0,929	CHE	4
Netherlands	NLD	0.702	0.735	CHE	9	0,955	CHE	3
Norway	NOR	0.704	0.766	CHE	8	0,919	CHE	10
Portugal	PRT	0.581	0.692	CHE	12	0,839	CHE	17
Spain	ESP	0.65	0.783	CHE	7	0,83	CHE	18
Sweden	SWE	0.591	0.643	CHE	15	0,92	CHE	8
Switzerland	CHE	1	1	CHE	1	1	CHE	1
United Kingdom	GBR	0.649	0.727	CHE	11	0,893	CHE	13
United states	USA	0.756	0.847	CHE	2	0,893	CHE	13
Average		0.657	0.732			0.894		
Minimum		0.457	0.605			0.723		

Fig. 11 DEA results (Model 2), 2009–2013

These results also suggest that governments are performing more efficiently in the health and education sections than in administrative and infrastructure sections despite the fact that they apply a higher level of expenditure in administrative functions.

		Model 1	Model 2	Model3	Model 4	Model 5	Model 6
Inputs		Total public expenditure	Total public expenditure	Government Consumption	Education Expenditure	Health Expenditure	Public investment
Outputs		PSP	PSP Opportunity PSP Musgravian	PSP Administration	PSP Education	PSP Health	PSP infrastructure
Countries on the frontier		CHE	CHE	CHE	FIN, JPN, LUX, NLD	IRL, JPN, LUX, CHE	DEU, CHE
Average scores	**Input**	0.732	0.732	0.56	0.812	0.839	0.673
	output	0.769	0.894	0.808	0.933	0.992	0.876
Minimum score	**Input**	0.605	0.605	0.422	0.586	0.684	0.493
	Output	0.431	0.723	0.492	0.854	0.972	0.644
Total countries		20	20	20	20	20	20
Efficient countries		1	1	1	4	4	2

Fig. 12 Summary results of different DEA models

Due to the significant distance between the Switzerland's efficiency score and the other countries especially the least efficient ones, we decided to conduct the DEA once again without considering Switzerland in the sample in order to acquire a more precise image of the differences in the efficiency scores (Fig. 19 and Fig. 29).

Figure 13 shows the results of the recalculations of DEA for Model 1 excluding Switzerland from the sample. These results denote the increase in the average efficiency scores of the countries for both input and output oriented analysis. Model 1 as depicted in Fig. 14, suggests that Canada, Japan, Luxembourg and the United States are performing on the efficiency frontier. Again, France and Greece are obtaining respectively the least input and output oriented efficiency scores in both models. The countries on average could have de-creased the level of the public expenditure by 14.6% and still performed efficiently. More details on the DEA results when excluding Switzerland are presented in Figs. 24–28, 30 and 32.

Model 1- 1 Input (Normalized Total Spending), 1 Output (Total PSP scores)								
COUNTRY	Code	CRT	INPUT ORIENTED			OUTPUT ORIENTED		
			VRT	PEERS	RANK	VRT	PEERS	RANK
Austria	AUT	0.736	0.769	CAN.USA	13	0,936	LUX	6
Belgium	BEL	0.671	0.751	USA.JPN	15	0,866	LUX	9
Canada	CAN	1	1	CAN	1	1	CAN	1
Denmark	DNK	0.612	0.722	JPN	18	0,819	LUX	14
Finland	FIN	0.643	0.751	JPN	15	0,828	LUX	13
France	FRA	0.631	0.715	USA.JPN	19	0,854	LUX	11
Germany	DEU	0.767	0.864	JPN.USA	9	0,859	LUX	10
Greece	GRC	0.353	0.744	JPN	17	0,46	LUX	19
Ireland	IRL	0.764	0.933	JPN	6	0,793	LUX,CAN	15
Italy	ITA	0.494	0.8	JPN	12	0,597	LUX	18
Japan	JPN	0.869	1	JPN	1	1	JPN	1
Luxembourg	LUX	0.958	1	LUX	1	1	LUX	1
Netherlands	NLD	0.82	0.87	CAN.USA	8	0,918	LUX	7
Norway	NOR	0.93	0.949	LUX.CAN	5	0,994	LUX	5
Portugal	PRT	0.515	0.816	JPN	11	0,61	LUX	17
Spain	ESP	0.674	0.917	JPN	7	0,711	LUX	16
Sweden	SWE	0.691	0.759	USA.JPN	14	0,882	LUX	8
United Kingdom	GBR	0.750	0.859	USA.JPN	10	0,845	LUX	12
United states	USA	0.925	1	USA	1	1	USA	1
MEAN		0.726	0.854			0.841		
MINIMUM		0.353	0.715			0.460		

Fig. 13 DEA results (Model 1) excluding Switzerland, 2009–2013

Although Afonso et al. (2005) applied a FDH approach in order to assess the public spending efficiency and considered a bigger country-sample than what we did, we take the opportunity to compare our results from DEA, with more recent data, with the results they achieved from implementing FDH. By looking at Fig. 15, we observe an improvement in the efficiency scores of Canada, Finland, Germany, Italy, Netherlands, Norway, Sweden and Switzerland during that 10-year period.

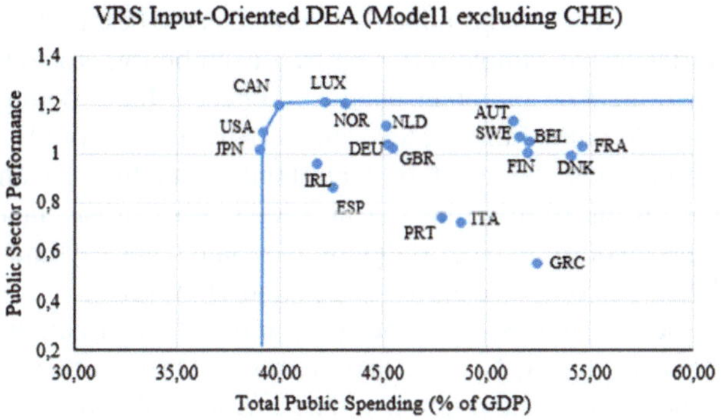

Fig. 14 Production Possibility Frontier (Model 1) excluding Switzerland

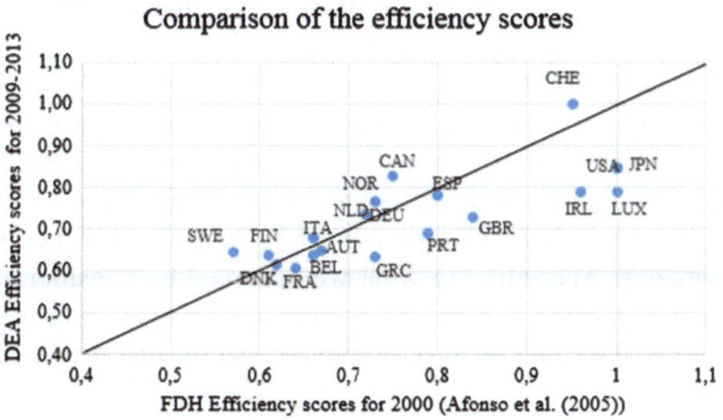

Fig. 15 Comparison of the efficiency scores of 2000 [obtained by Afonso et al. (2005)] and 2009–2013

5 Conclusion

We assessed the public spending efficiency for 20 OECD countries for the period 2009–2013 by applying a non-parametric approach called Data Envelopment Analysis (DEA). The reason of preferring DEA to Free Disposal Hull (FDH) is that FDH is more restrictive on the data, therefore, it considers more DMUs performing efficiently. However, some of these DMUs that are considered efficient under FDH may not be considered efficient under DEA due to the convexity assumption imposed by DEA.

In order to assess the efficiency, first, we constructed the composite indicators of Public Sector Performance (PSP) and Public Sector Efficiency (PSE) and then implemented the DEA approach for six different models by considering the level of the public spending as the input and the PSP scores as the output of our analysis.

The derived PSP scores suggest that Switzerland is the best performer among all the other countries in the sample followed by Luxembourg, Norway and Canada. The bottom performers on the other hands are Greece, Italy, Portugal and Spain. France, Denmark, Belgium, Finland, Sweden and Austria also could have performed the same by decreasing the level of their total expenditure. Comparing these results with the results from Afonso et al. (2005) we can say that Switzerland, Canada, United Kingdom, France, Belgium, Germany, Norway and United States had improved their performance during this period of 10 years.

PSE results indicate that Switzerland is the most efficient country followed by Luxembourg Canada, Japan, Norway and Germany. On the other hand Greece is considered as the least efficient country. These results also propose that being a good performer doesn't necessarily mean that the country is spending in an efficient manner. We can mention at France and Sweden those of which are relatively good performers but not efficient countries. Switzerland, Canada, Germany and Belgium showed an improvement in the scores of their public performance efficiency when comparing the results with the PSE results obtained by Afonso et al. (2005).

The results of the implemented DEA for model 1 that assesses the efficiency of the public spending as a whole, show that the only country in this sample that is performing on the efficiency frontier is Switzerland and all the other countries on average could decreased the expenditure level by 26.8% and still attained the same level of performance. According to what we observed by considering Switzerland as an outlier and excluding it from the sample and recalculating the DEA scores, countries could got the same level of outputs by decreasing the level of the public spending by 14.6%.

In summary, our results suggest that countries with a higher level of expenditures perform less efficiently than countries that have a lower level of public spending. However, following Mandl et al. (2008) we recommend individual analyses for each country to complement our analysis due to the different traditions and cultures in institutional settings, aspects of political economy, etc. Also applying a parametric analysis (e.g., Stochastic Frontier Analysis (SFA)) for checking the robustness of the results could be used in future work.

Acknowledgements UECE is supported by the Fundacào para a Ciência e a Tecnologia (Portuguese Foundation for Science and Technology).

We thank an anonymous referee and participants at the 2016 Public Sector Economics Conference organised by the Institute of Public Finance, Zagreb. The opinions expressed herein are those of the authors and do not necessarily reflect those of the authors' employers.

Appendix

See Figs. 16, 17, 18, 19, 20, 21, 22, 23, 24, 25, 26, 27, 28, 29, 30, 31, 32.

Sub Index	Variable	Source	Series
Opportunity Indicators			
Administration	Corruption	Transparency International's Corruption Perceptions Index (CPI) (2009-2013)	Average (5y) corruption on a scale from 10 (Perceived to have low levels of corruption) to 0 (highly corrupt)
	Red Tape	World Economic Forum: The Global competitiveness Report (2010-2015)	Average (5y) Burden of government Regulation on a scale from 7 (not burdensome at all) to 1 (extremely burdensome),(2009-2013)
	Judicial Independence	World Economic Forum: The Global competitiveness Report (2010-2015)	Average (5y) judicial independence on a scale from 7 (entirely independent) to 1 (heavily influenced),(2009-2013)
	Property Rights	World Economic Forum: The Global competitiveness Report (2010-2015)	Average (5y) property rights on a scale from 7 (very strong) to 1 (very weak), (2009-2013)
	Shadow Economy	Friedrich Schneider (2015)	%of official GDP. Reciprocal value 1/x. Average (5y) shadow economy (2009-2013)
Education	School Enrolment Secondary, gross (%)	World Bank, World Development Indicators (2009-2013)	Average (5y) Ratio of total enrolment in secondary education, (2009-2013)
	Quality of Educational System	World Economic Forum: The Global competitiveness Report (2010-2015)	Average (5y) quality of educational system on a scale from 7 (very well) to 1 (not well at all), (2009-2013)
	PISA scores	PISA Report, (2012)	Simple average of mathematics, reading and science scores
Health	Infant Mortality	World Bank, World Development Indicators (2009-2013)	Per 1000 lives birth in a given year. We used the Infant Survival Rate in our computations which is equal to: (1000-IMR)/1000. Average (5y) ISR
	Life Expectancy	World Bank World Development Indicators (2009-2013)	Average (5y) life expectancy at birth, Total (years)
Public Infrastructure	Infrastructure Quality	World Economic Forum: The Global Competitiveness Report (2010-2015)	Average (5y) infrastructure quality on a scale from 7 (extensive and efficient) to 1 (extremely underdeveloped), (2009-2013)
Standard Musgravian Indicators			
Distribution	Gini Index	Eurostat, OECD (2009-2013)	Average (5y) Gini Index on a scale from 100 (Perfect Inequality) to 0 (perfect equality), (2009-2013) Transformed to 100-Gini for better comparison
Stabilization	Coefficient of Variation of GDP Growth	C.V= Standard Deviation/Mean	Based on GDP at constant prices (percent change) Reciprocal value 1/x
	Standard Deviation of Inflation	IMF World Economic Outlook (WEO database) 2015	Inflation, average consumer prices (percent change). Reciprocal value 1/x of the standard deviation
Economic Performance	GDP per capita	IMF World Economic Outlook (WEO database) 2015	GDP based on PPP per capita GDP, current International dollar
	GDP Growth	IMF World Economic Outlook (WEO database) 2015	Average (10y) GDP, constant prices (percent change)
	Unemployment	IMF World Economic Outlook (WEO database) 2015	Average (10y) unemployment rate, percent of total labour force Reciprocal value 1/x

Fig. 16 Detailed list of output components

Sub Index	Variable	Source	Series
Administration	Government Consumption	The World Bank (2004-2013)	Average (10y) general government final consumption expenditure (% of GDP) at current prices
Education	Public Education	UIS Statistics (2004-2013)	Average (10y) expenditure on education (% of GDP)
Health	Public Health	OECD database (2004-2013)	Average (10y) expenditure on health % of GDP
Public Infrastructure	Public Investment	European Commission, AMECO (2004-2013)	Average (10y) General government gross fixed capital formation (% of GDP) at current prices
Distribution	Expenditure on Social Protection	European Commission, AMECO (2004-2013)	Average (10y) aggregation of the social transfers other than in kind (% of GDP) and Subsidies (% of GDP) at current prices
Stabilization\ Economic Performance	Government Total Expenditure	European Commission, AMECO (2004-2013)	Average (10y) of Total Expenditure (% Of GDP)

Fig. 17 Detailed list of input components (Expenditure Categories)

Country	Government Consumption	Education	Health	Public Investment	Transfers and Subsidies	Total Spending
Austria	19.53	5.43	7.45	2.97	20.20	51.31
Belgium	23.09	6.09	7.38	2.22	18.75	52.04
Canada	20.68	4.96	6.88	3.09	11.40	39.91
Denmark	25.92	8.10	8.28	3.17	18.48	54.07
Finland	22.77	6.27	5.93	3.77	17.86	51.97
France	23.21	5.55	8.21	4.02	20.01	54.63
Germany	18.61	4.61	7.97	2.13	17.62	45.21
Greece	20.48	3.83	5.94	4.24	17.68	52.48
Ireland	17.53	5.25	5.67	3.38	12.71	41.81
Italy	19.62	4.34	6.67	2.89	19.07	48.80
Japan	19.25	3.63	7.35	3.33	13.41	39.02
Luxembourg	16.32	3.55	5.87	4.11	16.64	42.12
Netherlands	24.79	5.30	8.31	3.91	12.01	45.19
Norway	20.25	6.83	7.19	3.91	14.78	43.14
Portugal	20.14	5.09	6.49	3.64	16.36	47.82
Spain	18.89	4.45	6.13	3.99	14.64	42.54
Sweden	25.19	6.53	7.52	4.32	15.76	51.57
Switzerland	10.83	5.14	6.48	2.96	13.35	32.95
United Kingdom	20.70	5.34	7.02	2.73	14.14	45.44
United States	15.79	5.28	7.36	3.81	13.76	39.16
Average	20.18	5.28	7.01	3.43	15.93	46.06
Maximum	25.92	8.10	8.31	4.32	20.20	54.63
Minimum	10.83	3.55	5.67	2.13	11.40	32.95

Fig. 18 Public Expenditure (% of GDP) 2004–2013. Sources: The World Bank, European Commission (AMECO), OECD database, UIS Statistics

Country	Opportunity Indicators					Musgravian Indicators				Total Public Sector Performance	
	Administration	Education	Health	Infrastructure	PSP Opportunity	Distribution	Stability	Economic Performance	PSP Musgravian	**Equal weights**	Different weights
Austria	1.13	0.97	1.00	1.09	1.05	1.03	1.33	1.29	1.22	**1.13**	1.16
Belgium	0.89	1.08	1.00	1.02	1.00	1.05	1.23	1.02	1.10	**1.05**	1.07
Canada	1.10	1.05	1.00	1.02	1.04	0.97	1.84	1.24	1.35	**1.20**	1.25
Denmark	1.08	1.06	0.99	1.05	1.05	1.03	0.86	0.91	0.94	**0.99**	0.97
Finland	1.17	1.12	1.00	1.12	1.10	1.06	0.72	0.93	0.90	**1.00**	0.97
France	0.96	0.98	1.01	1.11	1.01	1.00	1.28	0.88	1.05	**1.03**	1.04
Germany	1.03	1.01	1.00	1.08	1.03	1.01	1.15	0.99	1.05	**1.04**	1.04
Greece	0.61	0.86	1.00	0.79	0.81	0.95	-0.01	-0.04	0.30	**0.56**	0.47
Ireland	1.05	1.09	1.00	0.85	1.00	1.00	0.66	1.10	0.92	**0.96**	0.94
Italy	0.64	0.88	1.01	0.74	0.82	0.97	0.45	0.46	0.63	**0.72**	0.69
Japan	1.10	0.98	1.01	1.05	1.04	0.95	1.03	1.02	1.00	**1.02**	1.01
Luxembourg	1.19	0.95	1.00	1.05	1.05	1.02	1.18	1.91	1.37	**1.21**	1.26
Netherlands	1.15	1.10	1.00	1.07	1.08	1.06	1.25	1.13	1.14	**1.11**	1.12
Norway	1.06	1.03	1.00	0.90	1.00	1.10	1.51	1.62	1.41	**1.20**	1.27
Portugal	0.78	0.94	0.99	1.06	0.94	0.94	0.28	0.38	0.53	**0.74**	0.67
Spain	0.77	0.95	1.01	1.02	0.94	0.95	0.72	0.68	0.78	**0.86**	0.83
Sweden	1.10	1.00	1.01	1.04	1.04	1.08	1.01	1.22	1.10	**1.07**	1.08
United Kingdom	1.09	0.99	1.00	0.95	1.01	0.97	1.14	1.00	1.04	**1.02**	1.03
United States	1.11	0.95	0.99	1.00	1.01	0.87	1.36	1.26	1.16	**1.09**	1.11
Average	1.00	1.00	1.00	1.00	1.00	1.00	1.00	1.00	1.00	**1.00**	1.00
Maximum	1.19	1.12	1.01	1.12	1.10	1.10	1.84	1.91	1.41	**1.21**	1.27
Minimum	0.61	0.86	0.99	0.74	0.81	0.87	-0.01	-0.04	0.30	**0.56**	0.47

Fig. 19 Public Sector Performance (PSP) Indicators excluding Switzerland, 2009–2013

COUNTRY		CRS	INPUT ORIENTED			OUTPUT ORIENTED		
			VRS	PEERS	RANK	VRS	PEERS	RANK
Austria	AUT	0.498	0.557	CHE	8	0,895	CHE	5
Belgium	BEL	0.336	0.474	CHE	16	0,71	CHE	16
Canada	CAN	0.465	0.529	CHE	13	0,879	CHE	7
Denmark	DNK	0.364	0.422	CHE	20	0,863	CHE	11
Finland	FIN	0.447	0.478	CHE	15	0,935	CHE	3
France	FRA	0.36	0.47	CHE	17	0,766	CHE	15
Germany	DEU	0.483	0.587	CHE	5	0,823	CHE	14
Greece	GRC	0.263	0.535	CHE	12	0,492	CHE	20
Ireland	IRL	0.521	0.621	CHE	4	0,839	CHE	12
Italy	ITA	0.283	0.557	CHE	8	0,508	CHE	19
Japan	JPN	0.500	0.568	CHE	7	0,879	CHE	7
Luxembourg	LUX	0.634	0.667	CHE	3	0,952	CHE	2
Netherlands	NLD	0.400	0.439	CHE	18	0,911	CHE	4
Norway	NOR	0.453	0.54	CHE	10	0,839	CHE	12
Portugal	PRT	0.335	0.54	CHE	10	0,621	CHE	17
Spain	ESP	0.352	0.574	CHE	6	0,613	CHE	18
Sweden	SWE	0.376	0.432	CHE	19	0,871	CHE	9
Switzerland	CHE	1	1	CHE	1	1	CHE	1
United Kingdom	GBR	0.457	0.524	CHE	14	0,871	CHE	9
United states	USA	0.614	0.692	CHE	2	0,887	CHE	6
Average		0.457	0.560			0.808		
Minimum		0.263	0.422			0.492		

Fig. 20 DEA results (Model 3) 2009–2013 Model 3—1 Input (Normalized Government Consumption), 1 Output (Administration PSP scores)

COUNTRY		CRS	INPUT ORIENTED			OUTPUT ORIENTED		
			VRS	PEERS	RANK	VRS	PEERS	RANK
Austria	AUT	0.663	0.663	JPN	16	0,881	FIN	16
Belgium	BEL	0.661	0.825	NLD	10	0,975	FIN	6
Canada	CAN	0.786	0.926	NLD	7	0,975	NLD	6
Denmark	DNK	0.488	0.586	NLD	20	0,955	FIN	10
Finland	FIN	0.657	1	FIN	1	1	FIN	1
France	FRA	0.657	0.657	JPN	17	0,889	FIN	15
Germany	DEU	0.817	0.882	NLD	9	0,962	NLD	9
Greece	GRC	0.831	0.931	LUX	6	0,857	NLD	18
Ireland	IRL	0.76	0.948	NLD	5	0,982	NLD	5
Italy	ITA	0.756	0.817	LUX	11	0,854	NLD	20
Japan	JPN	1	1	JPN	1	1	JPN	1
Luxembourg	LUX	0.998	1	LUX	1	1	LUX	1
Netherlands	NLD	0.774	1	NLD	1	1	NLD	1
Norway	NOR	0.557	0.615	NLD	18	0,919	FIN	11
Portugal	PRT	0.689	0.698	LUX	14	0,867	NLD	17
Spain	ESP	0.796	0.798	LUX	12	0,915	NLD	12
Sweden	SWE	0.568	0.598	NLD	19	0,901	FIN	13
Switzerland	CHE	0.769	0.924	NLD	8	0,974	NLD	8
United Kingdom	GBR	0.690	0.709	NLD	13	0,9	FIN	14
United states	USA	0.662	0.67	LUX	15	0,855	NLD	19
Average		0.729	0.812			0.933		
Minimum		0.488	0.586			0.854		

Fig. 21 DEA results (Model 4) 2009–2013 Model 4—1 Input (Normalized Education Expenditure), 1 Output (Education PSP scores)

COUNTRY		CRS	INPUT ORIENTED			OUTPUT ORIENTED		
			VRS	PEERS	RANK	VRS	PEERS	RANK
Austria	AUT	0.760	0.760	IRL	16	0,986	JPN	14
Belgium	BEL	0.764	0.767	IRL	15	0,982	JPN	17
Canada	CAN	0.823	0.827	LUX/IRL	10	0,988	CHE/JPN	11
Denmark	DNK	0.679	0.684	IRL	20	0,979	JPN	19
Finland	FIN	0.954	0.956	IRL	6	0,994	CHE/LUX	7
France	FRA	0.694	0.741	CHE/LUX	17	0,992	JPN	9
Germany	DEU	0.710	0.711	IRL	18	0,985	JPN	16
Greece	GRC	0.952	0.954	IRL	7	0,994	LUX/CHE	7
Ireland	IRL	1	1	IRL	1	1	IRL	1
Italy	ITA	0.856	0.932	LUX/CHE	8	0,996	JPN/CHE	6
Japan	JPN	0.782	1	JPN	1	1	JPN	1
Luxembourg	LUX	0.968	1	LUX	1	1	LUX	1
Netherlands	NLD	0.682	0.690	LUX/IRL	19	0,987	JPN	13
Norway	NOR	0.789	0.802	LUX/IRL	13	0,988	CHE/JPN	11
Portugal	PRT	0.866	0.873	IRL	9	0,982	CHE/JPN	17
Spain	ESP	0.929	0.993	LUX/CHE	5	0,999	CHE/LUX	5
Sweden	SWE	0.757	0.805	LUX/CHE	12	0,991	JPN	10
Switzerland	CHE	0.884	1	CHE	1	1	CHE	1
United Kingdom	GBR	0.806	0.807	IRL	11	0,986	JPN/CHE	14
United states	USA	0.760	0.770	IRL	14	0,972	JPN	20
Average		0.821	0.839			0.992		
Minimum		0.679	0.684			0.972		

Fig. 22 DEA results (Model 5) 2009–2013 Model 5—1 Input (Normalized Health Expenditure), 1 Output (Health PSP scores)

COUNTRY	Code	CRS	INPUT ORIENTED			OUTPUT ORIENTED		
			VRS	PEERS	RANK	VRS	PEERS	RANK
Austria	AUT	0.729	0.775	CHE/DEU	5	0,943	CHE	5
Belgium	BEL	0.907	0.959	DEU	3	0,937	CHE/DEU	6
Canada	CAN	0.657	0.69	DEU	7	0,883	CHE	13
Denmark	DNK	0.657	0.672	DEU	9	0,907	CHE	9
Finland	FIN	0.589	0.684	CHE/DEU	8	0,967	CHE	3
France	FRA	0.547	0.616	CHE/DEU	12	0,958	CHE	4
Germany	DEU	1	1	DEU	1	1	DEU	1
Greece	GRC	0.368	0.503	DEU	19	0,679	CHE	19
Ireland	IRL	0.496	0.63	DEU	11	0,73	CHE	18
Italy	ITA	0.508	0.737	DEU	6	0,644	CHE/DEU	20
Japan	JPN	0.623	0.64	DEU	10	0,904	CHE	10
Luxembourg	LUX	0.503	0.518	DEU	18	0,901	CHE	11
Netherlands	NLD	0.539	0.545	DEU	15	0,919	CHE	7
Norway	NOR	0.457	0.545	DEU	15	0,778	CHE	17
Portugal	PRT	0.576	0.585	DEU	13	0,913	CHE	8
Spain	ESP	0.506	0.534	DEU	17	0,88	CHE	14
Sweden	SWE	0.474	0.493	DEU	20	0,892	CHE	12
Switzerland	CHE	0.775	1	CHE	1	1	CHE	1
United Kingdom	GBR	0.687	0.78	DEU	4	0,833	CHE/DEU	16
United states	USA	0.517	0.559	DEU	14	0,859	CHE	15
Average		0.606	0.673			0.876		
Minimum		0.368	0.493			0.644		

Fig. 23 DEA results (Model 6) 2009–2013 Model 6—1 Input (Public Investment), 1 Output (Infrastructure PSP scores)

COUNTRY	Code	CRS	INPUT ORIENTED			OUTPUT ORIENTED		
			VRS	PEERS	RANK	VRS	PEERS	RANK
Austria	AUT	0.773	0.796	JPN.CAN. NLD	14	0,984	NLD, LUX	8
Belgium	BEL	0.726	0.753	CAN.JPN	16	0,931	NLD, LUX	15
Canada	CAN	1	1	CAN	1	1	CAN	1
Denmark	DNK	0.722	0.746	NLD.JPN	17	0,96	NLD,FIN	10
Finland	FIN	0.791	1	FIN	1	1	FIN	1
France	FRA	0.693	0.712	CAN.JPN	19	0,934	NLD,FIN	14
Germany	DEU	0.851	0.859	CAN.JPN	10	0,954	NLD	11
Greece	GRC	0.577	0.741	JPN	18	0,736	FIN	19
Ireland	IRL	0.897	0.933	JPN	8	0,946	NLD,JPN	12
Italy	ITA	0.629	0.798	JPN	13	0,752	NLD,FIN	18
Japan	JPN	1	1	JPN	1	1	JPN	1
Luxembourg	LUX	0.958	1	LUX	1	1	LUX	1
Netherlands	NLD	0.895	1	NLD	1	1	NLD	1
Norway	NOR	0.965	1	NOR	1	1	NOR	1
Portugal	PRT	0.735	0.814	JPN	12	0,865	NLD,FIN	17
Spain	ESP	0.824	0.912	JPN	9	0,884	NLD,JPN	16
Sweden	SWE	0.760	0.760	CAN.JPN	15	0,963	LUX, NLD	9
United Kingdom	GBR	0.835	0.858	JPN.CAN	11	0,935	NLD	13
United states	USA	0.972	0.999	JPN.CAN	7	0,987	JPN,CAN	7
Average		0.821	0.878			0.938		
Minimum		0.577	0.712			0.736		

Fig. 24 DEA results (Model 2) excluding Switzerland 2009–2013 Model 2—1 Input (Normalized Total Spending), 2 Output (Opportunity and Musgravian PSP scores)

COUNTRY	Code	CRS	INPUT ORIENTED			OUTPUT ORIENTED		
			VRS	PEERS	RANK	VRS	PEERS	RANK
Austria	AUT	0.789	0.813	LUX.USA	7	0.944	LUX	5
Belgium	BEL	0.525	0.684	USA	15	0.743	LUX	15
Canada	CAN	0.728	0.764	USA	12	0.923	LUX	7
Denmark	DNK	0.570	0.609	USA	19	0.905	LUX	10
Finland	FIN	0.703	0.71	LUX.USA	14	0.980	LUX	3
France	FRA	0.566	0.68	USA	16	0.805	LUX	14
Germany	DEU	0.757	0.848	USA	4	0.864	LUX	13
Greece	GRC	0.410	0.771	USA	11	0.515	LUX	19
Ireland	IRL	0.823	0.901	USA	3	0.884	LUX	12
Italy	ITA	0.445	0.805	USA	8	0.535	LUX	18
Japan	JPN	0.784	0.82	USA	6	0.925	LUX	6
Luxembourg	LUX	1	1	LUX	1	1	LUX	1
Netherlands	NLD	0.633	0.646	LUX.USA	17	0.961	LUX	4
Norway	NOR	0.713	0.78	USA	10	0.885	LUX	11
Portugal	PRT	0.528	0.784	USA	9	0.651	LUX	16
Spain	ESP	0.556	0.836	USA	5	0.644	LUX	17
Sweden	SWE	0.596	0.627	USA	18	0.920	LUX	8
United Kingdom	GBR	0.723	0.763	USA	13	0.917	LUX	9
United states	USA	0.964	1	USA	1	1	USA	1
Average		0.674	0.781			0.842		
Minimum		0.410	0.609			0.515		

Fig. 25 DEA results (Model 3) excluding Switzerland 2009–2013 Model 3—1 Input (Normalized Government Consumption), 1 Output (Administration PSP scores)

COUNTRY	Code	CRS	INPUT ORIENTED			OUTPUT ORIENTED		
			VRS	PEERS	RANK	VRS	PEERS	RANK
Austria	AUT	0.663	0.663	JPN.LUX	15	0.879	NLD,FIN	15
Belgium	BEL	0.661	0.825	NLD.JPN	9	0.968	FIN,NLD	7
Canada	CAN	0.786	0.926	NLD.JPN	7	0.975	NLD,JPN	6
Denmark	DNK	0.488	0.586	NLD.JPN	19	0.946	FIN	9
Finland	FIN	0.663	1	FIN	1	1	FIN	1
France	FRA	0.657	0.657	JPN	16	0.887	NLD,FIN	14
Germany	DEU	0.817	0.882	NLD.JPN	8	0.962	NLD,JPN	8
Greece	GRC	0.841	0.931	LUX	6	0.867	NLD,JPN	16
Ireland	IRL	0.775	0.984	NLD.JPN	5	0.994	NLD,JPN	5
Italy	ITA	0.756	0.817	LUX	10	0.854	NLD,JPN	19
Japan	JPN	1	1	JPN	1	1	JPN	1
Luxembourg	LUX	0.998	1	LUX	1	1	LUX	1
Netherlands	NLD	0.774	1	NLD	1	1	NLD	1
Norway	NOR	0.562	0.635	NLD.JPN	17	0.92	FIN	10
Portugal	PRT	0.689	0.698	LUX	13	0.867	NLD,JPN	16
Spain	ESP	0.796	0.798	LUX	11	0.915	NLD,JPN	11
Sweden	SWE	0.568	0.598	NLD.JPN	18	0.893	FIN	13
United Kingdom	GBR	0.690	0.709	NLD.JPN	12	0.899	NLD,FIN	12
United states	USA	0.669	0.670	LUX	14	0.864	NLD	18
Average		0.729	0.809			0.931		
Minimum		0.488	0.586			0.854		

Fig. 26 DEA results (Model 4) excluding Switzerland 2009–2013 Model 4—1 Input (Normalized Education Expenditure), 1 Output (Education PSP scores)

COUNTRY	Code	CRS	INPUT ORIENTED			OUTPUT ORIENTED		
			VRS	PEERS	RANK	VRS	PEERS	RANK
Austria	AUT	0.760	0.760	IRL	15	0.986	JPN	14
Belgium	BEL	0.764	0.767	IRL	14	0.982	JPN	17
Canada	CAN	0.823	0.828	LUX.IRL	9	0.990	ESP,JPN	10
Denmark	DNK	0.679	0.684	IRL	19	0.979	JPN	18
Finland	FIN	0.954	0.956	IRL	5	0.994	LUX,ESP	7
France	FRA	0.694	0.747	ESP	16	0.992	JPN	8
Germany	DEU	0.710	0.711	IRL	17	0.985	JPN	15
Greece	GRC	0.953	0.954	IRL	7	0.995	ESP,LUX	6
Ireland	IRL	1	1	IRL	1	1	IRL	1
Italy	ITA	0.856	0.956	ESP.JPN	5	0.998	ESP,JPN	5
Japan	JPN	0.782	1	JPN	1	1	JPN	1
Luxembourg	LUX	0.968	1	LUX	1	1	LUX	1
Netherlands	NLD	0.682	0.690	LUX.IRL	18	0.987	JPN	12
Norway	NOR	0.789	0.803	LUX.IRL	12	0.989	JPN,ESP	11
Portugal	PRT	0.866	0.873	IRL	8	0.985	ESP,JPN	15
Spain	ESP	0.929	1	ESP	1	1	ESP	1
Sweden	SWE	0.757	0.810	LUX.ESP	10	0.991	JPN	9
United Kingdom	GBR	0.806	0.808	IRL	11	0.987	ESP,JPN	12
United states	USA	0.760	0.770	IRL	13	0.972	JPN	19
Average		0.817	0.848			0.990		
Minimum		0.679	0.684			0.972		

Fig. 27 DEA results (Model 5) excluding Switzerland 2009–2013 Model 5—1 Input (Normalized Health Expenditure), 1 Output (Health PSP scores)

COUNTRY	Code	CRS	INPUT ORIENTED			OUTPUT ORIENTED		
			VRS	PEERS	RANK	VRS	PEERS	RANK
Austria	AUT	0.728	0.858	FIN.DEU	4	0.991	FIN,DEU	3
Belgium	BEL	0.915	0.969	DEU	3	0.943	FIN,DEU	9
Canada	CAN	0.658	0.697	DEU	8	0.925	FIN,DEU	11
Denmark	DNK	0.655	0.674	DEU	9	0.950	FIN,DEU	6
Finland	FIN	0.590	1	FIN	1	1	FIN	1
France	FRA	0.549	0.838	FIN.DEU	5	0.991	FIN	3
Germany	DEU	1	1	DEU	1	1	DEU	1
Greece	GRC	0.369	0.504	DEU	18	0.705	FIN	18
Ireland	IRL	0.498	0.633	DEU	11	0.765	FIN,DEU	17
Italy	ITA	0.506	0.738	DEU	7	0.674	FIN,DEU	19
Japan	JPN	0.628	0.646	DEU	10	0.947	FIN,DEU	8
Luxembourg	LUX	0.507	0.521	DEU	17	0.937	FIN	10
Netherlands	NLD	0.544	0.549	DEU	14	0.955	FIN	5
Norway	NOR	0.457	0.549	DEU	14	0.804	FIN	16
Portugal	PRT	0.580	0.590	DEU	12	0.949	FIN,DEU	7
Spain	ESP	0.505	0.534	DEU	16	0.911	FIN	13
Sweden	SWE	0.473	0.496	DEU	19	0.92	FIN	12
United Kingdom	GBR	0.69~0	0.785	DEU	6	0.868	FIN,DEU	15
United states	USA	0.522	0.564	DEU	13	0.893	FIN	14
Average		0.599	0.692			0.901		
Minimum		0.369	0.496			0.674		

Fig. 28 DEA results (Model 6) excluding Switzerland 2009–2013 Model 6—1 Input (Public Investment), 1 Output (Infrastructure PSP scores)

Country	**Opportunity Indicators**						**Musgravian Indicators**					**Total Public Sector Efficiency**	
	Administrat ion	Education	Health	Infrastructur e	PSE	Opportunity	Distribution	Stability	Economic Performanc e	PSE	Musgravian	**Equal weights**	Different Weights
Austria	1.19	0.95	0.94	1.27	1.09		0.82	1.22	1.17	1.07		**1.08**	1.08
Belgium	0.79	0.94	0.94	1.58	1.07		0.90	1.11	0.91	0.97		**1.03**	1.00
Canada	1.10	1.12	1.02	1.15	1.10		1.37	2.16	1.45	1.66		**1.34**	1.47
Denmark	0.86	0.69	0.84	1.15	0.89		0.90	0.75	0.79	0.81		**0.85**	0.84
Finland	1.06	0.94	1.18	1.03	1.05		0.95	0.64	0.84	0.81		**0.95**	0.89
France	0.85	0.93	0.86	0.96	0.90		0.80	1.10	0.75	0.88		**0.89**	0.89
Germany	1.14	1.16	0.88	1.75	1.23		0.92	1.19	1.03	1.05		**1.15**	1.11
Greece	0.62	1.18	1.18	0.64	0.91		0.86	-0.01	-0.04	0.27		**0.63**	0.48
Ireland	1.24	1.09	1.24	0.87	1.11		1.27	0.73	1.23	1.08		**1.10**	1.09
Italy	0.67	1.07	1.06	0.89	0.92		0.82	0.43	0.44	0.57		**0.77**	0.68
Japan	1.18	1.42	0.97	1.09	1.17		1.13	1.23	1.22	1.20		**1.18**	1.19
Luxembourg	1.51	1.42	1.20	0.88	1.25		0.98	1.31	2.13	1.47		**1.35**	1.40
Netherlands	0.96	1.10	0.84	0.94	0.96		1.42	1.29	1.17	1.29		**1.10**	1.18
Norway	1.08	0.80	0.97	0.80	0.91		1.19	1.64	1.76	1.53		**1.18**	1.32
Portugal	0.80	0.98	1.07	1.01	0.96		0.92	0.28	0.37	0.52		**0.77**	0.67
Spain	0.84	1.13	1.15	0.88	1.00		1.04	0.79	0.75	0.86		**0.94**	0.91
Sweden	0.90	0.81	0.94	0.83	0.87		1.10	0.92	1.10	1.04		**0.94**	0.98
United Kingdom	1.09	0.98	1.00	1.20	1.07		1.10	1.17	1.03	1.10		**1.08**	1.09
United States	1.46	0.95	0.94	0.90	1.06		1.02	1.62	1.50	1.38		**1.20**	1.27
Average	1.02	1.04	1.01	1.04	1.03		1.03	1.03	1.03	1.03		**1.03**	1.03
Maximum	1.51	1.42	1.24	1.75	1.25		1.42	2.16	2.13	1.66		**1.35**	1.47
Minimum	0.62	0.69	0.84	0.64	0.87		0.80	-0.01	-0.04	0.27		**0.63**	0.48

Fig. 29 Public Sector Efficiency (PSE) Indicators excluding Switzerland, 2009–2013

		Model 1	Model 2	Model3	Model 4	Model 5	Model 6
Inputs		Total public expenditure	Total public expenditure	Government Consumption	Education Expenditure	Health Expenditure	Public investment
Outputs		PSP	PSP Opportunity PSP Musgravian	PSP Administration	PSP Education	PSP Health	PSP infrastructure
Countries on the frontier		CAN, JPN, LUX, USA	CAN, FIN, JPN, LUX, NLD, NOR	LUX, USA	FIN, JPN, LUX, NLD	IRL, JPN, LUX, ESP	FIN, DEU
Average scores	**Input**	0.854	0.878	0.781	0.809	0.848	0.692
	output	0.841	0.938	0.842	0.931	0.990	0.901
Minimum scores	**Input**	0.715	0.712	0.609	0.586	0.684	0.496
	Output	0.46	0.736	0.515	0.854	0.972	0.674
Total countries		19	19	19	19	19	19
Efficient countries		4	6	2	4	4	2

Fig. 30 Summary results of different DEA models excluding Switzerland

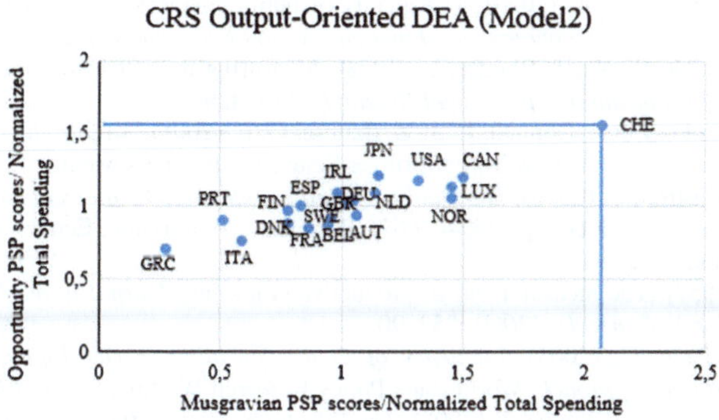

Fig. 31 Production Possibility Frontier (Model 2)

Fig. 32 Production Possibility Frontier (Model 2) excluding Switzerland

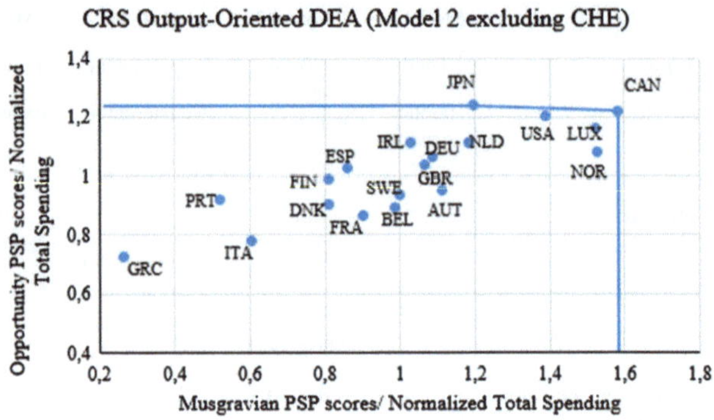

References

Afonso, A., & Fernandes, S. (2008). Assessing and explaining the relative efficiency of local government. *Journal of Socio-Economics, 37*(5), 1946–1979.

Afonso, A., & St. Aubyn, M. (2005). Non-parametric approaches to education and health efficiency in OECD countries. *Journal of Applied Economics, VIII*(2), 227–46.

Afonso, A., Schuknecht, L., & Tanzi, V. (2005). Public sector efficiency: an international comparison. *Public Choice, 123*(3–4), 321–47.

Afonso, A., Romero, A., & Monsalve, E. (2013). *Public Sector Efficiency: Evidence for Latin America*. Inter-American Development Bank, 80478, Inter-American Development Bank. Department of Economics, ISEG-UL, Working Paper no. 19/2013/DE/UECE.

Banker, R. D., Charnes, A., & Cooper, W. W. (1984). Some models for estimating technical and scale inefficiencies in data envelopment analysis. *Management Science, 30*(9), 1078–1092.

Barrios, S., & Schaechter, A. (2008). The quality of public finances and economic growth. 337 European Commission, *Economic and Financial Affairs Economic Papers*.

Charnes, A., Cooper, W. W., & Rhodes, E. (1978). Measuring the efficiency of decision making units. *European Journal of Operational Research, 2*(6), 429–44.

Coelli, T. J., Rao, D. S. P., ODonnell, C. J., & Battese, G. E. (2005). *An introduction to efficiency and productivity analysis*. New York: Springer Science & Business Media.

Deroose, S., & Kastrop, C. (2008). *The Quality of Public Finances: Findings of the Economic Policy Committee-Working Group (2004–2007)*. European Commission, Economic and Financial Affairs Economic Papers.

Farrell, M. J. (1957). The measurement of productive efficiency. *Journal of the Royal Statistical Society. Series A (General), 120*(3), 253–90.

Herrera, S., & Pang, G. (2005). *Efficiency of Public Spending in Developing Countries: An Efficiency Frontier Approach*. World Bank Policy Research Working Paper 3645.

Mandl, U., Dierx, A., & Ilzkovitz, F. (2008). *The Effectiveness and Efficiency of Public Spending*. European Commission, Economic and Financial Affairs Economic Papers.

Scheubel, B. (2015). Public sector efficiency revisited the quality of public policy during the crisis and beyond. Mimeo.

St. Aubyn, M., Pina, A., Arcia, F., & Pais, J. (2009). Study on the efficiency and effectiveness of public spending on tertiary education. *Economic Papers no. 390*.

Sutherland, D., Robert, P., Joumard, I., & Nicq, C. (2007). *Performance Indicators for Public Spending Efficiency in Primary and Secondary Education*. OECD Economics Department, Working Paper 546.

Government Debt, Fiscal Rules and Singular Growth Dynamics

Paulo Brito

Abstract A fiscal rule controlling the government surplus as a function of the deviation of the actual debt ratio from a target level is introduced in an otherwise benchmark endogenous growth model in which productive government expenditures are financed by taxes and government debt. This generates a feedback mechanism from the government debt ratio to expenditure that can generate impasse-singular dynamics, in the sense that rates of growth can become locally infinitely valued. We characterize locally the different impasse-singular dynamics that can exist and discuss their consequences for the existence and characterization of general equilibrium endogenous growth paths, for different parameterizations of the fiscal rule. We present some consequences of impasse-singular dynamics generated by particular fiscal rules, which are not present in regular models: existence of multiple over-determinate balanced growth paths (BGP), existence of constraints in the domain of existence of determinate equilibrium paths converging to a regular BGP, and the existence of singular BGP's.

1 Introduction

Within different contexts, and with several justifications, there is a widespread adoption of fiscal rules (see Kinda et al. (2013) for a survey) introducing targets on the primary surplus, on the government debt relative to GDP, or both. Rules can be explicit or implicit. A clear example of the first type are the rules agreed upon within the Treaty on Stability, Coordination and Governance in the Economic and Monetary Union, in the chapter known as the Fiscal Compact, that bind the fiscal policy of most members of the European Union. There is also empirical evidence that there are fiscal reaction functions relating the primary surplus to the level of debt, even for countries as the U.S whose fiscal policy is not bound by international treaties (see Bohn (1998, 2008) and Ghosh et al. (2013)).

P. Brito
Universidade de Lisboa, ISEG and UECE, 1249-078 Lisboa, Portugal
e-mail: pbrito@iseg.ulisboa.pt

B. Bökemeier, A. Greiner (eds.), *Inequality and Finance in Macrodynamics*, Dynamic Modeling and Econometrics in Economics and Finance 23,
DOI 10.1007/978-3-319-54690-2_3

Budget rules, both in the form of a budget balance or of any other form of fiscal target, endogenize one or more fiscal policy instruments and introduce a feedback mechanism relating government expenditures or taxes to the level of government debt.

In the macroeconomic theory literature the case which has been more studied is related to the consequences of budget balance rules over transition dynamics. Papers in this strand of literature (see Schmitt-Grohe and Uribe (1997), or a recent contribution Nishimura et al. (2015)), find that a budget balance rule can be destabilizing and a source of (permanent) indeterminacy. In all those papers the government deficit is potentially financed by taxes only. A budget rule introduces a pro-cyclical mechanism, through the effect of endogenously set taxes distorting the rate of return of capital, which can generate a self-fulfilling future change in taxes.

In that strand of models the government expenditures are not productive, there is not economic growth, and the government debt dynamics is absent. Since Arrow and Kurz (1970) it is know that public capital is a relevant source of growth. In most actual economies the government can be seen as a productive sector specializing in the provision of goods and services which have an intrinsic aggregate nature. It is, therefore, a generator of externalities, and consequently it is a major source of economic growth. In the endogenous growth literature Barro (1990) offers a benchmark model in which the government expenditures are productive (see Irmen and Kuehnel (2009) for a survey on the voluminous literature following that paper). However, government financing, by taxes or debt, has also an aggregate effect over the cost of capital that can partially or totally offset the positive effect of externalities' production. Fiscal rules introduce another transmission mechanism for fiscal policy which tends to work through the expectations of future changes in the rate of return of the economy.

Contributions dealing with productive government expenditures, economic growth, debt financing and fiscal rules can be separated into two strands, depending on the nature of the fiscal rule: Futagami et al. (2008) consider an adjustment equation where the variation of the debt over capital ratio is adjusted against a benchmark while Greiner (2008), Kamiguchi and Tamai (2012) and Greiner and Fincke (2015, pp. 169–173) consider a rule in which the primary surplus is adjusted to the actual ratio of government debt over GDP relative to a target. This literature has two shortcomings: either the rules are unrealistic (they feature simply a feedback mechanism for government debt), or the dynamic implications of the feedback mechanism introduced by the rules were not fully addressed. Indeed, in some of these papers (for instance, in Greiner and Fincke (2015)) singular dynamics, taking the form of impasse-singularities, is present.

In this paper we extend an otherwise benchmark Barro (1990) model, in which government expenditures are productive and are financed by distortionary taxes and government debt, by introducing a (symmetric) fiscal rule according to which the primary surplus is a linear function of the deviation of the debt to GDP ratio relative to a target level (as in Bohn (1998) and Greiner and Fincke (2015)). We are not concerned with the determination or the setting of an optimal target but with the existence and characterization of the impasse-singularities that different fiscal rules can generate.

A general analysis for two-dimensional dynamic general equilibrium (DGE) models in the presence of impasse-singularities is provided in Brito et al. (2017a). In that paper it is shown that the vector field in the dynamic system representing the DGE is non-Lipschitzian which implies that the existence of solutions for all $t \in [0, \infty)$, for any given initial value for the non-anticipating variables, is not guaranteed. This means that the existence of DGE paths should be explicitly checked before any attempt to characterize them. There are several symptoms of the presence of impasse-singularities: existence of infinitely-valued eigenvalues of the Jacobian of the dynamic system, multiplicity of steady-states with the same type of local dynamics (e.g. two saddle-points as in Greiner and Fincke (2015)), or the existence of changes in the determinacy properties of the DGE paths along the transition path (if it exists). In the presence of impasse-singularities the local analysis in the neighborhood of a steady-state can be misleading. In Brito et al. (2017a) it is proved that if there is a unique DGE path, and there is an impasse-singularity, only four cases can happen: black-hole singularities and no transitional dynamics, white-hole singularities and transitional dynamics confined to a subset of the state space, or worm-hole dynamics and trajectories cross the impasse surface but change their determinacy properties, from locally determinate to locally indeterminate, or from locally indeterminate to locally determinate.

In this paper we use the method presented in Brito et al. (2017a)[1] and go beyond by extending it along three dimensions: first, we deal with an endogenous growth model in which the transition dynamics can exist along a BGP and the steady states refer to the detrended variables; second, we have multiplicity of steady states; and, third, we also have multiple impasse-singularities. In Brito et al. (2017a) the steady state refers to a stationary DGE path, and there is at most one isolated steady state and one isolated impasse-singularity. The method consists in studying local dynamics not only in the neighborhood of the steady state but also in the neighborhood of the impasse-singularity. However, as we proved in that paper, those two types of local analysis although providing necessary conditions for the existence and the characterization of DGE paths are not sufficient: there are existence conditions of a global nature and a complete characterization of the dynamics, in particular the existence of temporary indeterminacy, also depends on global type of conditions.

We present a thorough local bifurcation analysis in the space of two fiscal rule parameters: the debt target and the speed of adjustment to that target. We then select some specific parameter values, and present the global phase diagrams for cases that cannot occur in regular DGE models, in which there are multiple over-determinate

[1]In both papers we deal with implicit differential equations by using the approach and terminology presented in Zhitomirskii (1993) and Llibre et al. (2002) and extending to cases in which there are steady states. See also Riaza (2008).

BGP's, or there are bounded transitional determinate DGE dynamics converging to a regular BGP path, or to singular BGP path.

The rest of the paper is organized in the following way: in Sect. 2 we present the model, in Sect. 3 we characterize analytically local dynamics both at the impasse surface and at the BGP, Sect. 4 presents the phasc diagrams for a sample of possible DGE scenarios, and Sect. 5 closes the paper.[2]

2 The Model

The representative private consumer-producer maximizes the intertemporal utility functional

$$\int_0^\infty \frac{C(t)^{1-\sigma}}{1-\sigma} e^{-\rho t} dt$$

where C is consumption subject to the budget constraint

$$\dot{K} + \dot{B} = (1 - \tau(t))(r(t)B(t) + Y(t)) - C(t), \tag{1}$$

where K and B are the stocks of physical capital and government debt, τ is the income-tax rate, r is the rate of interest on government bonds, and Y is the income from production.[3] We assume that the rate of time preference, ρ, is positive, and the intertemporal elasticity of substitution is below unity, $\sigma \geq 1$. The initial levels for the two stocks are given and are non-negative in present value terms $\lim_{t\to\infty}(K(t) + B(t))e^{-\int_0^t r(s)ds} \geq 0$.

There is only one good in this economy, which is produced by a technology using public goods provided by the government. We posit a Cobb-Douglas production function

$$Y(t) = K(t)^{\alpha} G(t)^{1-\alpha},$$

where G is the government expenditure, and $\alpha \in (0, 1)$ is the income share of private capital. A non-arbitrage condition holds such that the gross rate of interest on government bonds is instantaneously equalized to the marginal productivity of capital,

$$r(t) = \alpha K(t)^{\alpha-1} G(t)^{\alpha-1}.$$

---

[2]The proofs are assembled at the Appendix.

[3]Greiner and Fincke (2015, p. 169–173) assume that only the income generated by private sources is taxed. That is, there are no taxes over the return on the private investment on government bonds. This assumption changes significantly the types of dynamics that can occur.

From household optimality conditions we derive the Euler equation

$$\frac{\dot{C}}{C(t)} = \frac{(1-\tau(t))r(t)-\rho}{\sigma} \tag{2}$$

and the transversality condition $\lim_{t\to\infty} C(t)^{-\sigma}(K(t)+B(t))e^{-\rho t}=0$.

The government finances expenditures by taxing all types of income and issuing bonds such that the budget constraint

$$\dot{B} = r(t)B(t) - S(t) \tag{3}$$

holds, where B is the stock of outstanding government bonds (indeed debt) and the primary surplus is

$$S(t) = \tau(t)(r(t)B(t) + Y(t)) - G(t). \tag{4}$$

We consider the fiscal rule similar to the one introduced in Greiner (2008), Kamiguchi and Tamai (2012) and Greiner and Fincke (2015, pp. 169–173),

$$S(t) = \phi(B(t) - \bar{B}(t)) \tag{5}$$

where ϕ parameterizes the speed of the adjustment of the actual level of debt to a time-varying target $\bar{B}(t)$. We assume that the debt target is set such that its ratio with the GDP is constant

$$\frac{\bar{B}(t)}{Y(t)} = \bar{b}, \text{ for all } t \in [0,\infty).$$

We assume both parameters defining the policy, ϕ and $\bar{b}$, are strictly positive. A higher ϕ speeds up the adjustment of the actual level of the government debt to the target. We also assume that the tax rate is fixed parametrically and that the government implements the rule by controlling government expenditures.

Consolidating the government and the household budget constraints we get the good's market clearing equation

$$\dot{K} = Y(t) - C(t) - G(t). \tag{6}$$

In order to render them clear we formally define both dynamic general equilibrium (DGE) and equilibrium balanced growth (BGP) paths:

Definition 1 (DGE) A DGE is a trajectory $(K(t), C(t), B(t), G(t), r(t))_{t\in[0,\infty)}$ satisfying, **for all** $t \in [0,\infty)$, Eqs. (2), (6) and (3) together with the fiscal rule (5), the transversality and initial conditions, $K(0) = K_0$ and $B(0) = B_0$, where K_0 and B_0 are given.

Definition 2 (BGP) A BGP path is a particular DGE path such that the rates of growth for K, C, B and G are constant and r is constant.

In the rest of the paper we denote the ratios of government debt and government expenditures over the GDP by $b \equiv B/Y$ and $g \equiv G/Y$, respectively.

Setting $G = gY$ in the production function we can rewrite it as a linear function of the capital stock $Y = A(g)K$, where $A(g) \equiv g^{\frac{1-\alpha}{\alpha}}$. Therefore, the average productivity of capital is an increasing function of g and its curvature depends on α. By assuming $\alpha > 1/2$ we set it as concave.

This implies that the interest rate, $r = r(g) = \alpha A(g)$, is also an increasing (and concave if $\alpha > 1/2$) function of the ratio of government expenditures over the GDP. In addition, the rate of growth of consumption is also an increasing (and concave) function of the government expenditure over the GDP

$$\frac{\dot{C}}{C} = \gamma(g) \equiv \frac{(1-\tau)r(g) - \rho}{\sigma}. \tag{7}$$

2.1 Endogenous Debt Adjustment

If we eliminate the budget surplus, S, in Eqs. (4) and (5) we get a relationship between the government debt and the government expenditure, engineered by the fiscal rule, $\tau(t)(1 + r(t)b(t)) - g(t) = \phi(b(t) - \bar{b})$. As the debt ratio and the government expenditures are both endogenous, the government has one degree of freedom consisting in fixing the income tax rate exogenously. From now on we assume that it is a constant parameter: $\tau(t) = \tau$.

In order to get explicit expressions, it is convenient to express the endogeneity of the expenditure ratio regarding the debt ratio, introduced by the fiscal rule and the exogeneity of the tax rate, by the inverse relationship

$$b(t) = b(g(t)) \equiv \frac{g(t) - \bar{g}}{\tau r(g(t)) - \phi}. \tag{8}$$

In Eq. (8) we introduce two critical levels for the expenditure ratio: the level associated with zero government debt,

$$\bar{g} \equiv \tau + \phi\bar{b}, \tag{9}$$

and the level associated with an unbounded level for b,

$$\tilde{g} \equiv \left(\frac{\phi}{\alpha\tau}\right)^{\frac{\alpha}{1-\alpha}}, \tag{10}$$

which is attained when the tax collected from the interest on government bonds is equal to the rate of adjustment of the debt ratio, $\tilde{g} \equiv \{g : \tau r(g) = \phi\}$. The relationship of the expenditure ratio relative to those two critical values determines the creditor position of the government: if $\min\{\bar{g}, \tilde{g}\} < g < \max\{\bar{g}, \tilde{g}\}$ then, because $b(g) < 0$, the government is a creditor; if $g < \min\{\bar{g}, \tilde{g}\}$ or $g > \max\{\bar{g}, \tilde{g}\}$ then it is a debtor because $b(g) > 0$; and if $g = \bar{g}$ it is neither a debtor nor a creditor.

Two countervailing forces are at work: a high ratio of government expenditure, above the level consistent with the debt target $\bar{g}$, increases the level of debt, but on the other hand, as government expenditures are productive, they tend to increase tax returns on government bonds possibly above the speed of the adjustment set by the fiscal rule. The first factor is measured by the wedge $g - \bar{g}$ and the second factor is measured by the wedge $r(g) - r(\tilde{g})$, because $\tau r(g) - \phi = \tau(r(g) - r(\tilde{g}))$. Consequently, if $\bar{g} > \tilde{g}$ ($\bar{g} < \tilde{g}$) then function $b(g)$ is increasing (decreasing) with g. It is also increasing if $\bar{g} = \tilde{g}$ and $\alpha > 1/2$.

The next Lemma characterizes the debtor position of the government depending on the value of the two parameters defining the fiscal rule ϕ and $\bar{b}$[4]:

Lemma 1 (Government Debtor Position) *Let Eq.* (8) *represent the government debtor position as a function of g, b(g). Then the following cases are possible :*

1. *if* $\phi \geq \max\{\alpha\tau, (1-\tau)/\bar{b}\}$ *then* $\min\{\bar{g}, \tilde{g}\} \geq 1$ *then* $b(g) > 0$ *for all* $g \in (0, 1)$.
2. *if* $(1-\tau)/\bar{b} \leq \phi < \alpha\tau$ *then* $\tilde{g} < 1 \leq \bar{g}$ *and* $b(g) > 0$ *for* $0 < g < \tilde{g}$ *and* $b(g) < 0$ *for* $\tilde{g} < g < 1$;
3. *if* $\alpha\tau < \phi \leq (1-\tau)/\bar{b}$ *then* $\bar{g} \leq 1 < \tilde{g}$ *which implies that* $b(g) > 0$ *for* $0 < g < \bar{g}$ *and* $b(g) < 0$ *for* $\bar{g} < g < 1$;
4. *if* $\phi < \min\{\alpha\tau, (1-\tau)/\bar{b}\}$ *then* $\max\{\bar{g}, \tilde{g}\} < 1$ *then* $b(g) < 0$ *if* $\min\{\bar{g}, \tilde{g}\} < g < \max\{\bar{g}, \tilde{g}\}$ *and* $b(g) > 0$ *otherwise.*

Figure 1 illustrates Lemma 1, by showing the values of the policy parameters associated with the four cases.

In all four cases, if g is close to zero the government will be a debtor. When g is high the debtor position depends on the parameters: if both parameters are high the government should be a debtor (case $L1(1)$), but when they are both low it should be a creditor (case $L1(4)$). In the other two cases, for a high debt target and a low speed of attainment, $L1(2)$, or for a low debt target and a high speed, $L1(3)$, there is an interval, $g \in (\min\{\bar{g}, \tilde{g}\}, \min\{\max\{\bar{g}, \tilde{g}\}, 1\})$, such that the government should be a creditor and a debtor outside that interval.

[4]All the proofs are in the Appendix.

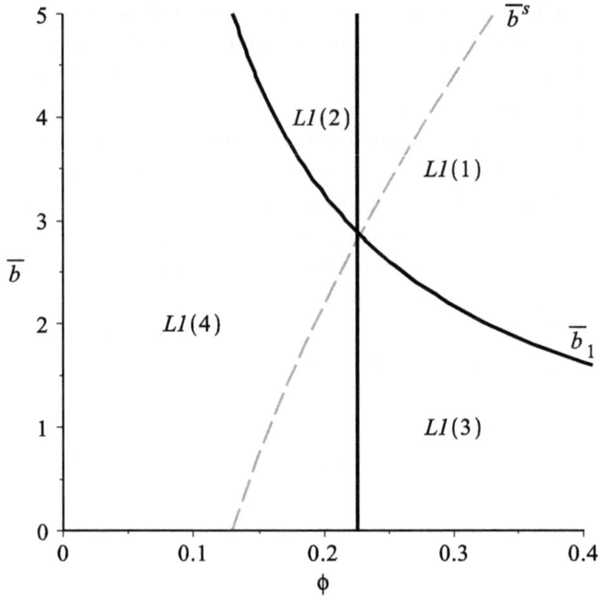

Fig. 1 Graphical illustration of Lemma 1 in the space $(\phi, \bar{b})$ for $\alpha = 0.65$ and $\tau = 0.35$. We set $\bar{b}^1 \equiv \dfrac{1-\tau}{\phi}$

2.2 *The DGE in Detrended Variables*

The rate of growth of the debt ratio can be obtained from two different sources. From the definition of b we get

$$\frac{\dot{b}}{b} = \frac{\dot{B}}{B} - \frac{\dot{Y}}{Y} = \gamma_B(g) - \gamma_K(g) - \left(\frac{1-\alpha}{\alpha}\right)\frac{\dot{g}}{g} \tag{11}$$

where, from Eq. (3), the rate of growth of the government debt is

$$\frac{\dot{B}}{B} = \gamma_B(g) = (1-\tau)r(g) + \frac{g-\tau}{b(g)}. \tag{12}$$

However, we can also time-differentiate the feedback mechanism introduced by the fiscal rule in Eq. (8) to get

$$\frac{\dot{b}}{b} = \frac{(\tau r(g) - \phi - (g-\bar{g})\tau r'(g))\, g}{(\tau r(g) - \phi)(g-\bar{g})}\left(\frac{\dot{g}}{g}\right). \tag{13}$$

Although we can eliminate either the rate of growth of the debt ratio or the expenditure ratio we only get an explicit differential equation if we eliminate the debt ratio, obtaining

$$\left(\frac{\alpha g-\phi(1-\alpha)b(g)}{\alpha(g-\bar{g})}\right)\frac{\dot{g}}{g}=\frac{g-\tau}{b(g)}+(\alpha(1-\tau)+g-1)A(g)+z,$$

where $z \equiv C/K$ denotes consumption detrended by the capital stock. From this equation we get the rate of growth of the government expenditure ratio

$$\frac{\dot{g}}{g}\equiv\gamma_g(g,z)=\frac{\psi(g,z)}{\delta(g)}.$$

The numerator contains three wedges,

$$\psi(g,z)\equiv\alpha\left(\tau r(g)-\phi\right)\left(g-\bar{g}\right)\left(z-\zeta(g)\right), \tag{14}$$

the speed wedge, $\tau r(g)-\phi$, the level wedge, $g-\bar{g}$, and a consumption wedge, $z-\zeta(g)$, where

$$\zeta(g)\equiv(1-g-\alpha(1-\tau))A(g)+(\tau-g)\frac{\tau r(g)-\phi}{g-\bar{g}}, \tag{15}$$

measures the potential for financing government expenditures out of consumption. The denominator is also a function of the level and speed wedges:

$$\delta(g)\equiv\alpha g(\tau r(g)-\phi)-\phi(1-\alpha)(g-\bar{g}). \tag{16}$$

The rate of growth for private capital is a function of g and z

$$\frac{\dot{K}}{K}=\gamma_K(g,z)\equiv(1-g)A(g)-z. \tag{17}$$

Using Eqs. (7) and (17) we find the rate of growth of the consumption-capital ratio

$$\frac{\dot{z}}{z}=\gamma_z(g,z)=\gamma(g)-\gamma_K(g,z)=z-z(g), \tag{18}$$

is equal to the wedge between consumption and potential consumption allowed for the level of the government expenditures,

$$z(g)\equiv 1-g-\gamma(g)=\frac{(\sigma(1-g)-\alpha(1-\tau))A(g)+\rho}{\sigma}. \tag{19}$$

Function $z(g)$ is usually interpreted as a Laffer-curve associating government expenditures to the rate of growth of the economy. It has an inverted-U shape, starting from $z(0) = \rho/\sigma > 0$, reaching a maximum at a point where $g = (1-\alpha)(\sigma - \alpha(1-\tau))/\sigma > 0$ and decreasing for higher values of g (possibly becoming negative for large g). There are two effects involved: first, productive government expenditures increase capital productivity thus having a positive effect on output and on the rate of capital accumulation, but second, they are financed by capital income taxation via the interest rate, which is also increasing in g. For low values of g the first effect dominates and for high values of g the second effect dominates.

The DGE representation in detrended variables is the path $(g(t), z(t))_{t\in[0,\infty)}$, such that every $(g(t), z(t)) \in \Omega$, where the domain of the detrended variables is $\Omega = (0,1) \times \mathbb{R}_{++}$, are solutions of the system

$$\dot{g} = g\,\frac{\psi(g,z)}{\delta(g)} \tag{20}$$

$$\dot{z} = z\,(z - z(g)) \tag{21}$$

together with the initial conditions and the transversality condition. We will only consider DGE paths which are equilibrium BGP paths or converge asymptotically to an equilibrium BGP path.

An equilibrium BGP is a DGE path satisfying $(C^*(t), K^*(t), B^*(t))_{t\in[0,\infty)}$ such that $C^*(t) = c^* e^{\gamma(g^*)t}$, $K^*(t) = k^* e^{\gamma_K(g^*,z^*)t}$ and $B^*(t) = b^* e^{\gamma_B(g^*,z^*)t}$ where (g^*, z^*) are steady states of the system (20)–(21). A sufficient condition for the verification of the transversality condition is that we should have $\rho + \sigma\gamma(g^*) - \gamma_K(g^*, z^*) > 0$ and $\rho + \sigma\gamma(g^*) - \gamma_B(g^*, z^*) > 0$.

The structure of this model is similar to the one found in Kamiguchi and Tamai (2012) and Greiner and Fincke (2015, pp. 169–173). In particular both papers include a function similar to $\delta(g)$ that could take a zero value. If this is the case, function $\gamma_g(g, z)$ is not Lipschitzian. This implies that the existence and uniqueness of solutions for $t \in [0,\infty)$ of system (20)–(21) is not guaranteed, and therefore a DGE path may not exist.

We call *impasse point* to $(g^s, z^s) \in \Omega$ such that $\delta(g^s) = 0$.[5] A point $(g, z) \in \Omega$ such that $\delta(g) \neq 0$ will be called a *regular point*.

The set of impasse points, or *impasse-set*, is the one-dimensional manifold over Ω,

$$\mathcal{S} = \{(g,z) \in \Omega : \delta(g) = 0\}. \tag{22}$$

If $\mathcal{S}$ is non-empty (that is, if impasse points exist) they can be reached in finite or in infinite time. At any impasse point the rate of growth γ_g becomes locally infinitely valued (and the Jacobian of the system (20)–(21) has locally infinitely-

[5]As function $\delta(g)$ can only be equal to zero at a specific value for the state variables and not for a particular value of the parameters we have impasse-singularities and not fast-slow singularities (see Brito et al. (2017a)).

valued eigenvalues). We call *singular steady state* to a steady state lying at the impasse surface. Observe that this case is non-generic, as it only occurs for specific parameter values. In the generic case in which the impasse point is not a steady state, it is reached at a particular finite time, following a sequence of regular points, but the solution may not be continued after it is reached. If there is no continuation any trajectory passing through such an impasse point cannot be a DGE path, and there is an existence problem which does not occur in regular DGE models. If there is continuation the determinacy properties of a crossing DGE path change in ways that also do not occur in regular models.

It is shown in Brito et al. (2017a) that there are two local necessary conditions for the existence of singular DGE paths: crossing trajectories and at least one steady state should both exist. There are also global conditions that should be met. We address the local conditions in the next section and the global conditions in Sect. 4. In addition to singular DGE paths, regular DGE paths that do not cross the singular surface $\mathcal{S}$, can also exist. However, they are bounded to a subset of space Ω and cannot have points in the whole space as in DGE models without singularities.

3 Local Analysis

The first subsection characterizes the impasse surface and the second the set of steady states of system (20)–(21).

3.1 Impasse-Surface

If $\mathcal{S}$ is non-empty it partitions the domain of (g, z), Ω, into three subsets $\mathcal{S}$, Ω_+, containing points where $\delta(g) > 0$, and Ω_-, where $\delta(g) < 0$.

In order to check for the existence and location of impasse points we need to define two critical values for $\bar{b}$

$$\bar{b}^s = \frac{1}{\phi}(\tilde{g} - \tau) = \frac{1}{\phi}\left(\left(\frac{\phi}{\alpha\tau}\right)^{\frac{\alpha}{1-\alpha}} - \tau\right) \tag{23}$$

$$\bar{b}_1^s = \frac{\phi(1 - \tau + \alpha\tau) - \alpha^2\tau}{(1-\alpha)\phi^2}, \tag{24}$$

which are different except at point $(\phi, \bar{b}) = (\alpha\tau, (1-\tau)/(\alpha\tau))$.

Lemma 2 (Singular Points and Partition of Ω) *Consider function $\delta(g)$ in Eq. (16). It is a convex function such that*

1. *if $\bar{b} > \bar{b}^s$ or if $\phi \geq \alpha\tau$ and $\bar{b}_1^s < \bar{b} < \bar{b}^s$ then the set $\mathcal{S}$ is empty and $\Omega = \Omega_+$;*

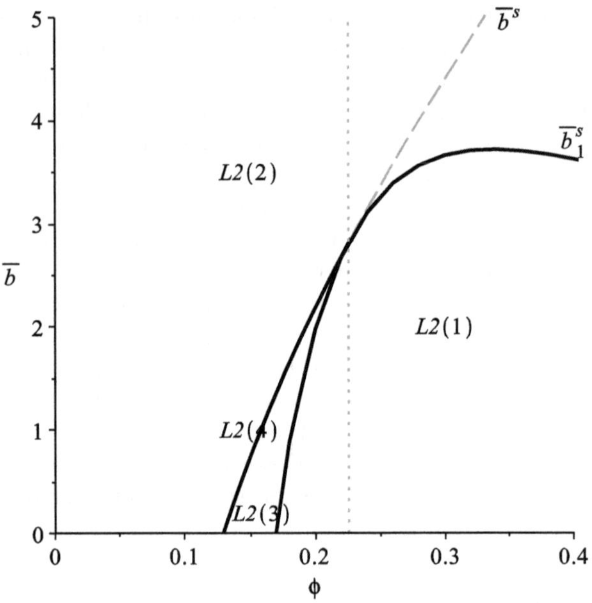

Fig. 2 Graphical illustration of Lemma 2 in the space $(\phi, \bar{b})$ for $\alpha = 0.65$ and $\tau = 0.35$

2. *if* $0 < \bar{b} < \bar{b}^s_1$ *then there is one unique singular point* $g^s \in (0, \min\{\tilde{g}, 1\})$ *and the set* Ω *is partitioned into two regular subsets* $\Omega_+ = \{(g,z) : 0 < g < g^s\}$ *and* $\Omega_- = \{(g,z) : g^s < g < 1\}$*, and the impasse subset* $\mathcal{S} = \{(g,z) : g = g^s\}$*;*
3. *if* $\phi < \alpha\tau$ *and* $\bar{b}^s_1 < \bar{b} < \bar{b}^s$ *then there are two singular points* g^s_1 *and* g^s_2*, such that* $0 < g^s_1 < \bar{g} < \tilde{g} < g^s_2 < 1$ *and the set* Ω *is partitioned into three regular subsets* $\Omega_{+,1} = \{(g,z) : 0 < g < g^s_1\}$*,* $\Omega_{+,2} = \{(g,z) : g^s_2 < g < 1\}$ *and* $\Omega_- = \{(g,z) : g^s_1 < g < g^s_2\}$*, and two impasse subsets* $\mathcal{S}_1 = \{(g,z) : g = g^s_1\}$ *and* $\mathcal{S}_2 = \{(g,z) : g = g^s_2\}$*;*
4. *if* $\phi < \alpha\tau$ *and* $\bar{b} = \bar{b}^s$ *then there is one unique singular point such that* $g^s = \bar{g} = \tilde{g} < 1$ *and the set* Ω *is partitioned into the regular set* $\Omega_+ = \{(g,z) : g \neq g^s\}$ *and the impasse set* $\mathcal{S} = \{(g,z) : g = g^s\}$*. This is a non-generic case.*

Figure 2 illustrates Lemma 2, by showing the values of the policy parameters associated with the four cases, labelled $L2(1)$, $L2(2)$, $L2(3)$ and $L2(4)$ (which corresponds to curve $\bar{b} = \bar{b}^s$ for $\phi < \alpha\tau$).

A necessary condition for the existence of singular points is $\bar{g} \leq \tilde{g}$ which is equivalent to $\bar{b} \leq \bar{b}^s$. In Fig. 2 this corresponds to the points below curve $\bar{b} = \bar{b}^s$. This means that in that figure the points above curve $\bar{b} = \bar{b}^s$ represent the combinations of fiscal rule parameters yielding $\bar{g} > \tilde{g}$.

From Lemma 2 we conclude that, if the debt target is set at relatively low (high) levels or the speed of convergence is high (low), there will (not) exist impasse points and the associated singular dynamics will be absent. Impasse points exist only in the generic cases $L2(1)$ and $L2(2)$ and in the non-generic case $L2(4)$. Comparing with Lemma 1, observe that the government can be a creditor only if there is an impasse surface and the variables (g, z) lie within set Ω_-. If the impasse set is empty (as in the $L2(1)$ case) the government is always a debtor.

From Lemmas 1 and 2 we can also extract some intuition for the speed of the adjustment of the government debt position for different values of g. The elasticity of the creditor position is

$$\beta(g) \equiv -\frac{b'(g)g}{b(g)} = \frac{1-\alpha}{\alpha} - \frac{\delta(g)}{(\tau r(g) - \phi)(g - \bar{g})},$$

where the second term has the same sign as function $b(g)$. We can easily infer the following: first, at an impasse point the elasticity $\beta(.)$ is equal to the share of public expenditures versus private capital in output, $(1-\alpha)/\alpha$; second inside subsets Ω_+, where the government is always a debtor, the elasticity is lower than $(1-\alpha)/\alpha$; third inside subset Ω_- the elasticity is higher (lower) if the government is a debtor (creditor). This means that trajectories that cross $\mathcal{S}$, if they exist, witness a large change in the reaction of the creditor position after a relatively small change in g, and the rule can introduce non-monotonous reactions in the government budget.

We now characterize the local dynamics for points within $\mathcal{S}$ when it is non-empty (that is when $\bar{g} \leq \tilde{g}$). There are four types of impasse points. First, if $\delta'(g^s)\psi(g^s, z^s) < 0$ point $(g^s, z^s) \in \mathcal{S}$ is an attractor point and we call it black-hole because this points attracts neighbouring trajectories which reach $\mathcal{S}$ in finite time and does not allow them to be continued afterwards. This means that trajectories hitting a black-hole cannot be DGE trajectories. Second, if $\delta'(g^s)\psi(g^s, z^s) > 0$ point $(g^s, z^s) \in \mathcal{S}$ is an attractor point and we call it white-hole because this points repels neighbouring trajectories allowing them to be continued in time. Although they can be DGE trajectories they cannot cross $\mathcal{S}$, which means that they remain confined to subset Ω_- or to subset Ω_+. Those points would not belong to DGE paths if crossing $\mathcal{S}$ is necessary to reach a steady state. Third, if $\delta'(g^s)\psi(g^s, z^s) = 0$, under the conditions $\psi(g^s, z^s) = 0$ and $\delta'(g^s) \neq 0$, we have one impasse-transversal point. An impasse transversal point can be seen as a fixed point of the de-singularized dynamic system

$$\dot{g} = g\,\psi(g, z) \tag{25}$$

$$\dot{z} = z\,\delta(g)\,(z - z(g)) \tag{26}$$

where the vector field has the same integral curves as system (20)–(21), but we introduce a time-reversal for the integral curves within set Ω_-, and remove the singularity.[6] From now on we denote the impasse-transversal points by (g^i, z^i) and

[6] Using the analogy with fast-slow singularities, we can see desingularization as a device to uncover the singularities by locally changing the time scale such that the speed of the trajectories close to the singular surface becomes finite.

denote the *set of impasse-transversal points* by

$$\Gamma_I = \{(g, z) \in \mathcal{S} : z = \zeta(g)\}$$

where $\zeta(g)$ is defined in Eq. (15). Every impasse-transversal point, $(g^i, z^i) = (g^s, \zeta(g^s))$, can be classified further depending on their local dynamic properties, which can be determined from the eigenvalues of the Jacobian of the de-singularized system evaluated at an impasse-transversal point

$$J^i = J(g^i, z^i) = \begin{pmatrix} g^i\,\psi_g(g^i, z^i) & g^i\,\psi_z(g^i, z^i) \\ z^i\,\delta'(g^i)\,(z^i - z(g^i)) & 0 \end{pmatrix}.$$

There are three types of impasse-transversal points: First, point (g^i, z^i) is a *transversal saddle* if J^i has two real eigenvalues, one negative and one positive. Through this point there are two trajectories crossing S, one coming from Ω_+ with a continuation in Ω_-, and another coming from Ω_- with a continuation in Ω_+. Second, point (g^i, z^i) is a *transversal node* if J^i has two real eigenvalues with the same sign. Transversal nodes have a basin of attraction belonging to set Ω_+, if the eigenvalues are negative, or to set Ω_-, if the eigenvalues are positive. There are infinitely many trajectories coming from that basin of attraction and having a continuation in the opposite subset. Third, point (g^i, z^i) is a transversal *focus* if J^i has two complex-conjugate eigenvalues. In this case there are no crossing trajectories. We say we have worm-hole dynamics, meaning that there are crossing trajectories, in the first two cases and these are the only cases in which there could exist crossing DGE paths.[7]

In Lemma 2 we proved that one or two impasse points can exist, i.e $\mathcal{S}$ is non-empty. Impasse-transversal points further introduce a partition over set $\mathcal{S}$ into a subset of attractor points, $\mathcal{S}^-$, and a subset of repeller points, $\mathcal{S}^+$, which determine the existence and the type of DGE paths in their neighborhood. In Lemma 3 we determine the existence and number of impasse-transversal points and the partitions they introduce over $\mathcal{S}$, and in Lemma 4 we characterize the types of impasse-transversal points (i.e, we partition further Γ_I).

Lemma 3 (Partition of the Impasse Set) *Let*

$$\bar{b}^i = \{\bar{b} : \delta(g, \bar{b}) = 0,\ \zeta(g, \bar{b}) = 0\}.$$

and assume $\bar{b} < \bar{b}_s$. Five cases are possible:

1. *If $\phi > \alpha\tau$ and $\bar{b}^i \le \bar{b} < \bar{b}_1^s$ then set $\mathcal{S}$ has only one component, $\mathcal{S}_1$, and contains only attractor points. Then $\mathcal{S} = \mathcal{S}_1^-$;*

[7] See Brito et al. (2017a) for further details. There is another type of impasse point, impasse-tangent points at which a tangent trajectory passes through $\mathcal{S}$. In the present model there are no impasse-tangent points.

2. *If* $0 < \bar{b} < \min\{\bar{b}^i, \bar{b}_1^s\}$ *then set* $\mathcal{S}$ *has only one component,* $\mathcal{S}_1$*, containing one impasse-transversal point,* $(g^i, z^i) = (g_1^s, \zeta(g_1^s))$*, dividing* $\mathcal{S}_1$*, into a subset of attractor points, if* $z > z^i$*, and a subset of repeller points, if* $z < z^i$*. Then* $\mathcal{S} = \mathcal{S}_1^- \cup \Gamma_I \cup \mathcal{S}_1^+$*;*
3. *If* $\phi < \alpha\tau$ *and* $\max\{\bar{b}_1^s, \bar{b}^i\} \leq \bar{b} \leq \bar{b}^s$ *then set* $\mathcal{S}$ *has two components,* $\mathcal{S}_1$*, containing only attractor points, and* $\mathcal{S}_2$*, containing only repeller points. Then* $\mathcal{S} = \mathcal{S}_1^- \cup \mathcal{S}_2^+$*;*
4. *If* $\phi < \alpha\tau$ *and* $\max\{0, \bar{b}_1^s\} \leq \bar{b} \leq \bar{b}^i$ *then set* $\mathcal{S}$ *has two components,* $\mathcal{S}_1$ *and* $\mathcal{S}_2$*. Set* $\mathcal{S}_1$*, contains one impasse-transversal point,* $(g^i, z^i) = (g_1^s, \zeta(g_1^s))$ *dividing* $\mathcal{S}_1$ *into a subset of repeller points, if* $z < z^i$*, and attractor points, if* $z > z^i$*, and set* $\mathcal{S}_2$ *only contains repeller points. Then* $\mathcal{S} = \mathcal{S}_1^- \cup \Gamma_{I,1} \cup \mathcal{S}_1^+ \cup \mathcal{S}_2^+$ *;*
5. *If* $\phi < \alpha\tau$ *and* $\max\{0, \bar{b}^i\} < \bar{b} < \bar{b}^s$ *then set* $\mathcal{S}$ *has two components,* $\mathcal{S}_1$ *and* $\mathcal{S}_2$ *and every subset contains one impasse-transversal point. Set* $\mathcal{S}_1$*, contains one impasse-transversal point,* $(g_1^i, z_1^i) = (g_1^s, \zeta(g_1^s))$ *dividing* $\mathcal{S}_1$ *into a subset of repeller points, if* $z < z_1^i$*, and a set of attractor points, if* $z > z_1^i$*, and set* $\mathcal{S}_2$ *also contains one impasse-transversal point,* $(g_2^i, z_2^i) = (g_2^s, \zeta(g_2^s))$*, dividing it into a subset of repeller points, if* $z > z_2^i$*, and attractor points, if* $z < z_2^i$*. Then* $\mathcal{S} = \mathcal{S}_1^- \cup \Gamma_{I,1} \cup \mathcal{S}_1^+ \cup \mathcal{S}_2^- \cup \Gamma_{I,2} \cup \mathcal{S}_2^+$*.*

Figure 3 illustrates Lemma 3, by depicting the values of the policy parameters associated with the five cases. There are no impasse-transversal points in $L3(1)$, in which there is a single impasse-surface, and $L3(3)$, in which there are two impasse-surfaces. There is one impasse transversal point in cases $L3(2)$ and $L3(4)$, and there

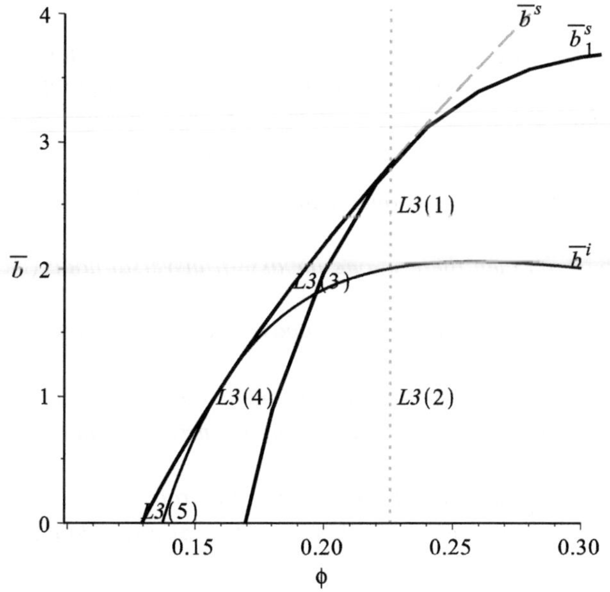

Fig. 3 Graphical illustration of Lemma 3 in the space $(\phi, \bar{b})$ for $\alpha = 0.65$ and $\tau = 0.35$

are two impasse-transversal points in case $L3(5)$. The last case occurs for relatively low values of the two parameters: low target value for the debt ratio and low speed of adjustment.

When there is a single impasse-transversal point it will belong to set $\mathcal{S}_1$ which is associated to relatively low levels of g and to a debtor position. When there are two impasse-transversal points they belong to two different branches of set $\mathcal{S}$, one for low levels of g and another for high values of g and both correspond to debtor positions.

This model produces different configurations as compared to Brito et al. (2017a), in which there is a single impasse surface and at most one impasse-transversal point, and to Brito et al. (2017b), in which there can be two impasse-transversal points but within a single impasse surface.

Lemma 4 (Characterization of Impasse-Transversal Points) *Let*

$$\bar{b}^i_d = \{\bar{b} : \ \delta(g^i_d(\bar{b}), \bar{b}) = 0\}$$

where

$$g^i_d \equiv \frac{\phi(1-\alpha)\left((1-\tau)(\sigma-1)\bar{g}+\sigma\tau^2\right)}{\rho+\alpha\tau+\phi\left(\sigma(1-\alpha\tau)-(1-\tau)\right)}, \tag{27}$$

and let

$$\bar{b}^i_\Delta = \left\{\bar{b} : \ \Delta J^i(\bar{b}) = 0\right\}.$$

where $\Delta J^i(\bar{b}) = \Delta J\left(g^i(\bar{b}), \zeta(g^i(\bar{b}))\right)$ is the discriminant of the Jacobian of the desingularized system evaluated at an impasse-transversal point. For reasonable parameter values there are two points $\bar{b}^i_{\Delta,1} < \bar{b}^i_{\Delta,2}$. Assume that $\sigma \geq 1$ and $\bar{b} < \bar{b}^s < \tau(\sigma\phi-\rho)/(\rho\phi)$. Then eight cases are possible:

1. *if $\bar{b}^i_b < \bar{b} < \bar{b}^i$ then there is one impasse-transversal saddle-point (g^i, z^i) such that $g^i < \bar{g}$;*
2. *if $\bar{b}^i_\Delta < \bar{b} < \bar{b}^i_d$ then there is one impasse-transversal node (g^i, z^i) such that $g^i < \bar{g}$;*
3. *if $0 < \bar{b} < \bar{b}^i_{\Delta,1}$ then there is one impasse-transversal focus (g^i, z^i) such that $g^i < \bar{g}$;*
4. *if $\bar{b}^i_b < \bar{b}^i < \bar{b} < \bar{b}^i_{\Delta,2}$ then there are two impasse-transversal points, (g^i_1, z^i_1) and (g^i_2, z^i_2), such that $g^i_1 < \bar{g} < \tilde{g} < g^i_2$, where the first is a saddle and the second a node;*
5. *if $\bar{b}^i_{\Delta,2} < \bar{b} < \bar{b}^s$ then there are two impasse-transversal points, (g^i_1, z^i_1) and (g^i_2, z^i_2), such that $g^i_1 < \bar{g} < \tilde{g} < g^i_2$, where the first is a saddle and the second a focus;*

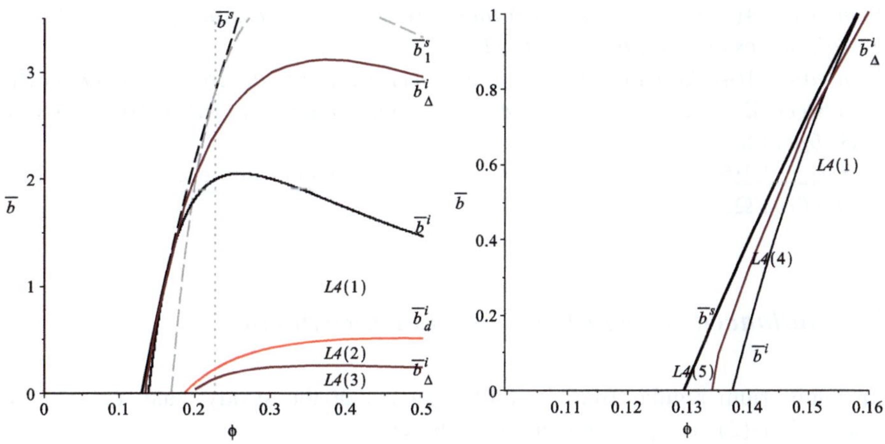

Fig. 4 Graphical illustration of Lemma 4 in the space $(\phi, \bar{b})$ for $\alpha = 0.65$, $\tau = 0.35$, $\sigma = 2$ and $\rho = 0.02$. The RHS panel zooms a small area in the LHS panel to the left of area $L4(1)$

6. *if* $\bar{b}^i < \bar{b}^i_{\Delta,1} < \bar{b} < \min\{\bar{b}^i_d, \bar{b}^i_{\Delta,2}\}$ *then there are two impasse-transversal points,* (g^i_1, z^i_1) *and* (g^i_2, z^i_2)*, such that* $g^i_1 < \bar{g} < \tilde{g} < g^i_2$*, where the two are transversal nodes;*
7. *if* $\bar{b}^i < \bar{b}^i_{\Delta,1} < \bar{b}^i_{\Delta,2} < \bar{b} < \bar{b}^i_d$ *then there are two impasse-transversal points,* (g^i_1, z^i_1) *and* (g^i_2, z^i_2)*, such that* $g^i_1 < \bar{g} < \tilde{g} < g^i_2$*, where the first is a node and the second is a focus;*
8. *if* $\bar{b}^i < \bar{b} < \bar{b}^i_{\Delta,1} < \bar{b}^i_{\Delta,2} < \bar{b}^i_d$ *there are two impasse-transversal points,* (g^i_1, z^i_1) *and* (g^i_2, z^i_2)*, such that* $g^i_1 < \bar{g} < \tilde{g} < g^i_2$*, where the two are transversal foci.*

Figure 4 illustrates Lemma 4, by depicting the values of the policy parameters associated with the five cases $L4(1)$ to $L4(5)$. Cases $L4(6)$, $L4(7)$ and $L4(8)$ only occur if curves $\bar{b} = \bar{b}^i_d$ and $\bar{b} = \bar{b}^i_\Delta$ cross curve $\bar{b} = \bar{b}^i$, which only happen for $\sigma \approx 1$. This also means that there are values for the parameters such that a *singularity induced bifurcation* point can exist, such that $b = b^s = \bar{b}^i_d$, implying that $g^*_s = \bar{g} = \tilde{g}$, where g^*_s is an impasse-transversal steady state.

While the existence and the partition of the impasse surface is independent of the elasticity of intertemporal substitution and the rate of time preference, these two parameters determine the type of impasse-transversal points and their local dynamic properties. That is, they determine the existence of crossing trajectories.

Referring to the cases in Lemma 4 and Fig. 4 we can have the following types of crossing trajectories:

Proposition 1 (Crossing Trajectories) *Crossing trajectories exist in the following cases:*

1. *in case* $L4(1)$ *there are two trajectories crossing the unique impasse surface one from* Ω_+ *towards* Ω_- *and another in the opposite direction;*

2. *in case L4(2) and in case L4(7) there are, in most cases, an infinite number of trajectories crossing from* Ω_- *to* Ω_+*;*
3. *in case L4(4) there are two crossing trajectories, in both directions over set* $\mathcal{S}_1$*, between* Ω_- *and* $\Omega_{+,1}$*, and there are an infinite number of trajectories crossing* $\mathcal{S}_2$ *from* Ω_- *to* $\Omega_{+,2}$*;*
4. *in case L4(5) there are trajectories in infinite number going from* Ω_- *to* $\Omega_{+,1}$ *and from* Ω_- *to* $\Omega_{+,2}$.

3.2 Balanced Growth Paths and Local Dynamics

An equilibrium regular BGP is a BGP such that the steady state, (g^*, z^*), of system (20)–(21) is a point belonging to the set

$$\Gamma_E = \{(g, z) \in (0, 1) \times (0, \infty) : \ \psi(g, z) = 0, \ \gamma_z(g, z) = 0, \ \delta(g) \neq 0\}$$

and verifying the transversality condition. An equilibrium singular BGP is a BGP such that the steady state (g_s^*, z_s^*) is a point belonging to the set

$$P_E = \left\{(g, z, \bar{b}) \in (0, 1) \times (0, \infty)^2 : \ \psi(g, z) = 0, \ \gamma_z(g, z) = 0, \ \delta(g) = 0\right\}$$

and verifying the transversality condition.[8]

We introduce three new critical values for the debt ratio target parameter, $\bar{b}$:

$$\bar{b}_1^* \equiv \frac{\tau(\sigma\phi - \rho)}{\rho\phi}, \tag{28}$$

$$\bar{b}_z^* \equiv \left\{\bar{b} : z(g, \bar{b}) = 0, \ \tau r(g) = \varphi(g, \bar{b})\right\} \tag{29}$$

and

$$\bar{b}_s^* \equiv \left\{\bar{b} : \ \delta(g_s^*(\bar{b}), \bar{b}) = 0\right\}, \tag{30}$$

where $g_s^* = g_d^i$ (see Eq. (27)).

The next result characterises the equilibrium BGP's as regards their number, their location in the partition introduced by the impasse surface(s) and the asset position of the government.

Lemma 5 (Number and Location of BGPs) *Let* $\bar{b}_s$ $\bar{b}_1^*$, $\bar{b}_z$ *and* $\bar{b}_s^*$ *be given by Eqs.* (23), (28), (29) *and* (30), *respectively, and assume that* $\rho < \alpha(1 - \tau)$. *Then*

[8]Observe that given the constraint introduced by the singularity condition, the set of singular steady states has a higher co-dimension in the joint space of the variables and the parameters.

the following cases are possible, concerning the number and the location of steady states (g^*, z^*)*:*

1. *if* $\bar{b} < \bar{b}_1^*$ *then there is one isolated regular BGP such that* $g^* < g_1^s$ *is in* Ω_+ *and the government is a creditor;*
2. *if* $\bar{b} > \max\{\bar{b}_z, \bar{b}_s^*\}$ *then there is one isolated regular BGP such that* $g_1^s < g^* < g_1^s$ *is in* Ω_+ *and the government is a debtor;*
3. *if* $\bar{b}_z < \bar{b} < \bar{b}_s^*$ *then there is one isolated regular BGP such that* $\bar{g} < g^* < \tilde{g}$ *is in set* Ω_- *and the government is a debtor;*
4. *if* $\bar{b} = \bar{b}_s^* \geq \bar{b}_z$ *then there is one isolated singular BGP such that* $g^* = g_s^*$ *is in* $\mathcal{S}$ *which corresponds to a debtor position;*
5. *if* $\bar{b}^s < \bar{b} < \max\{\bar{b}_1^*, \bar{b}_z^*\}$ *then there are two regular BGP's such that* $g_1^* < g_2^* < g_1^s$ *are both in* Ω_+ *and* g_1^* *corresponds to a debtor and* g_2^* *to a creditor position;*
6. *if* $\bar{b}_s^* < \bar{b} < \max\{\bar{b}^s, \bar{b}_z^*\}$ *then there are two regular BGP's such that* $g_1^* < g_1^s < g_2^*$*,* g_1^* *is in* Ω_+ *and corresponds to a debtor position , while* g_2^* *is in* Ω_- *and corresponds to a creditor position;*
7. *if* $0 < \bar{b} < \min\{\bar{b}_s^*, \bar{b}_z^*\}$ *then there are two regular BGP's such that* $g_1^s < g_1^* < g_2^*$*, where both are in* Ω_- *and* g_1^* *corresponds to a debtor position and* g_2^* *to a creditor position;*
8. *if* $\bar{b} = \bar{b}_s^* < \bar{b}_z^*$ *then there is one singular BGP, such that* $g_1^* = g_s^*$ *is in* $\mathcal{S}$ *and a regular BGP such that* $g_s^* < g_2^*$ *is in* Ω_- *and corresponds to a debtor position.*

Figure 5 presents a graphical illustration of Lemma 5 for different values of $\bar{b}$ and ϕ, and given values of the other parameters, including σ and ρ. A unique BGP exists for cases $L5(1)$ to $L5(4)$ and two BGP's can exist in the other cases. In cases $L5(4)$

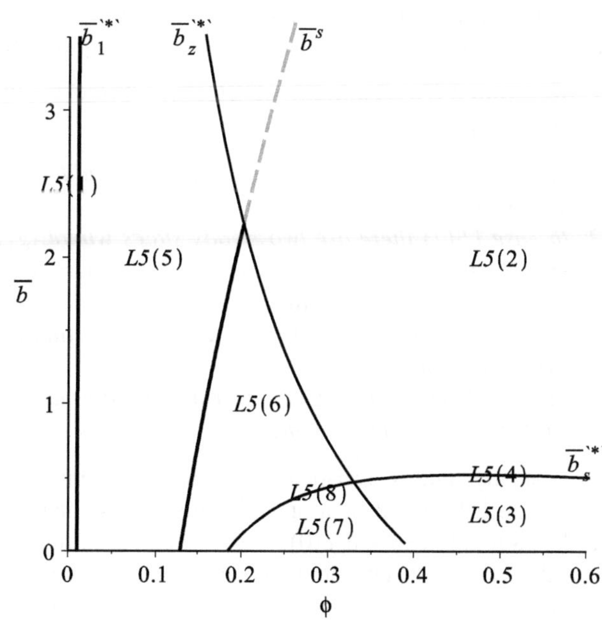

Fig. 5 Graphical illustration of Lemma 5 in the space $(\phi, \bar{b})$ for $\alpha = 0.65$ and $\tau = 0.35$, $\sigma = 2$ and $\rho = 0.02$

and $L5(8)$ there is a singular BGP, isolated in the first case and jointly with a regular BGP in the second case. Multiple equilibrium BGP's tend to occur for relatively low (high) values of the speed of the adjustment and high (low) levels for the debt target. Singular BGP's are located in the unique impasse surface $\mathcal{S}$ when they are isolated or are in surface $\mathcal{S}_1$ when they are not. They exist for relatively high levels for the speed of adjustment and for a wide interval of values for the debt target. Given benchmark parameter values for the tax rate, the intertemporal elasticity of substitution and the rate of time preference, a singular BGP can occur for a target debt ratio set around or below 50%, thus very close to the Fiscal Compact target.

When there are multiple steady states, the BGP associated to the higher value for g, g_2^*, displays a higher rate of growth, because the rate of growth is an increasing function of g (see Eq. (7)). However, as can be seen in the next section, this case corresponds to a very high expenditure ratio involving an almost confiscatory economy, which makes this BGP possibly dominated in utility terms by the adjacent BGP which has a lower expenditure ratio and a lower growth rate.

It is also possible to prove that $\sigma < 1 - \tau$ and $\bar{b} < 0$ are necessary conditions for getting a steady state level of debt $b^* = b(g^*) = \bar{b}$. Therefore, for the accepted values of the parameters, although the target level influences the dynamics it does not correspond to a BGP outcome.

Next we study the local dynamics at equilibrium BGP's.

Lemma 6 (Local Dynamics at a Regular BGP) *If $\bar{b} < \bar{b}_{\underline{s}}^*$ then the regular steady state (g_1^*, z_1^*), isolated or not, is a saddle-point. If $\bar{b} > \bar{b}_s^*$ then the steady state (g_1^*, z_1^*), isolated or not, is an unstable node or focus. If it exists, the steady state (g_2^*, z_2^*), is always an unstable node or focus.*

Therefore, we have the following cases concerning asymptotic local dynamics, and the asymptotic determinacy properties, of equilibrium BGP's:

Proposition 2 (Determinacy of BGP's) *Consider the cases in Lemma 5. The following types of BGP dynamics are possible:*

1. *in case $L5(3)$ the steady state (g^*, z^*) is a saddle point, then the BGP is locally determinate;*
2. *in case $L5(7)$ there are two steady states where (g_1^*, z_1^*) is a saddle point and (g_2^*, z_2^*) is locally unstable, then the first BGP is locally determinate and the second is over-determinate;*
3. *in cases $L5(1)$ and $L5(2)$ the unique steady state is unstable, then the BGP is over-determinate;*
4. *in cases $L5(5)$ and $L5(6)$ the two steady states are unstable, then the associated BGP's are over-determinate.*

A steady state is over-determined if the dimension of the local stable manifold in the neighborhood of a steady state is less than one. This implies that the a DGE path only exists if $g(0) = g^*$. Then the initial values of the pre-determinate variables cannot be chosen freely and should belong to a one-dimensional manifold. Therefore, over-determinate steady states can only be DGE paths if the initial point and the steady state for the detrended variables coincide. If this is the case then over-determinacy means that the DGE and the BGP paths coincide and there is no transitional dynamics.

Two BGPs with the same local stability properties cannot exist in regular DGE models in which there are no impasse-singularities (see case 4 in Proposition 2). This property also occurs in Greiner and Fincke (2015, pp. 169–173) and in Brito et al. (2017a). However, while in those papers there are two saddle-point steady states, in our case there are two unstable steady states.

Furthermore, cases $L5(4)$ and $L5(8)$ cannot exist in regular DGE models. If there is a singular steady state, and we consider the local dynamics in terms of the dynamic system (20)–(21), there will be a DGE path converging with infinite speed to a singular point. However we know that in this case the BGP path grows at a finite rate of growth $\gamma(g_s^*)$. In order to uncover the local dynamics we need to note that the value of the parameter leading to this case, $\bar{b}_s^*$, is equal to the value of the parameter associated to a zero determinant of the de-singularized vector field (25)–(26), $\bar{b}_s^* = \bar{b}_d^i$ (see Lemma 4). This means that it is a saddle-node of the desingularized vector field, implying there is convergence from the Ω_--side of the impasse-surface and divergence from the Ω_+-side. That is, the DGE path converges locally at a very high speed, and in an asymmetric way, to a BGP path growing at rate $\gamma(g_s^*)$. This case corresponds to a bifurcation point introduced by the existence of an impasse singularity.

4 Equilibrium Dynamics

Propositions 1 and 2 suggest that several types of DGE paths exist in our model: over-determinate DGE which are coincident with one or two BGP's, determinate DGE paths converging to a BGP without crossing an impasse surface, or determinate DGE paths converging to a BGP after crossing an impasse surface or to a BGP lying in the impasse surface (we call these singular DGE paths).

A determinate DGE path involving crossing trajectories can only exist if specific both local and global specific conditions are jointly satisfied. There are two local conditions: at least one steady state in which (g^*, z^*) is a saddle-point or a stable node or stable focus should exist in Ω_+ or Ω_-, and the impasse surface, $\mathcal{S}$, at the boundary of those two subsets, contains at least one impasse-transversal saddle-point or node. There are also two global conditions: the initial point $(g(0), z(0))$ and the steady state belong to different subsets, Ω_+ or Ω_-, and there is at least one (heteroclinic) trajectory joining the impasse-transversal point and the steady state, or joining the initial point and a singular steady-state.

The other type of DGE paths, over-determinate or determinate with no crossing, can be seen as cases in which either one of the local conditions (in the case of over-determinacy) or one of the global conditions (in the case of non-crossing DGE paths) fail.

At first look, it seems that the difference introduced by the existence of impasse-singularities, as regards regular models, is only related to the existence of the crossing DGE paths. However, this is not the case. The existence of impasse-singularities can also be noted in the existence of multiple over-determinate DGE paths and on the confinement of the existence of non-crossing DGE to a particular subset of the state space Ω.

Next we gather some representative DGE dynamics that illustrate the effect of the existence of impasse-singularities introduced by the fiscal policy rule. In those examples we constrain the analysis to cases in which an impasse surface exists, that is $\bar{b} \leq \bar{b}^s$. From Propositions 1 and 2 we know that, for specific subsets of the parameters values, at least one of the local conditions holds, although we were unable to find parameter values for which determinate DGE paths with crossing trajectories exist. However, the existence of singularities still constrains the dynamics in significant ways.

4.1 *Multiple Over-Determined BGP's*

Most of the equilibrium DGE dynamics involve over-determinacy, that is the existence of BGP's isolated or multiple, in which the detrended variables are unstable nodes or foci, the equilibrium dynamics is reduced to the BGP and there is no transitional dynamics. These cases are possible only if the initial values for the pre-determined variables, K and B, are in a given proportion such that the initial values of the detrended variables belong to the set Γ_E. If there is a single steady state $\Gamma_E = \{(g^*, z^*)\}$, the initial values should verify $g(0) = g^*$ and $z(0) = z^*$. If there is multiplicity of steady states, $\Gamma_E = \{(g_1^*, z_1^*), (g_2^*, z_2^*)\}$ then there are two initial points with associated different equilibrium BGP's.

Figure 6 presents an example associated to reasonable parameter values corresponding to case $L4(1)$ in Fig. 4 and $L5(6)$ in Fig. 5. There are two steady states, labelled Γ_{E_1} and Γ_{E_2}, both unstable, where the first is in set Ω_+ and corresponds to a debtor position and the second is in set Ω_- and corresponds to a creditor position. In addition, observe that there is one impasse-transversal saddle point, Γ_I, and there are two crossing trajectories starting at each steady state and passing through the impasse-transversal point. None of those trajectories are equilibria, because they do not converge to a steady state. There is another crossing between the isoclines at point $\tilde{g}$ but it does not satisfy the transversality condition.

Again, this type of phase diagram cannot exist in regular DGE models. If there are two steady states, in regular models, they have different local dynamic properties, because the curvatures of the isoclines at the two steady states should be different. The existence of an impasse surface allows for the different curvatures of the isoclines at the two steady states to be associated to the same type of local dynamics.

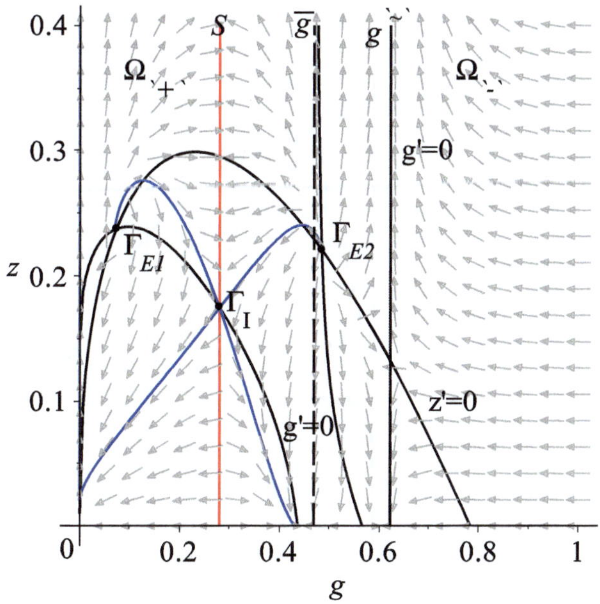

Fig. 6 Phase diagram for the case $L4(1)$ in Fig. 4 and $L5(6)$ in Fig. 5, and with $\alpha = 0.7$, $\sigma = 2$, $\rho = 0.02$, $\tau = 0.35$, $\phi = 0.2$ and $\bar{b} = 0.6$

4.2 Determinate BGP's

There are some combinations of parameters that allow the local conditions for the existence of a determinate DGE, involving local dynamics convergent to one BGP, for parameter values such that $\bar{b} \leq \bar{b}^s = \bar{b}^i_d$: cases $L4(2)$, $L4(6)$ and $L4(7)$ from Lemma 4 and $L5(3)$, $L5(4)$, $L5(7)$ and $L5(8)$ from Lemma 5.

However, for the cases represented in Figs. 4 and 5 crossing trajectories cannot exist, because a trajectory joining the impasse-transversal point (g^i_1, z^i_1) to (g^*_1, z^*_1) is not possible. In order to prove this, observe that the first point is an impasse-transversal node point where crossing trajectories flow from Ω_- to Ω_+ and the second is a saddle point belonging to Ω_-. Indeed, the stable manifold associated to (g^*_1, z^*_1) is the boundary of the basin of attraction to (g^i_1, z^i_1).

Therefore, although the necessary conditions for the existence of a crossing trajectory exist the global conditions fail. Figures 7, 8 and 9 present three related phase diagrams illustrating this case. The three phase diagrams share several common features. First, there is a unique steady state, $\Gamma_{E,1}$ or Γ_E, which is a saddle point, it is located in set Ω_- and corresponds to a long run debtor position (because it is to the left of the $g = \bar{g}$ line). Second, the stable manifold associated to this steady state is bounded to the left by the singular surface S but it is also bounded to the right.

However, the phase diagrams depicted in those figures differ in some aspects. The phase diagram in Fig. 7 in addition to the steady state $\Gamma_{E,1}$ has another steady state $\Gamma_{E,2}$ which is locally unstable, is still within set Ω_-, displays corresponds to a creditor position (because it is to the right of line $g = \tilde{g}$), and it has a

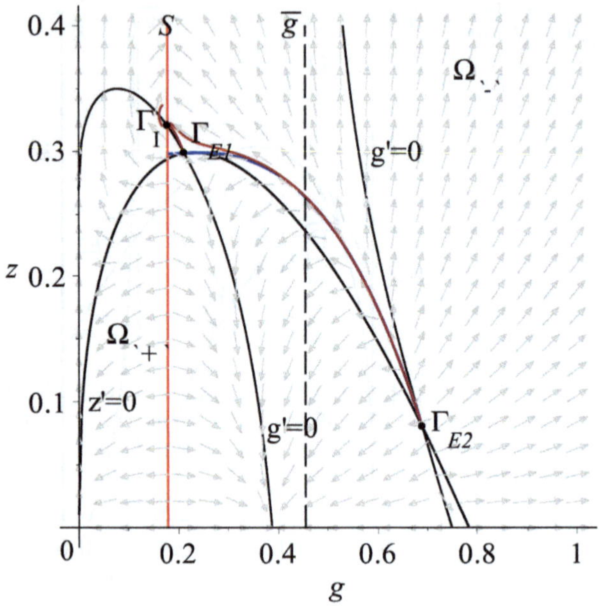

Fig. 7 Phase diagram for the case *L*4(2) in Fig. 4 and *L*5(7) in Fig. 5, and with $\alpha = 0.7$, $\sigma = 2$, $\rho = 0.02$, $\tau = 0.35$, $\phi = 0.35$ and $\bar{b} = 0.3$

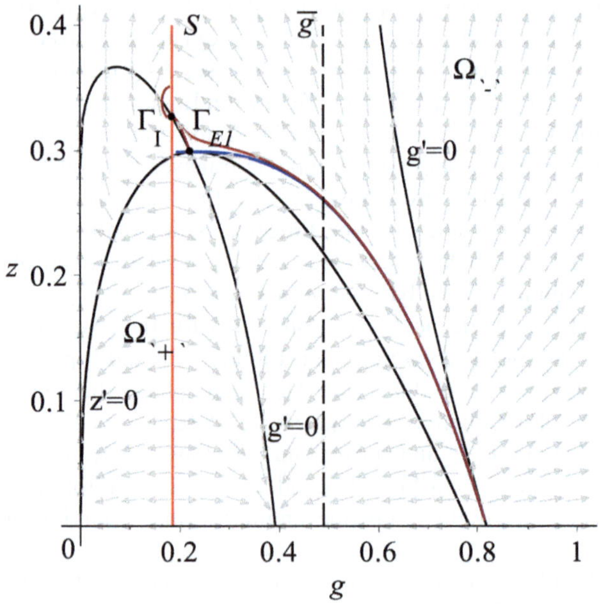

Fig. 8 Phase diagram for the case *L*4(2) in Fig. 4 and *L*5(3) in Fig. 5, and with $\alpha = 0.7$, $\sigma = 2$, $\rho = 0.02$, $\tau = 0.35$, $\phi = 0.4$ and $\bar{b} = 0.35$

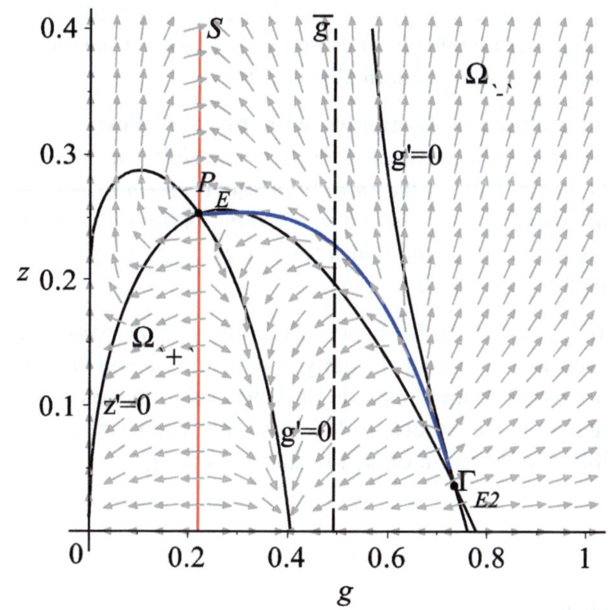

Fig. 9 Phase diagram for the case $L4(1)$ in Fig. 4 and $L5(8)$ in Fig. 5, and the BGP $g^* = g_1^*$ is a singular steady state. We set $\alpha = 0.65$, $\sigma = 1.8$, $\rho = 0.02$, $\tau = 0.35$, $\phi = 0.3$ and $\bar{b} = 0.4773478078$

heteroclinic trajectory connecting the two steady states. This means that either the initial values of the pre-determined variables are such that $(g(0), z(0)) = (g_2^*, z_2^*)$ and the BGP with rate of growth $\gamma(g_2^*)$ is over-determinate, or $(g(0), z(0)) \in \big((g_1^s, z(g_1^s)), (g_2^*, z_2^*)\big)$ and the DGE is determinate and converges to the BGP with rate of growth $\gamma(g_1^*)$. For other initial values DGE paths do not exist. We also depict a representative crossing trajectory, which is not a DGE path, departing from steady state $\Gamma_{E,2}$ passing through the impasse-transversal node point Γ_I, entering into set Ω_+ and terminating in finite time in an attractor point belonging to set $\mathcal{S}$.

The phase diagram in Fig. 8 displays only one regular steady state (g_1^*, z_1^*) in set Ω_- and corresponding to a debtor position. In this case the higher limit for the initial value of g is not given by another steady state but by a value for g such that $z(0) > 0$, because the stable manifold associated to the steady state is negatively sloped. Again, we depict a non-DGE crossing trajectory.

The phase diagram in Fig. 9 is similar to the case in Fig. 7 with the exception that the steady state (g_1^*, z_1^*) is an impasse-singular point, corresponding again to a debtor position. This steady state corresponds to a non-generic impasse-equilibrium saddle-node point. It is a singular saddle-node point because it is an unstable node from the perspective of Ω_+ and a saddle point from the perspective of Ω_-. This is a limit case: if ϕ is reduced or $\bar{b}$ augmented then the steady state g_1^* will move to the set Ω_+, and become a regular unstable point as in phase diagram in Fig. 6, or it will move to Ω_- and become a regular saddle-point as in phase diagram in Fig. 7.

In all cases we see the equilibrium expenditure ratio can only be higher than the steady state if the government is in or close to a net creditor position.

Therefore, the following is generically satisfied:

Proposition 3 (Equilibrium Dynamics with Fiscal Rules) *Assume that* $\bar{b} \leq \bar{b}_s^* = \bar{b}_d^i$. *Then a DGE path converging to an unique BGP with rate of growth* $\gamma(g_1^*)$ *only exists if the initial level of the ratio of the government expenditures,* $g(0)$, *lies between* g_1^s *and* $\min\{g_2^*, 1, g^z\}$, *where* g^z *is the point in the stable manifold such that* $z = 0$, *and the government ratio converges to a point* $g_1^s \leq g_1^* < \bar{g}$, *meaning that the government is asymptotically a debtor.*

It is interesting to note that for benchmark values of the parameters for the rate of time preference, intertemporal elasticity of substitution and distortionary tax rate those cases correspond to unrealistic high debt levels (above 60%). The long run debt ratio deviates from the target $\bar{b}$ if $g^* \neq \tau$, which is generically the case. Observe that, one possible configuration of steady states is that $g_1^* = \tau$ and $g_2^* = \bar{g}$, which implies that $b(g_2^*) = 0$.

5 Conclusion

By introducing a realistic fiscal rule, consisting in the use of the fiscal surplus to control the difference between the actual and a target ratio of the government debt to the GDP, in an otherwise benchmark growth model with productive government expenditures, we found that depending on the design of the rule, and on the initial level of the expenditure ratio, a DGE path converging to a determinate BGP can exist or not. The existence is not guaranteed because the rule introduces impasse-singularities into the model. The types of singularities that are specifically introduced by the rule considered in this paper generate several types of dynamics which are not present in regular models: multiple over-determinate BGP's, or confined determinate DGE dynamics converging to a regular or a singular BGP path. The type of dynamics can be very sensitive to small variations of the parameters defining the rule, which implies that the level of the target and the speed of the adjustment can affect the structural stability properties of the economy.

Acknowledgements I acknowledge helpful comments by Emmanuelle Augeraud-Veron and Alfred Greiner on previous versions of this paper. It was presented at the Conference "Financial and Real Interdependencies: volatility, inequalities and economic policies", Lisbon, May 30 2015, Católica Lisbon School of Business and Economics and at the 11th Annual Meeting of the Portuguese Economic Journal, Coimbra, July 2, 2016, Faculty of Economics University of Coimbra. Financial support from national funds by FCT is acknowledged. This article is part of the Strategic Project UID/ECO/00436/2013.

Appendix: Proofs

Proof of Lemma 1 Function (8) has the following obvious properties: $b(g) < 0$ if $\min\{\bar{g}, \tilde{g}\} < g < \max\{\bar{g}, \tilde{g}\}$, $b(g) = 0$ if $g = \bar{g}$, $b(g) > 0$ if $g < \min\{\bar{g}, \tilde{g}\}$ or $g > \max\{\bar{g}, \tilde{g}\}$, and $b(g) = \pm\infty$ if $g = \tilde{g}$. In order to determine the parameter values that map back those cases, observe that we have $\bar{g} < 1$ if and only if $\bar{b} < \bar{b}_1 \equiv (1-\alpha)/\phi$ and $\tilde{g} < 1$ if and only if $\phi < \alpha\tau$. Then $\max\{\bar{g}, \tilde{g}\} < 1$ if $\phi < \tau\alpha$ and $\bar{b} < \bar{b}_1$ and $\min\{\bar{g}, \tilde{g}\} > 1$ if and only if $\phi > \tau\alpha$ and $\bar{b} > \bar{b}_1$ and $\min\{\bar{g}, \tilde{g}\} \leq 1 \leq \max\{\bar{g}, \tilde{g}\}$ for the other cases. The rest of the proof is obvious. □

Proof of Lemma 2 Function $\delta(g)$, defined in Eq. (16), has the following first and second derivatives $\delta'(g) = \tau r(g) - \phi$ and $\delta''(g) = \tau r'(g)$. As $\delta'(g) \gtreqless 0$ if and only if $g \lesseqgtr \tilde{g}$ and $\delta''(g) > 0$ for all g, then it is a convex function of g reaching a minimum at point $\tilde{g}$. At this point we have $\delta(\tilde{g}) = \phi(1-\alpha)(\bar{g} - \tilde{g})$. Therefore $\delta(\tilde{g}) \lesseqgtr 0$ if and only if $\bar{g} \lesseqgtr \tilde{g}$. Consequently, three cases exist: (a) If $\bar{g} > \tilde{g}$ then $\delta(g) > 0$ for all $g > 0$ and impasse points do not exist; (b) if $\bar{g} = \tilde{g}$ then $\delta(g) = 0$ if $g^s = \bar{g} = \tilde{g}$ and $\delta(g) > 0$ otherwise, and there is a non-generic impasse point in the first case and there are no impasse points in the second; and (c) if $\bar{g} < \tilde{g}$ then there two impasse points, g_1^s and g_2^s, can exist satisfying $0 < g_1^s < \bar{g} < \tilde{g} < g_2^s$, such that $\delta(g) < 0$ if $g_1^s < g < g_2^s$, $\delta(g) = 0$ if $g = g_1^s$ or $g = g_2^s$ and $\delta(g) > 0$ otherwise.

Now, consider the cases (b) and (c) in which impasse points exist. A necessary condition for this is $\bar{g} \leq \tilde{g}$. We need to determine under which conditions the impasse points are admissible, i.e., belong to the domain $(0, 1)$. If $\bar{g} = \tilde{g}$ then the singular point satisfies $g_s < 1$ if and only if $\delta(1) < 0$ and $\delta'(1) > 0$, that is if $\phi < \tau\alpha$ and $\bar{b} < \bar{b}_1 = (1-\tau)/(\alpha\tau)$. If $\bar{g} < \tilde{g}$ then two cases are possible: (a) there are two admissible singular points satisfying $g_1^s < g_2^s < 1$ only if $\delta(1) > 0$ and $\delta'(1) > 0$; (b) there is only one admissible singular point satisfying $g_1^s < 1$ if $\delta(1) < 0$. However, $\delta(1) \lesseqgtr 0$ if and only if $\bar{b} \lesseqgtr \bar{b}_1$ and $\delta'(1) \lesseqgtr 0$ if and only if $\phi \gtreqless \alpha\tau$ which is equivalent to $\tilde{g} \gtreqless 1$. □

Proof of Lemma 3 Impasse-transversal points, (g^i, z^i), are determined by solving $\delta(g) = 0$ and $\psi(g, z) = 0$ within set Ω, i.e., for $g \in (0, 1)$ and $z > 0$. First, from Lemma 2 we already know that a condition for existence of impasse-points g^s such that $\delta(g) = 0$ is $\bar{g} \leq \tilde{g}$. As impasse points verify $\tau r(g^s) \neq \phi$ and $g^s \neq \bar{g}$ then $\psi(g, z) = 0$ if and only if $z^i = \zeta(g^s)$. Then, taking stock of Lemma 2 which determines the conditions for the existence ($\bar{b} \leq \bar{b}_s$) of one ($\phi > \alpha\tau$ and $\bar{b} < \bar{b}_1^s$) or two impasse points ($\phi > \alpha\tau$ and $\bar{b}^s > \bar{b} > \bar{b}_1^s$) we only have to check if $\zeta(g^s) > 0$. Solving the system $\delta(g, \bar{b}) = 0$ and $\zeta(g, \bar{b}) = 0$ we find, implicitly, two critical values $\bar{b}_i$, such that $\bar{b}_i = \{\bar{b} : \delta(g, \bar{b}) = 0,\ Z(g, \bar{b}) = 0\}$, where

$$Z(g) \equiv g^2 - \left((1-\alpha)(1+\bar{g}) + \tau^2\right) g + (1-\alpha)\left(\bar{g}(1-\alpha(1-\tau)) - \alpha\tau^2\right).$$

This is because $\zeta(g) \geq 0$ if and only if $Z(g, \bar{b}) \geq 0$. Two image singular points exist if, additionally, $\bar{b}^s > \bar{b} > \bar{b}^i$ and one impasse-transversal point exists if $\bar{b} < \bar{b}_i$. Comparing the values of those critical levels for $\bar{b}$, we find that there are five possible cases, when there exists at least one singular point: two impasse surfaces, $\mathcal{S}_1$ and $\mathcal{S}_2$, one transversal point in both surfaces, one transversal point in surface $\mathcal{S}_1$ and none in $\mathcal{S}_2$, or one surface $\mathcal{S} = \mathcal{S}_1$ and one or zero transversal points, or zero transversal points.

In order to determine the partition over impasse surfaces that are introduced by transversal points, we should observe that attractor (repeller) points verify $\delta'(g)\psi(g,z) < 0$ ($\delta'(g)\psi(g,z) > 0$). As $\delta'(g_1^s) < 0$ and $\delta'(g_2^s) > 0$ and because for any singular point $(\tau r(g_s)-\phi)(g_s-\bar{g}) > 0$ then $\text{sign}(\psi(g_s,z)) = \text{sign}(z-\zeta(g_s))$ then the singular point g_1^s is an attractor (repeller) if $z > z_1^i = \zeta(g_1^s)$ ($z < z_1^i = \zeta(g_1^s)$) and the singular point g_2^s is an attractor (repeller) if $z < z_2^i = \zeta(g_2^s)$ ($z > z_2^i = \zeta(g_2^s)$).

□

Proof of Lemma 4 The Jacobian of system (25)–(26), evaluated at an impasse-transversal point (g^i, z^i), is

$$J(g^i, z^i) = \begin{pmatrix} g^i\psi'_g(g^i, z^i) & g^i\psi'_z(g^i, z^i) \\ z^i\delta'(g^i)(z^i - z(g^i)) & 0 \end{pmatrix},$$

where

$$\psi'_z(g^i, z^i) = \alpha\tau(g^i - \bar{g})(\tau r(g^i) - \phi)$$
$$\psi'_g(g^i, z^i) = -\zeta'(g^i)\psi'_z(g^i, z^i).$$

As $z^i = \zeta(g^i)$ their algebraic properties depends only on g^i. Therefore, its trace, determinant and discriminant are

$$\text{tr}(J^i) = -g^i\zeta'(g^i)\psi'_z(g^i, \zeta(g^i))$$
$$\det(J^i) = g^iz^i\delta'(g^i)(\zeta(g^i) - z(g^i))\psi'_z(g^i, \zeta(g^i))$$
$$\Delta(J^i) = g^i\psi'(g^i)\left(g^i(\zeta'(g^i))^2 + 4\zeta(g^i)(\zeta(g^i) - z(g^i))\right).$$

We have $\psi'_z(g^i) = \psi'_z(g^i, \zeta(g^i)) > 0$ for any g^i. Therefore, $\text{sign}(\text{tr}(J^i)) \neq \text{sign}(\zeta'(g^i))$ and $\text{sign}(\det(J^i)) = \text{sign}(\delta'(g^i)(\zeta(g^i) - z(g^i)))$. The discriminant is positive if $z^i = \zeta(g^i) \leq z(g^i)$ and may be zero or negative only if $\zeta(g^i) < z(g^i)$.

After some algebra we get the determinant of Jacobian $J(g^i) \equiv J(g^i, \zeta(g^i))$

$$\det J(g^i) = \alpha z^i g^i(\tau r(g^i) - \phi)^2 D(g^i)(g^i - \bar{g})(g^i - g_d^i)$$

where

$$D(g^i) \equiv \frac{\rho\alpha\tau + \phi(\sigma(1 - \alpha\tau) - (1 - \tau))}{\alpha\tau\sigma g^i}$$

and g^i_d is in Eq. (27). If $\sigma > 1$ then $D(g^i) > 0$ and $g^i_d < \bar{g}$. This implies, because we have proved that $g^i_1 < \bar{g} < g^i_2$, $\det(J(g^i_2)) > 0$ and $\det(J(g^i_1)) > 0$ if $g^i_1 < g^i_d < \bar{g}$ and $\det(J(g^i_1)) \leq 0$ if $g^i_1 \leq g^i_d < \bar{g}$. Therefore (g^i_1, z^i_1) can be a transversal saddle, node or focus and (g^i_2, z^i_2) can only be a node or a focus. In order to discriminate between the two cases, we need to determine the sign of the discriminant of the Jacobian evaluated at every impasse-transversal point.

The sign of the discriminant $\Delta(J^i)$ depends on a quadratic polynomial in $(1/g)$. This means that there can be generically, zero or two critical values for g^i, say $g^i_{\Delta,1} < g^i_{\Delta,2}$. In the first case the two image-transversal points are nodes and in the second one they are foci. Evaluating the trace for impasse-transversal nodes is required for determining the direction of the crossing trajectories. If the trace is negative (positive) then there is crossing from the adjacent Ω_+ set to set Ω_- (from Ω_- to the adjacent Ω_+ set). We readily see that $\text{sign}(\text{tr}(J^i)) = -\text{sign}(\zeta'(g^i))$, where the sign of $\zeta'(g^i_1)$ is ambiguous but $\zeta'(g^i_2) > 0$. That is, when there is an impasse-transversal node at (g^i_2, z^i_2) there is crossing from Ω_- to $\Omega_{+,2}$.

If $g^*_i = \bar{g} = \tilde{g}$ then the two impasse-transversal points collapse to one and as $\text{tr}(J^*_i) = \det(J^*_i) = 0$ we have a singularity induced bifurcation point with co-dimension 2. □

Proof of Lemma 5 Equilibrium steady states values (g^*, z^*) must verify three conditions: first, they are fixed points of the system (20)–(21); second, they belong to set Ω; third, the transversality condition should be satisfied. A regular (singular) steady state additionally verifies $\delta(g) \neq 0$ ($\delta(g) = 0$).

First, we determine the fixed points of system (20)–(21) in the Ω domain. From Eq. (21), we find $\dot{z} = 0$ if and only if $z = 0$ or $z^* = z(g^*)$. The first point is not in set Ω and the second is monotonously related to g^* and is a BGP if $z^* > 0$. On the other hand the set of fixed points for g is $g^* = \{g \in (0,1) : \Psi(g) \equiv \psi(g, z(g)) = 0\}$ where

$$\Psi(g) = \alpha(\tau r(g) - \phi)\left(\varphi_d(g)\tau r(g) - \varphi_n(g)\right)/\sigma$$

and $\varphi_d(g) = \left((\tau\sigma + (1-\tau)(\sigma-1))\, g - \left(\tau^2\sigma + g(1-\tau)(\sigma-1)\right)\right)/\tau$, and $\varphi_n(g) \equiv (\sigma\phi - \rho)g - (\sigma\phi\tau - \rho\bar{g})$. It is useful to define $\varphi(g) = \varphi_n(g)/\varphi_d(g)$, that is

$$\varphi(g) = \frac{\tau(\sigma\phi - \rho)}{\tau\sigma + (1-\tau)(\sigma-1)}\left(\frac{g - g_1}{g - g_0}\right)$$

where

$$g_0 \equiv \frac{\tau^2\sigma + \bar{g}(1-\tau)(\sigma-1)}{\tau\sigma + (1-\tau)(\sigma-1)}, \text{ and } g_1 \equiv \frac{\phi\sigma\tau - \rho\bar{g}}{\sigma\phi - \rho}.$$

The fixed points of $\Psi(g) = 0$ are determined from two conditions: $\tau r(g) = \phi$ and $\tau r(g) = \varphi(g)$. Then one fixed point is always $g^*_1 = \tilde{g}$ (solving equation $\tau r(g) = \phi$).

Other fixed points are the solutions of equation $\tau r(g) = \varphi(g)$. We find three cases: (a) if $\sigma\phi < \rho$ and $g_1 > g_0 > 0$ then there is a second fixed point $g_2^* \in (g_0, g_1)$; (b) if $\sigma\phi > \rho$ and $g_0 > 0 \geq g_1$ then there is a second fixed point $g_2^* > g_0$; and (c) if $\sigma\phi > \rho$ and $g_0 > g_1 > 0$ then there are two more fixed points, g_2^* and g_3^*, such that $g_2^* < g_1 < g_0 < g_3^*$. But, as $g_1 = 0$ if and only if $\bar{b} = \bar{b}_1^*$, in Eq. (28), then the first two cases can be taken together if $\bar{b} < \bar{b}_1^*$.

Next, we have to check if the transversality condition holds for all those roots of equation $\Psi(g) = 0$. First, the rates of growth for consumption, capital stock and the government debt are derived in Eq. (7), (17) and (12). Using the definitions in Eq. (19) we find $\gamma_K(g, z) - \gamma(g) = z(g) - z$. Then $\gamma_K^* = \gamma_K(g^*, z^*) = \gamma^* = \gamma(g^*)$ along any fixed point of the system (20)–(21).

On the other hand

$$\gamma_B(g) - \gamma(g) = \frac{\varphi_d(g)}{\sigma(g - \bar{g})} \left(\tau r(g) - \varphi(g)\right).$$

Then $\gamma_B(g) - \gamma(g) = 0$ if and only if: (a) $\varphi_d(g) \neq 0$ and $\tau r(g) = \varphi(g)$; or (b) $g = \bar{g}$ and $\tau r(g) = \varphi(g) = \phi$, that is if $\bar{g} = \tilde{g}$. Observe that if $\tau r(g) = \phi$ and $\bar{g} \neq \tilde{g}$ then $\gamma_B(g) - \gamma(g) \neq 0$. Therefore: (1) a fixed point such that $z^* = z(g^*)$ and $\tau r(g^*) = \varphi(g^*)$ implies $\gamma_B^* = \gamma_K^* = \gamma^* = ((1 - \tau)r(g^*) - \rho)/\sigma$, and, then $\rho + \sigma\gamma^* - \gamma_K^* = \rho + \sigma\gamma^* - \gamma_B^* = \rho + (\sigma - 1)\gamma^* = (\rho + (\sigma - 1)(1 - \tau)r(g^*))/\sigma > 0$ if $\sigma \geq 1$, and the transversality condition holds; (2) a fixed point such that $z^* = z(g^*)$ and $\tau r(g^*) = \phi$ leads to two possible cases: if $g^* = \bar{g}$ then $g^* = \tilde{g} = \bar{g}$ implying $\gamma_B^* = \gamma_K^* = \gamma^* = ((1 - \tau)r(g^*) - \rho)/\sigma$ and the transversality condition is also satisfied; however, if $g^* \neq \bar{g}$ then $\gamma_B^* = (1 - \tau)\phi/\tau$ which is different from $\gamma_K^* = \gamma^* = ((1 - \tau)r(g^*) - \rho)/\sigma$, and because $\rho + \sigma\gamma^* - \gamma_B^* = \rho + (1 - \tau)\phi/\tau - \rho - (1 - \tau)\phi/\tau = 0$ the transversality condition is not satisfied.

At last we check if the fixed points verifying the transversality condition belong to the domain Ω: that is if $g^* \in (0, 1)$ and $z(g^*) > 0$. As $z(0) = \rho/\sigma > 0$ and, if we assume that $\rho < \alpha(1 - \tau))$, $z(1) = (\rho - \alpha(1 - \tau))/\sigma < 0$ and $z'(0) > 0$ and $z'(1) < 0$ then $g^* \in (0, 1)$ if $z(g^*) > 0$. As $z(g) > 0$ if and only if $(\sigma(1 - g)/\alpha - (1 - \tau))r(g) + \sigma\rho > 0$ then there is a unique critical level g_z such that $z(g) > 0$ if and only if $g < g_z$. Then by substituting in the equilibrium condition we find a critical value for a parameter $\bar{b}$, $\bar{b}_z^*$, such that $\varphi(\bar{b}_z^*, g_z) = \tau r(g_z)$. This value exists and is unique and we have $z(g^*) > 0$ if and only if $\bar{b} < \bar{b}_z^*$.

Now, we address the location of the fixed points, as regards the impasse surface(s). Taking $\tau r(g) = \varphi(g)$ and solving jointly with $\delta(g) = 0$ we obtain both singular fixed points and the constraint in the parameters for their existence. Factoring the expression $\tau r(g) > 0$ in both conditions, we get $\tau r(g) = \frac{\phi(g-(1-\alpha)\bar{g})}{\alpha g} = \varphi(g)$ that is equivalent to $\phi(g - (1 - \alpha)\bar{g})\varphi_d(g) = \alpha g \varphi_n(g)$ and g_s^*. However only the last one can be an equilibrium BGP. Observe that, when there are two BGP's only the one corresponding to the smaller g can be a singular BGP. From $\delta(g_s^*) = 0$ we get the critical value $\bar{b}_s^*$.

If we consider additionally the proofs of Lemmas 1, 2, and the constraint $z(g) > 0$ (and the fact that if $z(1) < 0$ then $\tau r(1) > \psi(1)$, and if $z(1) < 0$ then $\bar{g} < 1$), we have the following cases: (i) if f $\bar{b} > \bar{b}_{\text{L}}^*$ then there is a unique BGP and $\delta(g^*) > 0$ and $\tilde{g} < g^* < \bar{g} < 1$; (ii) if $\bar{b} \geq \bar{b}_z$ then there are two roots of $\Psi(g) = 0$ but only one, g_1^*, verifies $z(g) > 0$. In addition we always have $g_1^* < \min\{\bar{g}, \tilde{g}\}$ and the sign of $\delta(g_1^*)$ is positive (negative) if $\bar{b} > (<)\bar{b}_s^*$; (iii) if $\bar{b} = \bar{b}_s^*$ then $\delta(g_1^*) = 0$; (iv) if $\bar{b} < \min\{\bar{b}_z, \bar{b}_1^*\}$ then there are two roots of $\Psi(g) = 0$, satisfying the transversality condition and $z(g) > 0$. In addition the following relationship is verified $\min\{g_1^*, g_1^s\} \leq \max\{g_1^*, g_1^s\} < \bar{g} < g_2^* < \tilde{g} < g_2^s$. But $g_1^* = \min\{g_1^*, g_1^s\}$ if $\bar{b} > \bar{b}_s^*$, $g_1^s = \min\{g_1^*, g_1^s\}$ if $\bar{b} < \bar{b}_s^*$ and $g_1^* = g_1^s\}$ if $\bar{b} = \bar{b}_s^*$. The creditor position associated to the steady states is easily determined from the previous inequalities and from Lemma 1. □

Proof of Lemma 6 The Jacobian of system (20)–(21) evaluated at a regular BGP (g^*, z^*) is

$$J^* \equiv J(g^*, z^*) = \frac{1}{\delta(g^*)} \begin{pmatrix} g^* \psi_g'(g^*, z^*) & g^* \psi_z'(g^*, z^*) \\ -z^* z'(g^*)\delta(g^*) & -z^* \delta(g^*) \end{pmatrix}.$$

It has trace and determinant

$$\text{tr}(J^*) = \frac{z^* \delta(g^*) + g^* \psi_g'(g^*, z^*)}{\delta(g^*)}$$

$$\det(J^*) = \frac{z^* g^*}{\delta(g^*)} \left(\psi_g'(g^*, z^*) + \psi_z'(g^*, z^*) z'(g^*) \right).$$

However, because $\Psi'(g^*) = \psi_g'(g^*, z^*) + \psi_z'(g^*, z^*) z'(g^*)$ (see proof of Lemma 5) then the trace and the determinant become, equivalently

$$\text{tr}(J^*) = z^* + g^* \left(\frac{\Psi'(g^*)}{\delta(g^*)} - \frac{\psi_z'(g^*, z^*) z'(g^*)}{\delta(g^*)} \right)$$

$$\det(J^*) = z^* g^* \frac{\Psi'(g^*)}{\delta(g^*)}.$$

If $\bar{g} > \tilde{g}$ as $\delta(g^*) > 0$ and $\Psi'(g^*) > 0$ for any root of $\Psi(g) = 0$, with the exception of $g^* = \tilde{g}$, then all the equilibrium BGP are locally unstable nodes or foci, i.e. the dimension of the local stable manifold is zero. If $\bar{g} < \tilde{g}$ the two fixed points of $\Psi(g) = 0$, with the exception of $g^* = \tilde{g}$ always verify $\Psi'(g_1^*) > 0$ and $\Psi'(g_2^*) < 0$. In addition as $g_1^s < \bar{g} < g_2^*$ then we always have $\delta(g_2^*) < 0$ and therefore the point g_2^* is also always an unstable node or focus. On the other hand, if $g_1^s < g_1^*$ ($g_1^s > g_1^*$) then $\delta(g_1^*) < 0$ and g_1^* is a saddle point (unstable node or focus). □

References

Arrow, K., & Kurz, M. (1970). *Public Investment, the Rate of Return and Optimal Fiscal Policy*. Baltimore: The Johns Hopkins Press.

Barro, R. (1990). Government spending in a simple model of endogenous growth. *Journal of Political Economy, 98*, S103–S125.

Bohn, H. (1998). The behavior of U.S. public debt and deficits. *Quarterly Journal of Economics, 113*(3), 949–963.

Bohn, H. (2008). The sustainability of fiscal policy in the United States. In R. Neck, & J. E. Sturm (Eds.), *Sustainability of Public Debt* (pp. 15–49). Cambridge, MA: MIT Press.

Brito, P., Costa, L., & Dixon, H. (2017a). From Sunspots to Black Holes: Singular dynamics in macroeconomic models. In K. Nishimura, et al. (Eds.), *Sunspots and Non-linear Dynamics* (pp. 41–70). Berlin: Springer.

Brito, P., Costa, L., & Dixon, H. (2017b). Singular macroeconomic dynamics and temporary indeterminacy. (In preparation).

Futagami, K., Iwaisako, T., & Ohdoi, R. (2008). Debt policy rule, productive government spending, and multiple growth paths. *Macroeconomic Dynamics, 12*, 445–462.

Ghosh, A. R., Kim, J. I., Mendoza, E. G., Ostry, J. D., & Qureshi, M. S. (2013). Fiscal fatigue, fiscal space and debt sustainability in advanced economies. *Economic Journal, 123*, F4–F30.

Greiner, A. (2008). Does it pay to have a balanced government budget? *Journal of Institutional and Theoretical Economics, 164*(3), 460–476.

Greiner, A., & Fincke, B. (2015). *Public Debt, Sustainability and Economic Growth*. Berlin: Springer.

Irmen, A., & Kuehnel, J. (2009). Productive government expenditure and economic growth. *Journal of Economic Surveys, 23*, 692–733.

Kamiguchi, A., & Tamai, T. (2012). Are fiscal sustainability and stable balanced growth simultaneously attainable? *Metroeconomica, 63*, 443–457.

Kinda, T., Kolerus, C., Muthoora, P., & Weber, A. (2013). Fiscal rules at a glance. Technical report, International Monetary Fund.

Llibre, J., Sotomayor, J., & Zhitomirskii, M. (2002). Impasse bifurcations of constrained systems. *Fields Institute Communications, 31*, 235–256.

Nishimura, K., Seegmuller, T., & Venditti, A. (2015). Fiscal policy, debt constraint and expectation-driven volatility. *Journal of Mathematical Economics, 61*, 305–316.

Riaza, R. (2008). *Differential-Algebraic Systems: Analytical Aspects and Circuit Applications*. Singapore: World Scientific.

Schmitt-Grohe, S., & Uribe, M. (1997). Balanced-budget rules, distortionary taxes, and aggregate instability. *Journal of Political Economy, 105*(5), 976–1000.

Zhitomirskii, M. (1993). Local normal forms for constrained systems on 2-manifolds. *Boletim da Sociedade Brasileira de Matemática, 24*, 211–232.

Financial Liberalization, Inequality and Inclusion in Low-Income Countries

Davide Furceri, Jun Ge, and Prakash Loungani

Abstract This paper examines the distributional impact of capital account reforms and the linkage among liberalization, inequality and inclusion in low-income countries. Using a panel data for 29 low-income countries from 1970 to 2010, we find that capital account liberalization reforms are associated with statistically significant and persistent increase in income inequality in both short and medium term. We also highlight that the level of financial development has an important role in determining the response of inequality to liberalization: impact of capital account liberalization on inequality is larger in countries with lower level of credit market development and financial inclusion.

1 Introduction

Rising income inequality within countries has received renewed attention in recent years. Much of the focus has been on advanced and emerging economies; however, many low-income countries (LICs) also experienced growing income inequality from the late-1980s to the early-2000s and again more recently. Some observers have pointed out that these periods of rising inequality coincide with increasing openness to foreign capital in these countries (Goldin and Reinert 2013).

This paper analyzes the relationship between global financial integration and the rise in income inequality in LICs. In comparison to other country groups, low-income countries currently have greater restrictions on the capital account, and thus have more space to relax such restrictions in years to come. Thus, understanding

Disclaimer: This Working Paper should not be reported as representing the views of the IMF. The authors acknowledge the support from U.K.'s Department for International Development (DFID) under the project Macroeconomic Research in Low-Income Countries. The views expressed here are those of the authors and do not necessarily represent those of the IMF, IMF policy, or DFID.

D. Furceri • J. Ge • P. Loungani (✉)
International Monetary Fund, 700 19th Street NW, Washington, DC 20431, USA
e-mail: dfurceri@imf.org; jge@imf.org; ploungani@imf.org

B. Bökemeier, A. Greiner (eds.), *Inequality and Finance in Macrodynamics*, Dynamic Modeling and Econometrics in Economics and Finance 23,
DOI 10.1007/978-3-319-54690-2_4

the consequences of further opening up the capital account is likely to remain in the near future an important policy issue for LICs. Of course, policymakers are likely to embark on easing these restrictions if they expect some benefit in terms of increased growth. Having evidence on the likely distributional consequences informs policymakers about the efficiency-equity tradeoffs associated with financial liberalization.

We highlight a key channel through which capital account liberalization could impact income distribution in low-income countries. In low-income countries, with under-developed financial institutions and access to credit market limited to a few, liberalization may create a bias in favor of elites instead of facilitating the broader consumption smoothing by many that is promised by theory (Kose et al. 2009).

We use data on 29 low-income countries from 1970 to 2010 to study the linkages among liberalization, inequality and inclusion. Our main findings are:

1. episodes of capital account liberalization are followed by persistent increases in income inequality in low-income countries; specifically, liberalization is associated with a 3% increase in inequality in short term (2 years after the reform) and over a 6% increase in medium term (5 years after the reform);
2. the impact of capital account liberalization on inequality is larger in countries with lower level of credit market development and financial inclusion.

The reminder of the paper is organized as follows. Section 2 gives a review of current literature related to capital account liberalization, financial inclusion and inequality. Section 3 describes the data and shows the descriptive statistics of income inequality and capital account openness in low-income countries for the past three decades. Section 4 discusses the empirical methodology of the paper, including the methods to evaluate how the level of financial development and inclusion could influence the effects of capital account liberalization on inequality. Section 5 presents the main results of the analyses and robustness checks of our findings. Section 6 draws some policy implications.

2 Literature Review

There is a large literature investigating the relationship between financial liberalization and economic growth, for example studies by Henry (2007) as well as Quinn and Toyoda (2008), and some focused specifically on low-income countries, such as Dabla-Norris et al. (2010) and Delechat et al. (2008). Jaumotte et al. (2013), using data for 51 developed and developing countries between 1981 and 2003, report that financial globalization, measured by FDI over GDP which is de facto proxy for openness, is usually associated with an increase in inequality. Their estimation suggests that globalization contributes around 0.08% annual average increase of the Gini coefficient over 22-year sample period and the effects could be larger if the increase in Gini coefficient associated with financial liberalization was not offset by the reduction in Gini coefficient associated with trade liberalization. Dabla-Norris et al. (2015) also note that financial globalization drives higher income inequality in advanced and developing countries.

LaGarda et al. (2016) construct a large unbalanced panel of 141 countries from 1990 to 2013 to study the distributional effects of capital account liberalization. The authors find that capital account liberalization is likely to cause a statistically significant and persistent increase in inequality during a financial crisis, whereas the results are not so clear during economic expansions. Their results also suggest that social safety net, such as cash transfers could possibly reduce the negative effects of financial liberalization on inequality.

Batuo and Asongu (2015) focuses on the impacts of liberalization policies on income distribution in 26 African countries from 1996 to 2010, obtaining similar results to those in Jaumotte et al. (2013). Batuo and Asongu (2015) finds that financial liberalization in both de jure and de facto measures tend to increase income inequality in African countries, in contrast to trade liberalization which tends to reduce inequality. However, the effects diminish or even reverse under alternative econometric specifications.

There are also studies noting that the gains of opening up capital account are dependent on the strength of domestic financial institutions. It has been argued that countries could benefit from financial globalization only if they reach certain thresholds of financial development, especially with respect to the depth of the domestic credit market (Kose et al. 2011). Bumann and Lensink (2016) develop a theoretical model featuring agents with different investment abilities to examine impacts of capital account liberalization through two channels: a reduction in reserve requirements and an opening of domestic debt market to foreign funds. Their model predicts that financial liberalization will only improve the distribution of income in countries that have high financial depth and will worsen inequality in countries where financial depth is low. They then support the model prediction by conducting an empirical study based on inequality and liberalization indicators of 106 countries from 1973 to 2008.

3 Data Description

3.1 Measuring Income Inequality

In this paper, we use Gini coefficients from the Standardized World Income Inequality Database (SWIID) to measure inequality in 29 low-income countries from 1970 to 2010. By standardizing data from the United Nations University's World Income Inequality Database (WIID) and the Luxembourg Income Study (LIS), the SWIID offers comparable Gini estimates of gross income inequality for

Table 1 Selected country sample and indicators

	Gini coefficient		Capital account openness		Inclusion indicator
	Begin	End	Begin	End	Value
Albania	31.34	34.43	−0.11	−0.11	7.53
Bangladesh	47.80	57.50	−1.86	−1.16	23.33
Bolivia	49.24	55.79	−0.63	1.14	16.62
Cameroon	49.08	44.14	−0.11	−1.16	4.45
Cote d'Ivoire	55.44	46.33	−0.89	−1.16	–
Egypt, Arab Rep.	38.38	32.21	−1.86	2.46	3.65
El Salvador	59.51	44.84	−1.16	2.46	3.90
Fiji	50.36	53.47	−0.11	−1.16	–
Ghana	37.98	42.40	−1.86	−1.16	5.75
Guatemala	54.06	54.64	−1.16	2.46	13.72
Honduras	61.70	53.54	−0.89	−0.11	7.10
India	36.33	35.74	−1.16	−1.16	7.70
Indonesia	36.21	38.47	0.42	1.13	8.55
Kenya	72.17	48.69	−1.16	1.13	9.74
Madagascar	46.11	47.05	−1.16	−0.11	2.29
Morocco	67.70	41.47	−1.16	−1.16	4.30
Nepal	64.11	48.51	−1.16	−1.16	10.76
Nicaragua	55.65	51.39	−1.16	2.46	7.63
Pakistan	39.67	32.92	−1.86	−1.16	1.57
Paraguay	37.03	49.34	−1.16	0.87	12.88
Philippines	57.40	42.88	−0.37	0.08	10.51
Rwanda	25.15	46.35	−1.16	−1.16	8.40
Senegal	58.56	39.45	−0.89	−1.16	3.50
Sierra Leone	60.44	44.40	−1.16	−1.16	6.15
Sri Lanka	43.19	59.35	−1.86	0.08	17.74
Tanzania	55.32	35.47	−1.16	−1.16	6.61
Uganda	49.24	41.19	−1.16	2.46	8.87
Ukraine	41.99	31.93	−1.16	−1.16	8.15
Zambia	64.19	50.97	−1.16	2.46	6.13

173 economies since 1960.[1] Given its consistency, comparability and coverage, the SWIID is suited for our analysis on income inequality across low-income countries.

By construction, Gini coefficients are bounded between 0 and 100, with higher values indicating greater inequality. Within LICs, values of the Gini vary greatly across economies. Table 1 provides the values of the Gini at the start and end dates for all the countries we use. As summarized in Table 2, the Gini values in our sample

[1]For mode details, please see: Frederick Solt, 2009, "The Standardized World Income Inequality Database", Harvard Dataverse, V15.

Table 2 Descriptive statistics of low-income countries

Variable	N.Obs	Average	SD	Min	Max
Gini	803	45.87	9.46	23.57	73.40
D. Gini	773	−0.17	1.93	−8.04	19.57
Kaopen	1102	−0.47	1.26	−1.86	2.46
D.Kaopen	1098	0.02	0.41	−3.25	3.25
Equity index	476	0.41	0.40	0.00	1.00
Bond index	426	0.47	0.40	0.00	1.00
FDI index	476	0.41	0.38	0.00	1.00

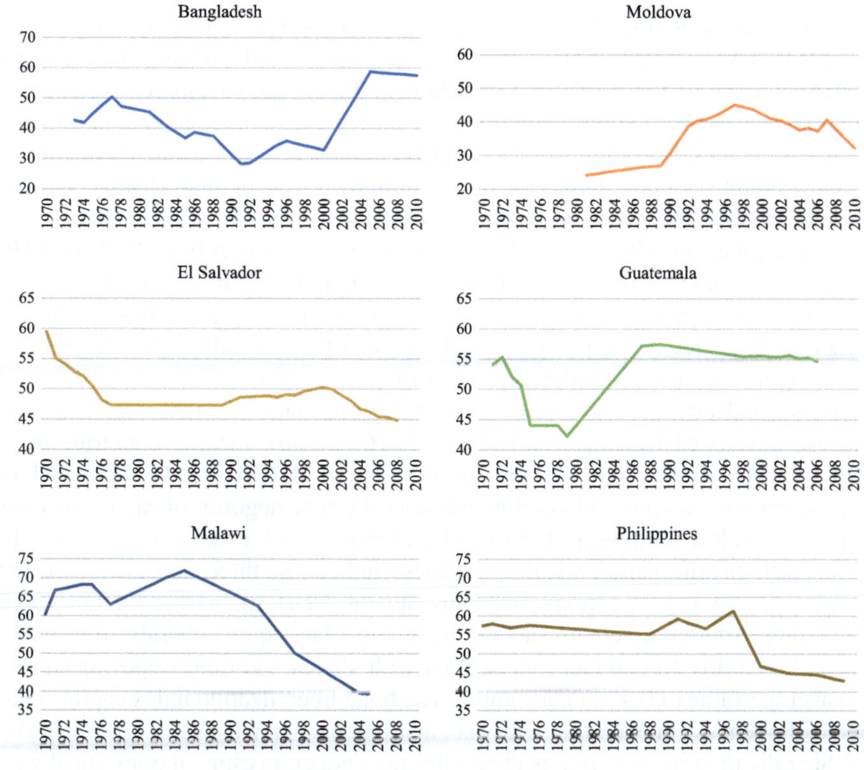

Fig. 1 Income inequality in selected low-income countries

range from 23.47 (Ukraine in 1991) to 73.4 (Kenya in 1971), with an average of 45.9 and a standard deviation of 9.5.

Though inequality in LICs increased on average, in the 1990s and the late-2000s, the evolution of inequality is different across countries. Selected cases are shown in Fig. 1. Inequality in Bangladesh has been rising since 1992, with a particularly sharp increase from 2000 to 2004. In 2010, Bangladesh's income inequality was much worse than 30 years prior in 1972. Countries like Moldova and Guatemala have experienced a period of increasing Gini indices beginning in the 1980s, but inequality has been gradually decreasing during the last decade. Nevertheless, their recent Gini indices indicate that inequality in these two countries is either the same as or worse than where it started. In contrast, some countries have lowered income inequality over time, such as El Salvador, Malawi and Philippines; however, even in these countries there are episodes of increased inequality for a few years.

3.2 Measuring Capital Account Openness

To measure financial liberalization, this paper uses two indices of openness that rely on the IMF's Annual Report on Exchange Arrangements and Exchange Restrictions (AREAER). The first index is from the Chinn and Ito (2008) database, which gives the largest coverage of country and time period for our sample, compared to other alternative indices. The Chinn-Ito (KAOPEN) index assesses a country's degree of capital account openness based on binary dummy variables that identify various regulatory restrictions on cross-border financial transactions reported in AREAER.[2] This paper covers the Chinn-Ito index for low income-countries from 1970 to 2010, with a maximum value of 2.46 (the least restricted) and a minimum of −1.86 (most restricted). The average value of Chinn-Ito index in the sample is −0.47 as shown in Table 2. Like income inequality, capital account openness, although it displays an increasing trend, varies considerably across different countries.

The second de jure measure of financial globalization is taken from a new database developed by Jahan and Wang (2016) for low-income countries from 1996 to 2013. Based also on the information in AREAER, the Wang-Jahan database provides a set of indices that measures the subcategories of capital account openness,[3] with a minimum of 0 (no liberalization) and a maximum of 1 (fully liberalized). In this paper, we use openness indices in three subcategories, (1) equity, (2) bond and (3) foreign direct investment (FDI), to analyze the impact of liberalizing different assets on income inequality. The equity liberalization index reflects the ability to conduct transactions with shares and other types of equity, with average value of 0.41 in our sample. The bond liberalization index captures the degree of openness in operating with debt instruments, with an average of 0.47. The FDI liberalization index captures the ability to conduct investment domestically or abroad through inward and outward flows of FDI, with an average of 0.41 (Table 2).

[2]For more details, please see: "A New Measure of Financial Openness", Journal of Comparative Policy Analysis, Volume 10, Issue 3 September 2008, pp. 309–322.

[3]For more detail, please see http://www.imf.org/external/datamapper/index.php?db=FM.

Table 3 Number of capital account liberalization reforms

	1970s	1980s	1990s	2000s	2010s
Low income countries	5	7	29	4	1

Table 4 Number of liberalization reforms by sub-index

(LICs)	1990s	2000s	2010s
Average equity	2	1	1
Average bond	4	10	2
Direct investment	8	7	1

3.3 *Identification Strategy*

For identifying capital account liberalization episodes, it could be ideal to obtain the dates when legislative decrees removed the restrictions. However, these dates are not easily accessible for low-income countries where legislative records are difficult to gather. Therefore, with the two sets of de jure indices, we constructed dummy variables to infer the dates of capital account liberalization episodes for every index. Following Furceri and Loungani (2015), we assume that a liberalization episode happens when, for a low-income country at a given time, the annual change in its openness index exceeds by two standard deviations the average annual change of all observations (covering also high-income and middle-income countries).

This criterion identifies 56 episodes of overall capital account liberalization in low-income countries from 1970 to 2010 (Table 3), with the greatest number of liberalization taking place in 1990s. In the period between 1996 and 2013, there are in total 4 liberalization episodes in equity market, 16 in bond market and 16 related to FDI across low-income countries (Table 4). Descriptive statistics also suggest that overall capital account liberalization tends to be associated with an increase in the Gini indices of 0.5% points (1.1%) in near-medium term (3 years) and of 1.5% points (3.2%) in medium term (5 years), compared with data before a liberalization reform as depicted in Fig. 2.

4 Methodology

4.1 *The Baseline*

The principal methodology used in this paper to examine the effects of capital account openness on income inequality for low-income countries follows Furceri and Loungani (2015), which estimates a univariate autoregressive equation and then derives its associated impulse response function (IRF). This approach is particularly suitable to evaluate the dynamic response of the dependent variable, through IRFs,

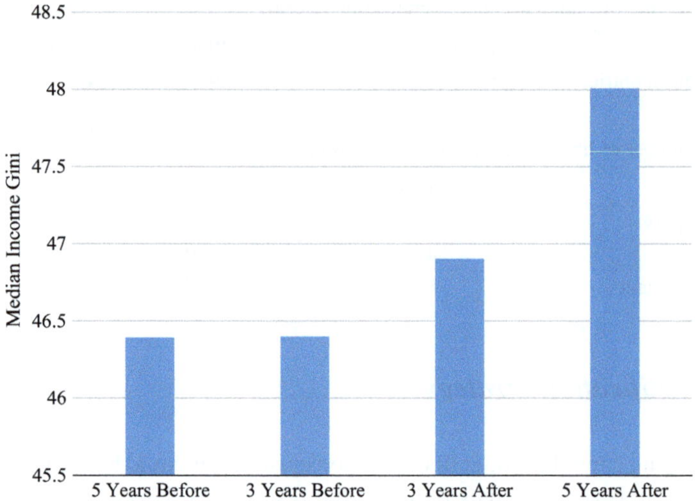

Fig. 2 The evolution of inequality before and after capital account liberalizations

after a reform that is a capital account liberalization episode in this study Furceri and Loungani (2015). The autoregressive equation estimated in this paper is as following:

$$g_{it} = a_i + \gamma_t + \sum_{j=1}^{l} \beta_j g_{i,t-j} + \sum_{j=0}^{l} \delta_j D_{i,t-j} + \sum_{j=1}^{l} \upsilon_j X_{i,t-j} + \epsilon_{it} \quad (1)$$

where g_{it} is the growth rate of inequality, measured by the annual change in the log of the Gini indices in country i at time t; a_i represent the country fixed effects included to control for unobserved heterogeneity across low-income countries and for the possible discrepancies between Gini indices derived from income and those derived from consumption in the original data; γ_t are the time-fixed effects to count for common global shocks.

In order to control for the normal dynamics of inequality and other serially-correlated factors that may influence short-term inequality, the equation also includes lagged inequality growth, $g_{i,t-j}$, for j periods. $D_{i,t-j}$ is a $(0, 1)$ dummy variable that takes the value 1 when a capital account liberalization episode takes place according to the criterion outlined in Sect. 3.3; D equals 0 otherwise. ε_{it} is the error term.

$X_{i,j-t}$ represents a set of control variables[4] that may influence the evolution of inequality as well as the effects of capital account liberalization on inequality in low-income countries. These control variables include: (1) current account reforms,

[4]For detailed description of variables and data sources, please refer to Data Appendix.

defined using Quinn and Toyoda (2008) index of current account openness; and (2) regulation reforms, defined using Economic Freedom of the World's (EFW) composite measure of credit regulations. By introducing these two structural reform indicators, we distinguish the effects of capital account liberalization from other policies which tend to be implemented at the same time. The set of control variables also contains: (3) GDP growth rate; (4) log GDP per capita; (5) square of log GDP per capita; (6) change in trade openness[5]; (7) changes in shares of government expenditure in GDP; (8) changes in the redistributive policies[6]; (9) changes in share of industry value added in GDP; (10) changes in agriculture share of GDP; (11) changes in dependency ratios; (12) changes in product, labor and credit market regulation.

With the data available, we constructed an unbalanced panel of annual observations from 1970 to 2010 for 29 low-income countries to estimate Eq. (1) with lag length of 2 (i.e. $j = 2$). Even though in theory, the estimation of coefficients might be biased when taking into consideration of lagged dependent variable, country fixed effects and our relatively small sample size (Nickell 1981), the problem is mitigated given the length of the time dimension in our sample. Based on the estimated coefficients of δ and β, impulse response functions of inequality for baseline are calculated to evaluate the impacts of capital account liberalization. The simultaneous response of inequality in the aftermath of a capital account liberalization episode is δ_0; and the cumulative response one-year ahead is computed as $\delta_0 + (\delta_1 + \beta_0\delta_0)$. We obtain the confidence band of the estimated IRFs using the clustered (at country level) heteroscedasticity robust standard errors associated with the estimated coefficients.

4.2 *Mechanisms: Financial Development and Inclusion*

In low-income countries, the influence of financial development warrants special attention because, in general, most low-income countries have weak financial institutions and non-inclusive access to credit market. It has been noted that more than half of the poorest 40% of the population in developing countries are without access to formal banking accounts (Sahay et al. 2015). Therefore, liberalizing capital account may exacerbate the bias in accessing financial products in favor of those who are already well-off, thus increasing income inequality, instead of reducing inequality through allowing for better consumption smoothing and lower volatility.

[5]Trade openness is defined as the sum of import and export over GDP.

[6]We use the difference between gross and net Gini indices as proxy of government's redistributive policies.

We then expand Eq. (1) to test for this hypothesis by estimating the following equation in which the impacts of capital account liberalization on inequality are allowed to vary across countries with different levels of financial development.

$$g_{it} = a_i + \gamma_t + \sum_{j=1}^{l} \beta_j g_{i,t-j} + \sum_{j=1}^{l} \upsilon_j X_{i,t-j} + \sum_{j=0}^{l} \delta_j^- D_{i,t-j} G(z_{it}) + \sum_{j=0}^{l} \delta_j^+ D_{i,t-j} (1 - G(z_{it})) + \epsilon_{it} \quad (2)$$

with

$$G(z_{it}) = \frac{exp(-\gamma z_{it})}{1 + exp(-\gamma z_{it})}, \quad \gamma > 0$$

In Eq. (2), z_{it} where $z_{it} \sim N(0, 1)$, indicates the level of financial development of country i; $G(z_{it})$ is the associated smoothing transition function of the degree of financial deepening.[7] Other variables remain the same as in Eq. (1). This approach, equivalent to Granger and Terasvirta (1993) smooth transition autoregressive model, enhances the stability and precision of the estimated IRFs of the inequality to capital account liberalization. In addition, this approach can deal with potential correlation of standard errors within countries through clustering at the country-level.[8]

We use two different variables[9] of financial development (z_{it}) in our analysis. The first variable is the credit market freedom indictor from Fraser Institute's EFW database, which assigns scores between 0 (low depth and level of openness) and 10 (high depth and level of openness) to countries according to the depth and openness of their credit markets. EFW's indicator serves as a measure for financial deepening. The second variable measures financial inclusion and access to credit developed by Demirguc-Kunt et al. (2015), defined as the ratio of adults who borrowed from a formal financial institution in the past years to the number of total adults.

[7] γ is chosen to be 1.5, see Abiad et al. (2015).

[8] The same approach has been used to model non-linearity in many economic issues, such as exchange rate dynamics (Sarno and Taylor 2002); sector performance in business cycle (Fok et al. 2005), money demand (Chen and Wu 2005) and fiscal multipliers (Auerbach and Gorodnichenko 2013).

[9] For detailed description of variables and data sources, please refer to Data Appendix.

5 Results

5.1 The Baseline

5.1.1 Baseline Results

Using an unbalanced panel, we estimate Eq. (1) for low-income countries. The estimated coefficients for capital account reform, current account reform, regulation reform and their respective lags are presented in Table 5, Column 1. The results show that capital account liberalization episode has contemporaneous (i.e. within the year) and persistent impacts on inequality in low-income countries, with the largest influence on income inequality occurring one-year after the capital account

Table 5 The effect of capital account liberalization of inequality in LICs (1970–2010), OLS

	(1)
Variables	Gini growth
Dep. variable (t−1)	0.255**
	(0.0863)
Dep. variable (t−2)	0.183***
	(0.0585)
Capital account reform (t)	0.0111*
	(0.00578)
Capital account reform (t−1)	0.0165**
	(0.00604)
Capital account reform (t−2)	0.0137**
	(0.00619)
Current account reform (t)	−0.00651
	(0.00761)
Current account reform (t−1)	−0.00453
	(0.00465)
Current account reform (t−2)	−0.000218
	(0.00724)
Regulation reform (t)	−0.00411
	(0.00588)
Regulation reform (t−1)	−0.00387
	(0.00483)
Regulation reform (t−2)	−0.0109*
	(0.00569)
Constant	−0.143
	(0.730)
Observations	493
R-squared	0.356

Robust standard errors in parentheses
*** $p< 0.01$, ** $p< 0.05$, * $p< 0.1$

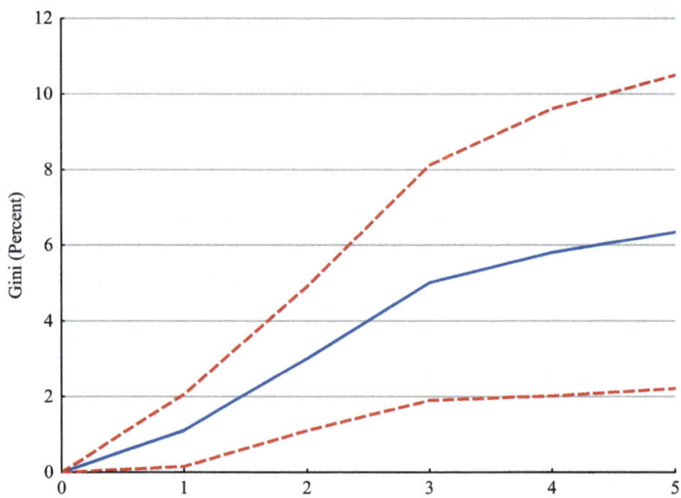

Fig. 3 The effect of capital account liberalization on inequality in LICs

liberalization. The persistent effects are further augmented by the high level of persistence in inequality itself. Interestingly, the results also reveal that both current account reforms and regulation reforms are associated with reduction in income inequality, even though their effects are not statistically significant.

The impulse response functions of inequality with respect to capital account liberalization are depicted in Fig. 3 with 90% confidence bands (dotted lines). The figure, in line with the results in Table 5, displays that capital account liberalization tends to increase inequality in low income countries and the effects are statistically significant and long-lasting. The estimated IRFs in Fig. 3 suggest that in short term, that is 2 years after the capital account reform, the liberalization episode may increase the Gini index by 3.0%, while in medium term, that is 5 years after the reform, the episode may increase the Gini index by 6.4% in low-income countries.

5.1.2 Robustness Check with Local Projection Method

There might be some concerns with IRFs computed using estimated coefficients of lagged inequality and of lagged capital account liberalization episode as the high correlation between lagged variables may cause t-statistic for individual coefficient to be weak. To mitigate the problem and check the robustness of our results, we use the local projection method developed by Jorda (2005) to estimate the effects of capital account reform on income inequality through following specification:

$$ln\left(Gini_{i,t+k}\right) - ln\left(Gini_{i,t}\right) = a_i^k + \gamma_t^k + \sum_{j=1}^{l} \beta_j g_{i,t-j} + \delta^k D_{i,t} + \upsilon^k X_{i,t} + \epsilon_{i,t}^k \quad (3)$$

where δ^k is the estimated response of inequality in each period $t + k$ to a capital account liberalization episode at time t; all other variables are the same as in Eq. (1). Equation (3) is then estimated for each $k = 1, \ldots, 5$. With this approach, we can directly trace the path of Gini indices in the aftermath of a capital account reform to obtain the dynamic response of inequality to the reform.

The estimated coefficients of equation (3) are presented in Table 6 and the estimated IRFs with 90% confidence band (dotted lines) are displayed in Fig. 4. As shown in the figure, while capital account liberalization does not have statistically

Table 6 The effect of capital account liberalization on inequality (1970–2010), local projection methods

	K=1	K=2	K=3	K=4	K=5
Capital account reform (t)	0.00891	0.0237**	0.0383***	0.0387**	0.0398***
	(0.00559)	(0.0101)	(0.0138)	(0.0145)	(0.0103)
Current account reform (t)	−0.00648	−0.0140	−0.0183	−0.00866	−0.00555
	(0.00672)	(0.00953)	(0.0130)	(0.0127)	(0.0149)
Regulation reform (t)	−0.00652	−0.0151**	−0.0262**	−0.0252	−0.0110
	(0.00412)	(0.00631)	(0.0107)	(0.0162)	(0.0268)
Constant	−0.157	−0.609	−0.981	−1.694	−2.145
	(0.636)	(1.448)	(2.407)	(3.169)	(3.607)
Observations	519	495	469	442	413
R-squared	0.364	0.484	0.539	0.622	0.694

Robust standard errors in parentheses
*** p< 0.01, ** p< 0.05, * p< 0.1

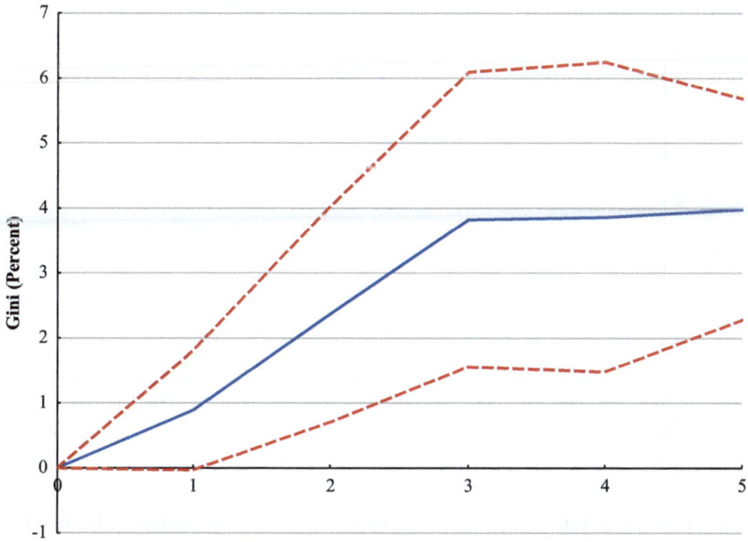

Fig. 4 The effect of capital account liberalization on inequality in LICs, local projections

significant effect in the first year immediately following the reform, it has strongly persistent and significant impacts on inequality afterwards, which confirms our baseline results. The IRFs are similar to the baseline estimation discussed in previous section. The results suggest that in low-income countries, capital account liberalization tends to increase Gini index by 2.4% in 2 years after the reform and by 4.0% in medium term (5 years after the reform), which is slightly smaller than the baseline.

5.1.3 Difference Between Equity, Bond and FDI Liberalization

To assess if there is difference in impacts of capital account liberalization on inequality when undergoing reforms in different asset markets, we re-estimated Eq. (1) by replacing Chinn-Ito index with Wang-Jahan indices of various subcategory of financial market, namely the equity market, the debt market and the FDI market. The results are presented in Fig. 5 and Table 7. Unfortunately, none of the coefficients using Wang-Jahan indices are statistically significant. That said, the point estimates are that, in the medium term, equity market liberalization tends to increase the inequality the most, by about 7%, followed by FDI liberalization which is likely to increase inequality by 4.0%. The results for FDI liberalization confirms previously estimations from previous study that FDI is associated with an increase in inequality (Jaumotte et al. 2013). In contrast, and surprisingly, opening up the debt market seems to reduce inequality by 1.0% in the medium term for low-income countries.

Fig. 5 The effect of different sub-category of capital account liberalization on inequality in LICs

Table 7 The effect of capital account liberalization on inequality (1970–2010), sub-category

	(1)	(2)	(3)
Variables	Gini growth	Gini growth	Gini growth
Dep. variable (t−1)	0.234**	0.204***	0.234**
	(0.112)	(0.0695)	(0.112)
Dep. variable (t−2)	0.173**	0.183***	0.168**
	(0.0700)	(0.0532)	(0.0708)
	Equity	Debt	FDI
Capital account reform (t)	0.0168	−0.000787	0.00316
	(0.0235)	(0.00696)	(0.0142)
Capital account reform (t−1)	0.0137	−0.00227	0.00927
	(0.0160)	(0.00964)	(0.0110)
Capital account reform (t−2)	0.0131	−0.00245	0.0138
	(0.0152)	(0.00930)	(0.0150)
Current account reform (t)	−0.0139**	−0.0130*	−0.0124*
	(0.00512)	(0.00660)	(0.00640)
Current account reform (t−1)	0.00162	−0.0201	0.00406
	(0.0112)	(0.0173)	(0.0116)
Current account reform (t−2)	−0.00629	−0.0114	−0.00404
	(0.00925)	(0.0114)	(0.0108)
Regulation reform (t)	0.000211	0.00527	0.000612
	(0.00638)	(0.0108)	(0.00629)
Regulation reform (t−1)	0.00293	0.00120	0.00178
	(0.00750)	(0.00765)	(0.00664)
Regulation reform (t−2)	0.000414	−0.00534	0.000215
	(0.0104)	(0.0119)	(0.0101)
Constant	0.817	0.479	0.834
	(1.856)	(2.422)	(1.746)
Observations	202	161	202
R-squared	0.562	0.632	0.563

Robust standard errors in parentheses
*** $p< 0.01$, ** $p< 0.05$, * $p< 0.1$

5.2 *Mechanisms: Financial Development and Inclusion*

To assess whether the level of financial development in low-income countries plays a role in determining the effects of capital account liberalization on income inequality, we first estimated Eq. (2) with EFW's credit market indicator. The IRFs obtained from the estimated coefficients of the two different levels of financial development are illustrated in Fig. 6, together with the baseline IRFs. The figure shows that in economies with high level of credit market openness, a capital account reform is only associated with 1.3% medium-term increase in Gini index in contrast to 9.6% increase in countries with low level of credit market openness. The results confirm

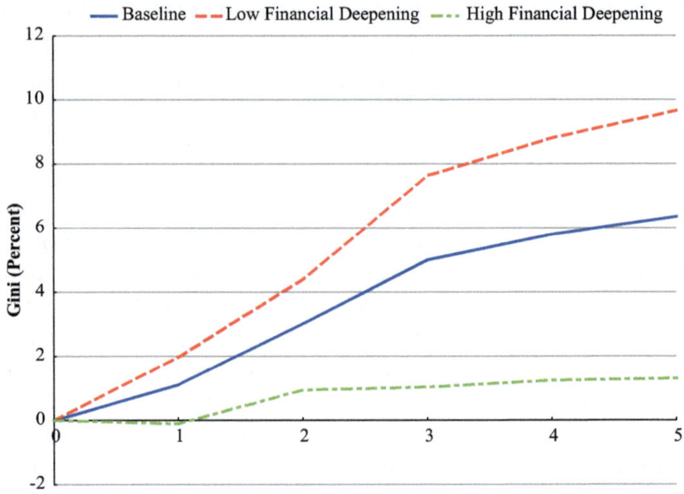

Fig. 6 The effect of capital account liberalization on inequality in LICs, financial depth

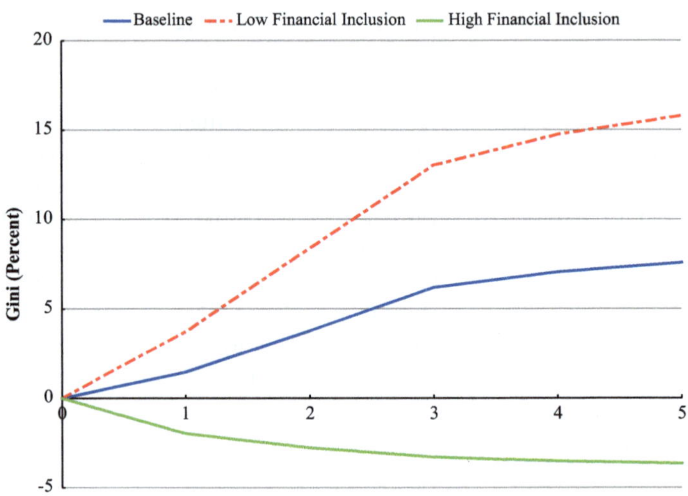

Fig. 7 The effect of capital account liberalization on inequality in LICs, financial inclusion

that the effects of capital account liberalization depend on the level of credit market development in low-income countries since for countries with low financial depth, the impact of capital account liberalization in the medium term is much larger than that of countries with high financial deepening.

We also re-estimate Eq. (2) with financial inclusion indicator and find that financial inclusion is crucial in determining the response of inequality to capital account liberalization episode in low-income countries. The respective IRFs of the high financial inclusion and low financial inclusion are displayed in Fig. 7. As

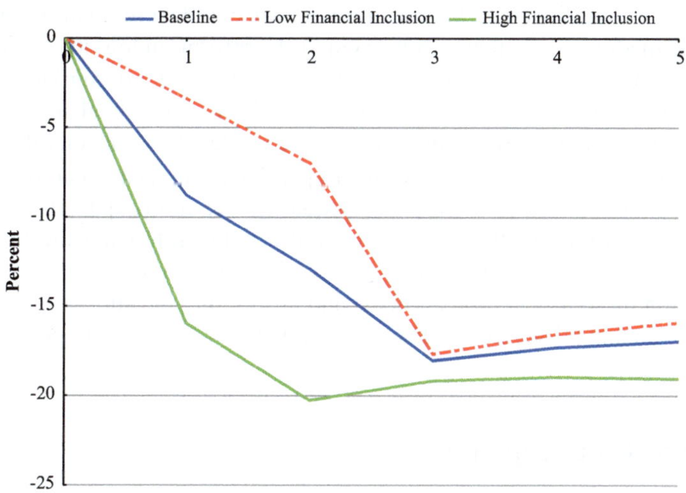

Fig. 8 The effect of capital account liberalization on poverty rates in LICs: the role of financial inclusion

demonstrated in the figure, in countries with low financial inclusion, Gini index typically experiences 15.7% increase in the medium term after a capital account reform, whereas in countries with high financial inclusion, inequality measured by Gini actually is decreased by 3.6%. In other words, for low-income countries, capital account liberalization helps to reduce the income inequality if the country has given people relatively more equal access to formal financial services.

In addition, we find that countries with high financial inclusion are likely to experience much larger reduction in poverty rates measured by poverty headcount ratio at $3.10 (2011 PPP) in short term, compared to countries with low financial inclusion, even though the effects tend to converge in the medium term after a capital account liberalization episode. The impacts of capital account reform on poverty are exhibited in Fig. 8.

6 Conclusion

Even though most of the literature has supported the idea that opening up capital account could bring considerable economic benefits, particularly on growth, to a country, the trend in inequality has casted doubt on whether these benefits can be distributed equally among the people, especially in low-income countries. This paper, using an unbalanced panel of 29 low-income countries over the period from 1970 to 2010, empirically examines the effects of capital account liberalization episode on income inequality and finds that capital account reform may lead to a long-lasting and statistically significant increase in inequality. Specifically, we estimate that a capital account liberalization reform could increase the Gini index by 6.5% in the medium term (that is, 5 years after the reform) in low-income countries.

In addition, our research provides evidence that the appropriate degree of capital account liberalization for a country depends on its financial and institutional development. For low-income countries, where the level of financial development and inclusion is relatively low, our results imply that a capital account liberalization reform is most likely to increase income inequality even more, up to 16% in medium term. When deciding to liberalize the capital account following the footsteps of high-income countries, policymakers in low-income countries should take into consideration these distributional effects and ensure that the supporting conditions are in place so that all segments of society can reap the benefits of opening up.

Appendix: Data Appendix

This appendix provides details on other variables used in this paper.

Current Account Openness Index

De jure index of current account openness is taken from Quinn and Toyoda (2008) dataset which is based on IMF's AREAER and is mostly available from 1980 to 2009 for 125 countries. The index measures a country's degree of compliance under IMF's Article VIII to free government restrictions in regard to international trade of goods and services. The higher the index, the more open the current account is.

Using the same method outlined in the paper to identify capital account liberalization, we assume current account liberalization reform occurs when, for a low-income country at a given time, the annual change in current account openness index exceeds by two standard deviations the average annual change of all observations covering 125 countries. As a result, with our low-income country sample, we are able to identify 10 current account liberalization episodes in 1980s, 21 in 1990s and 3 in 2000s.

Regulation Reform

The regulation index is from Fraser Institute's EFW (Economic Freedom of the World) database developed by Gwartney et al. (2016). It assesses the regulatory restraints of credit market, labor market and business regulations together for about 160 countries from 1970 to 2014, with lower number indicating more restrictions in these areas.

We identify an episode of regulatory reform if, for a given country at a given time, the annual change in the composite regulation index exceeds by two standard deviations the average annual change of all observations included in the EFW dataset. Following this methods, we find 8 regulatory reforms in 1980s, 21 in 1990s and 17 in 2000s for our low-income country sample.

Macroeconomic Variables

Macroeconomic variables: Gross domestic Product (GDP), GDP Growth Rate, Log GDP per Capita, Industry Value Added over GDP, Agriculture Value Added over GDP, Government Expenditure, Imports, Exports and Dependency Ratio are from World Bank's World Development Indicators (WDI), with annual data covering from 1970 to 2010. We calculated Trade Openness as the sum of exports and imports over GDP.

Redistributive Policies

To measure the redistributive policies of a government, we used the difference between market Gini indices, which measures the pre-tax and pre-transfer income inequality and net Gini indices, which measures inequality in post-tax and post-transfer income, as a proxy. Both measures of Gini indices are taken from Standardized World Income Inequality Database (SWIID) from 1970 to 2010.

Credit Market Freedom Indicator

Credit market freedom indictor (ownership of banks), based on the percentage of bank deposits held in privately owned banks, is from Fraser Institute's EFW as well, which measures conditions in the domestic credit market. The indicator is consisted of numeric variables between 0 and 10, covering around 160 countries from 1970 to 2014. When private held deposits ranges from 95 to 100% of total deposits, countries were given a score of 10 that indicates high financial depth and high level of credit market openness. When the private deposits were 10% or less of the total deposits, countries were assigned a rating of 0 to indicate low financial depth and low level of credit market openness.

Financial Inclusion

The financial inclusion indictor was taken from The Global Financial Inclusion database (Global Findex) developed by Demirguc-Kunt et al. (2015) at World Bank, covering around 140 economies. The inclusion indicator is defined as ratio of adults who borrowed from a formal financial institution in the past years to the number of total adults.

References

Abiad, A., Furceri, D., & Topalova, P. (2015). The macroeconomic effects of public investment: Evidence from advanced economies. *IMF Working Papers*, *15*(95), 1–26.

Auerbach, A. J., & Gorodnichenko, Y. (2013). Fiscal multipliers in recession and expansion. In A. Alesina, & F. Giavazzi (Eds.), *Fiscal policy after the financial crisis* (pp. 63–98). Chicago: University of Chicago Press.

Batuo, M. E., & Asongu, S. A. (2015). The impact of liberalisation policies on income inequality in African countries. *Journal of Economic Studies*, *42*(1), 68–100.

Bumann, S., & Lensink, R. (2016). Capital account liberalization and income inequality. *Journal of International Money and Finance*, *61*, 143–162.

Chen, S. -L., & Wu, J. -L. (2005). Long-run money demand revisited: Evidence from a non-linear approach. *Journal of International Money and Finance*, *24*(1), 19–37. doi:http://dx.doi.org/10.1016/j.jimonfin.2004.10.004.

Chinn, M. D., & Ito, H. (2008). A new measure of financial openness. *Journal of Comparative Policy Analysis: Research and Practice*, *10*(3), 309–322.

Dabla-Norris, E., Honda, J., Lahreche, A., & Verdier, G. (2010). FDI flows to low-income countries: Global drivers and growth implications. *IMF Working Papers*, *10*(132), 1–39.

Dabla-Norris, E., Kochhar, K., Suphaphiphat, N., Ricka, F., & Tsounta, E. (2015). Causes and consequences of income inequality: A global perspective. *IMF Staff Discussion Notes*, *15*(13), 1–39.

Delechat, C., Kovanen, A., & Wakeman-Linn, J. (2008, May). Sub-Saharan Africa: Private capital fueling growth. *IMF Survey Magazine*, *37*(5), 1–5.

Demirguc-Kunt, A., Klapper, L., Singer, D., & Oudheusden, P. V. (2015). The global findex database 2014: Measuring financial inclusion around the world. *World Bank Policy Research Working Papers*, WPS 7255, 1-97. http://documents.worldbank.org/curated/en/187761468179367706/The-Global-Findex-Database-2014-measuring-financial-inclusion-around-the-world.

Fok, D., van Dijk, D., & Franses, P. H. (2005). A multi-level panel STAR model for US manufacturing sectors. *Journal of Applied Econometrics*, *20*(6), 811–827. doi:10.1002/jae.822.

Furceri, D., & Loungani, P. (2015). Capital account liberalization and inequality. *IMF Working Papers*, *15*(243), 1–26.

Goldin, I., & Reinert, K.A. (2013, May). *Globalization for development: Meeting new challenges*. New York: Oxford University Press Inc.

Granger, C. W. J., & Terasvirta, T. (1993, December). *Modelling nonlinear economic relationships*. Oxford: Oxford University Press.

Gwartney, J., Lawson, R., & Hall, J. (2016). *Economic freedom of the World: 2016 annual report*. Vancouver: Fraser Institute.

Henry, P. B. (2007). Capital account liberalization: Theory, evidence, and speculation. *Journal of Economic Literature*, *45*(4), 887–935.

Jahan, S., & Wang, D. (2016). Capital account openness in low-income developing countries: Evidence from a new database. *IMF Working Paper*, *16*(252), 1–26.
Jaumotte, F., Lall, S., & Papageorgiou, C. (2013). Rising income inequality: Technology, or trade and financial globalization? *IMF Economic Review*, *61*(2), 271–309.
Jorda, O. (2005). Estimation and inference of impulse responses by local projections. *American Economic Review*, *95*(1), 161–182.
Kose, A. M., Prasad, E. S., & Taylor, A. D. (2011). Thresholds in the process of international financial integration. *Journal of International Money and Finance*, *30*(1), 147–179.
Kose, A. M., Prasad, E. S., & Terrones, M. E. (2009). Does financial globalization promote risk sharing? *Journal of Development Economics*, *89*(2), 258–270.
LaGarda, G., Gallagher, K., & Linares, J. (2016). Capital openness and income inequality: A cross-country study and the role of CCT in Latin America. MPRA Paper No.74181.
Nickell, S. (1981). Biases in dynamic models with fixed effects. *Econometrica*, *49*(6), 1417–1426.
Quinn, D. P., & Toyoda, A. M. (2008). Does capital account liberalization lead to growth? *Review of Financial Studies*, *21*(3), 1403–1449.
Sahay, R., Cihak, M., N'Diaye, P., Barajas, A., Mitra, S., Kyobe, A., et al. (2015). Financial inclusion: Can it meet multiple macroeconomic goals? *IMF Staff Discussion Notes*, *15*(17), 1–33.
Sarno, L., & Taylor, M. P. (2002). Purchasing power parity and the real exchange rate. *IMF Staff Papers*, *49*(1), 65–105.

On (Non-)Neutrality of Public Debt in Growing Economies

Alfred Greiner

Abstract In this paper we analyze effects of public debt on the long-run allocation of resources in a basic endogenous growth model with infinitely lived households. The government levies an income tax and issues government bonds to finance unproductive public spending. We demonstrate that in the case of flexible wages and elastic labour supply the balanced growth rate is the higher the smaller the ratio of public debt to GDP for a given income tax rate. When wages are rigid public debt is neutral in the sense that it does not affect the allocation of resources along the balanced growth path. Finally, in both cases the economy is stable only if the government puts a sufficiently high weight on stabilizing public debt.

1 Introduction

The financial and economic crises of 2008 that began as a sub-prime crisis in the USA in 2007 had plunged a great many economies throughout the world in deep recession. It seemed that the slump had been overcome by 2010 when some countries had reached their pre-crisis level of production. However, meanwhile the sub-prime crisis has turned into a public debt crisis because the bail-out of private financial institutions by governments led to an in part drastic increase of national debt ratios. In particular, in the Euro area some countries face severe problems and must be supported by others. This evolution drastically shows that public debt does affect real economies and with this paper we intend to contribute to the question of how public debt may affect the growth process of market economies in the long-run.

When one looks at the economics literature that studies the relationship between endogenous growth and public deficits and debt, one realizes that the way public spending is modelled plays an important role. For example, in an OLG model of endogenous growth Yakita (2008) demonstrates that there exists an upper bound for the level of public debt beyond which a sustainable debt policy is excluded that is the larger the higher the stock of productive public capital. Bräuninger (2005) also

A. Greiner
Department of Business Administration and Economics, Bielefeld University, P.O. Box 100131, 33501 Bielefeld, Germany
e-mail: agreiner@wiwi.uni-bielefeld.de

B. Bökemeier, A. Greiner (eds.), *Inequality and Finance in Macrodynamics*,
Dynamic Modeling and Econometrics in Economics and Finance 23,
DOI 10.1007/978-3-319-54690-2_5

finds that public debt policy is unsustainable once a critical value of public debt is crossed which, however, in his model only depends on the stock of private capital since he does not consider productive public spending. In the approach by Greiner (2008), who considers an endogenous growth model with infinitely lived agents and productive public capital, it is demonstrated that a higher ratio of public debt to GDP requires resources that reduce public spending so that higher debt leads to lower growth. But, once unemployment is allowed for, this result changes as has been pointed out by Greiner and Flaschel (2010). Then, an increase in deficit financed public spending may generate a higher long-run growth rate. In Greiner (2011) public spending is incorporated into the utility function and yields utility for the household sector. With this assumption it is shown that a debt policy that leads to higher long-run growth does not necessarily also imply higher welfare. Finally, Futagami et al. (2008) show that public debt policy may not only affect economic growth but can also be decisive as regards transition dynamics to the balanced growth path and the emergence of multiple growth paths.

With this paper we want to contribute to the research that analyzes how public debt affects economic growth in the long-run. We are interested in effects that come from public debt by itself so that we assume that public spending is a mere waste of resources. That means government spending is neither productive nor does it yield any utility. Further, we keep the income tax rate fixed so that any growth effects must indeed be attributed to variations of public debt.

When the government raises the stock of public debt, it has to run higher primary surpluses in the future to repay the higher debt. Then, the crucial question is which fiscal parameter the government uses to increase the primary surplus. When a distortionary tax is levied, the higher public debt will reduce future investment and growth. When the government raises a lump-sum tax or reduces non-distortionary transfers to the private sector, public debt will be neutral giving the Ricardo equivalence theorem. When the government reduces productive public spending as a result of higher debt, economic growth will also decline. However, when public spending is non-productive the effects of a decline in government spending with respect to economic growth are less obvious and it is this latter case that is analyzed in this contribution.[1]

Often endogenous growth models do not take into consideration unemployment when they are used to get insight into real world phenomena. The reason is that those models adopt a medium- to long-term perspective and markets, in particular labour markets, that are rigid in the short-run may become flexible in a longer-run perspective. But, if one looks at real world economies, one can see that unemployment in many European countries seems to present a persistent phenomenon. In the economics literature one can find different approaches that integrate unemployment in models of economic growth (Arico 2003, gives a survey of how unemployment can be integrated into endogenous growth models).

[1]We abstract from monetary aspects. A model showing how monetary and fiscal policy determine growth and inflation can be found in Greiner and Fincke (2015), Chap. 3.

Pissarides (1990), for example, identifies a link between economic growth and the labour market through a capitalization effect that states that firms are more willing to create new jobs in times of high growth. A different approach has been proposed by Aghion and Howitt (1994) who model the Schumpeterian idea of creative destruction meaning that innovations render existing technologies obsolete which then leads to reallocations of labour across firms. If specific skills are needed that not any job searcher disposes of, persistent unemployment may occur. Yet a different mechanism has been elaborated in the contribution by Acemoglu (1997). There, strategic interactions between firms affect their investment in new technologies and in human capital accumulation. This model can give rise to multiple equilibria where the equilibria with little employment is interpreted as an unemployment trap.

These considerations suggest that unemployment may indeed be relevant when one analyzes models of economic growth. Therefore, we distinguish between two cases in our analysis of growth effects of public debt. First, we consider the model with full employment and, then, we allow for wage rigidities that give rise to unemployment.

In the rest of the paper we proceed as follows. In the next section we present the structure of our growth model where we assume that wages are flexible and analyze the resulting model. Section 3, then, posits that the economy is characterized by wage rigidity and studies implications of that assumption with respect to long-run growth effects of public debt and deficit policies. Section 4, finally, summarizes the main results and concludes the paper.

2 The Growth Model with Flexible Wages and Elastic Labour Supply

To start with we describe the structure of our model for the case of flexible wages and elastic labour supply.

2.1 Model Structure

Our economy consists of three sectors: A household sector which receives labour income and income from its saving, a productive sector and the government. First, we describe the household sector.

2.1.1 The Household

The household sector consists of many identical households of mass one and where each household has measure zero. One household is representative for the whole household sector and maximizes the discounted stream of utility. As regards the utility function we adopt the function presented by Benhabib and Farmer (1994), where utility arises from per-capita consumption, $C(t)$, and where the household has disutility from labour, $L(t)$. The household maximizes its utility over an infinite time horizon subject to its budget constraint, taking factor prices as given. Thus, the maximization problem of the household can be written as[2]

$$\max_{C,L} \int_0^\infty e^{-\rho t} \left(\ln C - L^{1+\gamma}/(1+\gamma)\right) dt, \tag{1}$$

subject to

$$(1-\tau)(wL + rK + r_B B) = \dot{W} + C + \delta K. \tag{2}$$

$\rho \in (0, 1)$ is the household's rate of time preference, δ is the depreciation rate of capital and $\gamma \geq 0$ gives the inverse of the elasticity of labour supply. The variable w denotes the wage rate and r is the return to capital K and r_B the return to government bonds B. Wealth consists of public debt and of physical capital, $W := B+K$. Finally, $\tau \in (0, 1)$ is the constant income tax rate. The dot gives the derivative with respect to time.

A no-arbitrage condition requires that the return to capital equals the return to government bonds yielding $r_B = r - \delta/(1-\tau)$. This means that the household is indifferent whether its savings are used for investment or for financing government expenditures because both types of assets yield the same return. Thus, the budget constraint of the household can be written as

$$(1-\tau)(wL + rW) = \dot{W} + C + \delta W. \tag{3}$$

To solve this problem we formulate the current-value Hamiltonian which is written as

$$\mathcal{H} = \ln C + \lambda\left((1-\tau)(wL + rW) - C - \delta W\right) \tag{4}$$

with λ the shadow price of private wealth. Necessary optimality conditions are given by

$$C = w(1-\tau)L^{-\gamma} \tag{5}$$

$$\dot{C} = C((1-\tau)r - \rho) \tag{6}$$

If the transversality condition $\lim_{t\to\infty} e^{-\rho t} W/C = 0$ holds, which is the standard household's no-Ponzi game condition, that is fulfilled for a time path on which assets grow at the same rate as consumption, the necessary conditions are also sufficient.

[2]From now on we omit the time argument t if no ambiguity arises.

2.1.2 The Productive Sector

The productive sector is represented by one firm which behaves competitively and which maximizes static profits. The production function of the firm is given by,

$$Y = AK^{1-\alpha}\bar{K}^{\xi}L^{\alpha}, \tag{7}$$

with $(1-\alpha) \in (0,1)$ the capital share and $\alpha \in (0,1)$ the labour share, Y output and $\bar{K}$ represents the average economy-wide level of capital. The parameter $A > 0$ reflects total productivity and we assume constant returns to capital in the economy, i.e. $(1-\alpha) + \xi = 1$.

Using $(1-\alpha) + \xi = 1$ and that $K = \bar{K}$ in equilibrium, profit maximization gives

$$r = (1-\alpha)AL^{\alpha} \tag{8}$$

$$w = \alpha AL^{\alpha-1}K \tag{9}$$

2.1.3 The Government

The government in our economy receives tax revenues from income taxation and has revenues from issuing government bonds it uses to finance non-distortionary public spending, C_p, that is neither welfare enhancing nor productive. The period budget constraint of the government describing the accumulation of public debt is given by:

$$\dot{B} = r_B B(1-\tau) - \left(\tau Y - C_p\right) = r_B B(1-\tau) - S, \tag{10}$$

where S is the primary surplus, i.e. government surplus exclusive of net interest payments.

The inter-temporal budget constraint of the government is fulfilled if

$$B(0) = \int_0^{\infty} e^{-\int_0^{\mu}(1-\tau)r_B(\nu)d\nu} S(\mu)d\mu \Leftrightarrow \lim_{t\to\infty} e^{-\int_0^{t}(1-\tau)r_B(\mu)d\mu} B(t) = 0 \tag{11}$$

holds, which is the no-Ponzi game condition.

Now, assume that the government runs into debt today. Then, in order to meet the inter-temporal budget constraint, it has to repay that amount in the future in present value terms. This implies that the primary surplus must rise as public debt increases. Thus, the primary surplus is a function of public debt and of other variables. It should be noted that the government has some discretionary scope in setting the primary surplus so that assuming that public debt is the only determinant of the primary surplus would be too short-sighted. We posit that it is the level of GDP that determines the primary surplus, besides public debt, which seems to be reasonable.

Therefore, we assume that the government sets the primary surplus according to the following rule,

$$S(t) = \phi(t)\,Y(t) + \beta(t)\,B(t). \tag{12}$$

The parameter $\beta(t)$ determines how strong the primary surplus reacts to changes in public debt and will be denoted as reaction coefficient and $\phi(t)$ determines whether the level of the primary surplus rises or falls with an increase in GDP.

In Greiner (2008, 2011) it is shown that a positive coefficient β on average is necessary and sufficient for sustainability of public debt in our model. Since an average value for $\beta(t)$ that is strictly positive is necessary and sufficient for sustainability we set the reaction coefficient equal to that constant average value in our model and denote it by β. The parameter $\phi(t)$ will be bounded by above or below by a certain value since the effects of GDP on the determination of the primary surplus cannot go to plus or minus infinity. Hence, setting $\phi(t) = \phi =$ constant can be justified.

Here, it should also be mentioned that considerable effort has been devoted to estimating (12) empirically. For example, Bohn (1998) has obtained positive values for the reaction coefficient β for the USA by estimating (12) with OLS. Greiner et al. (2007) have performed the same estimation with data for euro area countries and also obtained statistically significant and positive values for β. In Fincke and Greiner (2011) time varying parameters have been estimated and it has been shown that the average coefficient is in most countries positive but time dependent. Setting β and ϕ equal to their average values, respectively, the evolution of public debt can be written as

$$\dot{B} = (r(1-\tau) - \delta - \beta)\,B - \phi\,Y, \tag{13}$$

with $\beta \in \mathbb{R}_{++}$ and $\phi \in \mathbb{R}$.

We should also like to point out that, in principle, public debt could be negative implying that the government would be a creditor. In this contribution, however, we assume throughout the paper that public debt is positive since this is definitely the more relevant and realistic case.

In the next subsection we define equilibrium conditions and the balanced growth path.

2.1.4 Equilibrium Conditions and the Balanced Growth Path

Before we analyze our model we give the definition of an equilibrium and of a balanced growth path. An equilibrium allocation for our economy is defined as follows.

Definition 3 An equilibrium is a sequence of variables $\{C(t), K(t), B(t)\}_{t=0}^{\infty}$ and a sequence of prices $\{w(t), r(t)\}_{t=0}^{\infty}$ such that

(a) equations (3), (5) and (6) hold,
(b) equations (8) and (9) hold and
(c) equations (12) and (13) hold.

In Definition 4 we define a balanced growth path.

Definition 4 A balanced growth path (BGP) is a path such that the economy is in equilibrium and such that consumption and capital grow at the same strictly positive constant growth rate, i.e. $\dot{C}/C = \dot{K}/K = g,\ g > 0,\ g = constant$, and either

(i) $\dot{B} = 0$ (balanced budget) or
(ii) $\dot{B}/B = g_B$, with $0 < g_B < g,\ g_B =$ constant, (slight deficit) or
(iii) $\dot{B}/B = \dot{C}/C = \dot{K}/K = \dot{G}/G = g$ (permanent deficit).

In Definition 4 we specify three different scenarios that imply different public debt and deficit policies. The first rule (i) is the balanced budget rule where the government has at each point in time a balanced budget. The second rule (ii) describes a situation where the government runs deficits so that the growth rate of public debt is positive in the long-run. But public debt grows at a smaller rate than capital, consumption and output[3] so that this scenario can be called slight deficit. The third rule, rule (iii), is called permanent deficit and describes a situation which is characterized by public deficits such that public debt grows at the same rate as all other economic variables in the long-run.

2.2 *Analysis of the Model*

To study our model, we note that it is completely described in equilibrium by the following equations,[4]

$$\frac{\dot{C}}{C} = (1-\tau)\,(1-\alpha)\,\psi\ (C/K)^{-\alpha/((1-\alpha)+\gamma)} - (\rho+\delta),\ C(0) > 0, \tag{14}$$

$$\frac{\dot{K}}{K} = \psi\ (C/K)^{-\alpha/((1-\alpha)+\gamma)}\ (1-\tau+\phi) - (C/K) + \beta\,(B/K) - \delta,\ K_0 > 0, \tag{15}$$

$$\frac{\dot{B}}{B} = \psi\ (C/K)^{-\alpha/((1-\alpha)+\gamma)}\ ((1-\tau)(1-\alpha) - \phi(K/B)) - \delta - \beta, B_0 > 0, \tag{16}$$

[3]Of course, GDP grows at the same rate as capital and consumption on a BGP.

[4]Without loss of generality we can set $A = 1$ in the analytical model what we do in the following.

with $\psi = (\alpha(1-\tau))^{\alpha/((1-\alpha)+\gamma)}$. The initial conditions with respect to capital and public debt are assumed to be given while consumption can be chosen by the household at time $t = 0$. It should be noted that (15) gives the resource constraint of the economy.

To analyze our economy around a BGP we define the new variables $c := C/K$ and $b := B/K$. Differentiating these variables with respect to time leads to a two dimensional system of differential equations given by,

$$\dot{c} = c\Big(c - c^{-\alpha/((1-\alpha)+\gamma)}\,\psi\,((1-\tau)(1-(1-\alpha))+\phi) - \rho - \beta\, b\Big), \tag{17}$$

$$\dot{b} = b\Big(c - c^{-\alpha/((1-\alpha)+\gamma)}\,\psi\,((1-\tau)(1-(1-\alpha))+\phi) - \phi\,\psi\, c^{-\alpha/((1-\alpha)+\gamma)}\, b^{-1}\Big) - b\,(\beta + \beta\, b). \tag{18}$$

A solution of $\dot{c} = \dot{b} = 0$ with respect to c, b gives a BGP for our model and the corresponding ratios $b^{\star}, c^{\star}$ along the BGP where the * is used to denote BGP values.

Before we analyze how the deficit policy of the government affects the allocation of resources and, thus, the growth rate in the long-run we first state in Proposition 1 that there exists a unique balanced growth path for this model.

Theorem 1 *Assume that the households' rate of time preference and the depreciation rate of capital are sufficiently small. Then, there exists a unique balanced growth path.*

Proof See Appendix. □

Proposition 1 shows that there exists a balanced growth path with sustained growth if the rate of time preference and the depreciation rate of capital are not too large. The latter requirement must be made because it can only be shown that a rest point of (17)–(18) exists but not that it is always associated with strictly positive growth which can be seen from (14). However, this is not a strict assumption. From an economic point of view it just states that the time preference of households plus the depreciation rate must be smaller than the after-tax return to capital so that ongoing growth is feasible.

Having established Proposition 1 we next deal with effects of public debt on the balanced growth path. Proposition 2 gives the result.

Theorem 2 *The balanced growth rate is the higher the smaller the ratio of public debt to GDP.*

Proof See Appendix. □

Proposition 2 demonstrates that higher public debt relative to GDP has a negative impact on the balanced growth rate. Looking at Eq. (14), which gives the balanced growth rate, this outcome is not so obvious because that equation is independent of public debt and also independent of the two parameters ϕ and β that determine public debt policy.[5] Further, it is true that the inter-temporal budget constraint of the government requires that a higher level of public debt be accompanied by higher primary surpluses. But, for a given income tax rate, higher primary surpluses are obtained through higher tax revenues in the future due to GDP growth and by reductions in public spending that is non-distortionary in our model. Hence, changes in public debt per se do not affect the allocation of resource and the question remains where this negative impact of public debt on economic growth comes from.

To answer this question we note that we assume inter-temporally optimizing households that decide how much labour they supply and how much they save and consume. When public debt and deficits are high, a large part of the household savings is used for unproductive purposes, namely for the public debt service and for unproductive public spending, so that the marginal value of additional saving, i.e. the shadow price of wealth λ, is low. This makes the household reduce its labour supply and saving, thus crowding out investment. It is in this way that public debt reduces saving and investment even if it does not cause any direct distortions.

Proposition 2 implies that governments can spur long-run growth by reducing the public debt to GDP ratio. In principle, in growing economies the government has two possibilities to do so. It can either run a balanced budget or it can run slight deficits such that public debt grows less than GDP. In both cases public debt relative to GDP will fall and in the long-run these two scenarios lead to the same growth performance which is stated in the following lemma.

Lemma 7 *The balanced budget scenario and a slight deficit yield the same balanced growth rate.*

Proof See Appendix. □

Given Lemma 7 we know that in the long-run it is irrelevant whether the government runs slight deficits or a balanced budget as long as GDP grows. Hence, the choice between a balanced budget and a slight deficit only has transitory affects and the debt ratio will be reduced faster when the government chooses to balance its budget compared to the case of a slight deficit.

To finish this section we state that a balanced budget and a slight deficit imply a higher long-run growth rate compared to permanent public deficits. We state this result as a corollary to Proposition 2.

Corollary 1 *The balanced budget scenario and a slight deficit yield a higher balanced growth rate than the scenario with permanent public deficits.*

Proof Follows immediately from Lemma 7 and Proposition 2. □

From our analysis up to now we know that a there exists a BGP for our model and how debt policy of the government affects the long-run growth rate. However, we

[5]Recall that the distortionary income tax rate τ is kept fixed.

have not yet studied the dynamics of our model. The next proposition shows under which conditions our model economy is stable in the sense that it asymptotically converges to the BGP.

Theorem 3 *Assume that the government runs a balanced budget or a slight deficit. Then, the model is saddle point stable.*

In the case of permanent deficits the model is saddle point stable if and only if the reaction of the primary surplus to higher public debt exceeds the difference between the net interest rate on public debt and the balanced growth rate.

Proof See Appendix. □

Proposition 3 states that the economy is saddle point stable when the government runs a balanced budget or only slight deficits. In this case, there exists a unique value of initial consumption, determined by the (shadow) price of capital, relative to capital such that the economy converges to the BGP in the long-run.

In the case of permanent public deficits the economy is stable only if the government puts a sufficiently high value on stabilizing public debt, i.e. if the reaction coefficient β is sufficiently large. Sufficiently large in this context means concretely that the reaction coefficient exceeds the difference between the net interest rate on public debt and the balanced growth, which has a nice and economically intuitive interpretation. Assume that public debt rises by one unit. Then, the additional interest burden just equals the net interest rate on debt which tends to make public debt explosive. On the other hand, on the BGP all economic variables and, thus, the tax revenue, too, rise at the balanced growth rate, which tends to stabilize the debt accumulation process. Since the difference between the net interest rate and the growth rate is positive, the increase in public debt makes this process unstable, unless the primary surplus rises at a magnitude at least as large as this difference. It should be noted that this increase in the primary surplus is achieved by a reduction of public spending in our setting where we keep the tax rate fixed.

These considerations demonstrate that the debt policy of the government is decisive as regards stability of the economy. If the government does not put a sufficiently high weight on stabilizing public debt, debt becomes explosive, implying that the government cannot fulfill its inter-temporal budget constraint, which makes the whole economy collapse.

3 The Model with Wage Rigidity

Next, we study our model for the case of wage rigidity that generates unemployment. We first describe how the model changes as we alter the basic assumption with respect to the functioning of the labour market, where we follow Greiner (2013).

3.1 The Model Setup

As in the previous section we again consider a decentralized economy with a household sector, a productive sector and the government.

3.1.1 The Household Sector

The household now maximizes over consumption alone and supplies L units of labor inelastically of which L^d is demanded by the productive sector. The rest $L - L^d$ is unemployed and the household receives unemployment benefits of $\kappa \cdot w$ per unemployed labor, which are not taxed, with w again the wage rate and $0 < \kappa < (1 - \tau)$. The latter assumption with respect to κ states that one unit of unemployed labour does not receive benefits exceeding the net wage income of one unit of employed labour. Total labor supply L is constant over time and the utility function is again logarithmic, $U(C) = \ln C$. Thus, the maximization problem of the household can be written as,

$$\max_{C} \int_{0}^{\infty} e^{-\rho t} \ln C \, dt, \tag{19}$$

subject to

$$(1 - \tau)\left(wL^d + rW\right) + \kappa w(L - L^d) = \dot{W} + C + \delta W. \tag{20}$$

The solution of this optimization problem again gives the growth rate of consumption which is equal to Eq. (6), except that the interest rate r is now different as will be shown next.

3.1.2 The Productive Sector and the Labor Market

The productive sector is represented by one firm which behaves competitively and which maximizes static profits. The production function of the firm is

$$Y = AK^{1-\alpha}\bar{K}^{\xi}\left(L^d\right)^{\alpha} \tag{21}$$

with the notation as in Sect. 2.1.

Profit maximization yields the interest rate and labour demand as

$$r = (1 - \alpha)(Y/K) \tag{22}$$

$$L^d = \alpha^{1/(1-\alpha)} A^{1/(1-\alpha)} (w/K)^{1/(\alpha-1)} \tag{23}$$

where we used $\bar{K} = K$ in equilibrium in (22) and (23).

The reason for rigid wages are labor market imperfections due to trade unions setting the wage rate. As concerns the wage setting we follow Raurich et al. (2006) and assume that there is a large number of unions that set the wage rate in order to maximize the following expression:

$$\max_w \left((1-\tau)w - w^s\right)^\gamma L^d(\cdot) \tag{24}$$

with $L^d(\cdot)$ given by (23), w^s denoting the reference wage and the coefficient $\gamma \in (0,1)$ gives a measure of the wage gap weight in the unions' objective function. The solution to the optimization problem yields the wage rate as

$$w = \frac{w^s}{(1-\tau)\,(1-\gamma(1-\alpha))} \tag{25}$$

As concerns the reference wage w^s there exist several approaches in the literature (see for example Layard et al. 1991; Blanchard and Katz 1999; Blanchard and Wolfers 2000; Collard and de la Croix 2000, or Raurich et al. 2006). We follow Raurich et al. (2006) and suppose that the reference wage equals workers' cumulated past average labor income where income further back in time contributes less than more recent labor income. In particular, we assume exponentially declining weights put on average labor income further back in time. Hence, the reference wage can be written as

$$w^s = \theta \int_{-\infty}^{t} e^{-\theta(t-s)} z(s) ds, \tag{26}$$

with z the workers' average income given by $z = (1-\tau)wL^d/L + \kappa w(L-L^d)/L$. The parameter $\theta > 0$ determines the weight attributed to more recent income. The higher θ, the larger the weight given to more recent levels of average income compared to income further back in time.

Differentiating (25) with respect to time and using (26) yields the growth rate of the wage rate as

$$\frac{\dot{w}}{w} = \theta \, \frac{(1-\tau)L^d/L + \kappa(L-L^d)/L}{(1-\tau)\,(1-\gamma(1-\alpha))} - \theta \tag{27}$$

3.1.3 The Government

The period budget constraint of the government in our economy describing the accumulation of public debt in continuous time is now given by

$$\dot{B} = r_B B(1-\tau) - \left(\tau Y - C_p - \kappa w(L-L^d)\right) = r_B B(1-\tau) - S, \tag{28}$$

where S again denotes the primary surplus of the government which it sets according to the rule given by Eq. (12) so that its inter-temporal budget constraint is fulfilled along the BGP.

3.1.4 Equilibrium Conditions and Balanced Growth

An equilibrium is in principle defined as in Sect. 2.1 but it differs a bit due to the fact that wages are not flexible now. Definition 5 determines an equilibrium.

Definition 5 An equilibrium is a sequence of variables $\{C(t), K(t), B(t)\}_{t=0}^{\infty}$ and a sequence of prices $\{w(t), r(t)\}_{t=0}^{\infty}$ such that

(a) equations (6) and (20) hold,
(b) equations (22), (23) and (27) with $L^d < L$ hold and
(c) equations (12) and (13) hold.

As regards long-run growth we define a balanced growth path analogously to the last section, that is a BGP is given if all variables grow at the same rate except for public debt possibly that may be constant or grow less than the other economic variables.

The growth rate of consumption is now obtained as

$$\frac{\dot{C}}{C} = (1-\tau)(1-\alpha)\alpha^{\alpha/(1-\alpha)}A^{\alpha/(1-\alpha)}(w/K)^{-\alpha/(1-\alpha)} - (\rho+\delta) \tag{29}$$

where we used the optimality conditions (22) and (23).

The economy wide resource constraint is derived by combining the budget constraint of the household with that of the government as

$$\frac{\dot{K}}{K} = \frac{Y}{K} - \frac{C}{K} + \beta\frac{B}{K} + (\phi-\tau)\frac{Y}{K} + \kappa(L-L^d)\frac{w}{K} - \delta\,, \tag{30}$$

Hence, the economy is completely described by Eqs. (13), (27), (29) and (30), with the return to capital, r, given by (22) and with L^d given by (23) and output determined by (21).

To get further insight into our model economy we define the new variables $c := C/K$, $x := w/K$ and $b := B/K$. Differentiating these variables with respect to time gives,[6]

$$\dot{c} = c\left((Y/K)((1-\alpha)(1-\tau)-1-(\phi-\tau)) - \rho + c - \beta b - \kappa(L-L^d)x\right) \tag{31}$$

$$\dot{x} = x\left(\theta\,\frac{(1-\tau)L^d/L + \kappa(L-L^d)/L}{(1-\tau)\,(1-\gamma(1-\alpha))} - \theta - \beta\, b - (\phi-\tau)(Y/K) - Y/K+\right.$$

[6]Again, we set $A = 1$ in the following analysis without loss of generality.

$$c + \delta - \kappa(L - L^d)\,x) \tag{32}$$

$$\dot{b} = b\left(\frac{Y}{K}\left((1-\alpha)(1-\tau) - \frac{\phi}{b} - 1 - (\phi - \tau)\right) - \beta + c - \beta b - \kappa(L - L^d)x\right) \tag{33}$$

with $Y/K = (L^d)^\alpha$ and $L^d = \alpha^{1/(1-\alpha)} x^{1/(\alpha-1)}$.

4 Analysis of the Model

A rest point of (31)–(33) gives a situation where all endogenous economic variables grow at the same rate, except public debt in case the government runs a balanced budget or a slight deficit. The next proposition clarifies the question of existence of a BGP.

Theorem 4 *Assume that the rate of time preference and the depreciation rate are sufficiently small and that $L^d < L$. Then, there exists a unique balanced growth path for*

$$\frac{\rho + \delta}{\theta} < 1 - \frac{\kappa}{(1-\tau)\,(1-\gamma(1-\alpha))}.$$

For

$$\frac{\rho + \delta}{\theta} > 1 - \frac{\kappa}{(1-\tau)\,(1-\gamma(1-\alpha))}$$

there exists no balanced growth path or there are two balanced growth paths.

Proof See Appendix. □

Proposition 4 gives conditions such that a unique BGP exists. It can also be seen that the model may produce two BGPs or no BGP at all. Whether there is a unique BGP or whether there exist two BGPs depends on the rate of time preference and on labour market conditions. The government affects the outcome indirectly by its income tax policy and by determining unemployment payments since these variables influence the reservation wage of trade unions. However, public debt and deficit policies have no effect on that outcome. The latter also holds for the long-run allocation of resources which is the contents of Proposition 5.

Theorem 5 *Assume that there exists at least one BGP for the model with wage rigidity. Then, the debt policy of the government does not affect the balanced growth rate.*

Proof See Appendix. □

Proposition 5 demonstrates that public debt in our model is neutral in the sense that it does not affect the allocation of resources in the long-run and, thus, has no effect on economic growth. The reason for this result is that the marginal product of capital that determines the incentive to invest depends on labour demand in the case of rigid wages which, for its part, is a negative function of the wage rate relative to the capital stock, $w/K = x$. Since public debt policy does not affect that variable, it has no effect on the balanced growth rate. The wage rate is determined by trade unions maximizing the wage sum relative to a reference wage sum according to Eq. (24). Hence, the wage policy of trade unions does not only determine labour income but also economic growth.

Next, we study stability of the model with wage rigidity. As in the case of wage flexibility s stronger reaction of the primary surplus to higher public debt tends to stabilize the economy. Proposition 6 gives the result assuming that the BGP is unique.

Theorem 6 *In the case of permanent deficits, there exists one negative real eigenvalue for β sufficiently small. For β sufficiently large, there are two negative eigenvalues (or two eigenvalues with negative real parts).*

Proof See Appendix. □

This proposition shows that for a balanced government budget or for a slight deficit the economy is saddle point stable. Given a fixed initial ratio of public debt to capital and also a fixed value of the wage rate relative to capital, there exists a uniquely determined value for initial consumption such that the economy converges to the BGP in the long-run.

If the government runs permanent deficits that also holds provided the reaction of the government to public debt is sufficiently large, where sufficiently large means that it exceeds at least the rate of time preference of the household. In fact, numerical examples suggest that two negative eigenvalues occur if and only if $\beta > \rho$ holds, although that cannot be proven for the analytical model. If that does not hold, i.e. if the reaction of the government to public debt is very small, there is only one negative eigenvalue. In this case, convergence to the balanced growth path is not given for a fixed initial public debt and wage rate.

To illustrate Proposition 6 we compute a numerical example. To do so we set the labour share to 60 percent, i.e. $\alpha = 0.6$, and $A = 0.5$. Total labour supply is $L = 1.25$ and unemployment benefits are 50 percent, $\kappa = 0.5$. The income tax rate is 10 percent, $\tau = 0.1$. The parameter θ reflecting the weight given to cumulated past income in the determination of the reference wage is set to 1. The inverse of the labour supply elasticity is set to $\gamma = 0.1$, the rate of time preference is $\rho = 0.08$ and the depreciation rate of capital is set to $\delta = 0.1$. Interpreting one time period as 2 years implies that the annual rate of time preference is about 3.9 percent and the annual depreciation rate about 4.9 percent. The balanced growth rate associated with these parameter values is $g = 0.0199$ which gives an annual growth rate of about 1 percent. The long-run unemployment rate amounts to 4.7 percent.

From Propositions 5 and 6 we know that the balanced growth rate is independent of public debt policy in the case of wage rigidity but public debt policy affects the dynamics. To see this we set $\phi = -0.01$ and $\beta = 0.12$ and $\beta = 0.0801$ which is larger than the rate of time preference which is $\rho = 0.08$. For these parameters there are two negative real eigenvalues. If we set $\phi = 0.01$ and $\beta = 0.07$ and $\beta = 0.01$ there is only one negative real eigenvalue such that this examples suggests that for values of β larger (smaller) than the time preference ρ there are two (one) negative eigenvalues.[7]

5 Conclusion

In this paper we have analyzed how public debt influences the long-run allocation of resources in a basic endogenous growth model with income taxation and unproductive public spending. Assuming that the inter-temporal budget constraint of the government must hold which implies that higher deficits must go along with higher primary surpluses, we could show that the long-run growth rate is the smaller the higher the debt to GDP ratio when wages are flexible. This holds for a fixed income tax rate and with non-distortionary public spending. The reason is that higher government debt implies a smaller shadow price of wealth that makes the household reduce its saving and labour supply.

In the case of rigid wages that give rise to unemployment public debt turns out to be neutral. This means that the long-run growth rate is independent of the debt and deficit policy of the government. The economic mechanism behind this outcome is that labour demand crucially determines the marginal product of private capital and, thus, the incentive to invest. If public spending is unproductive, variations in public debt that imply shifts in public spending have no effect on labour demand and on the return to private capital and, therefore, they do not influence long-run growth.

Further, we have seen that the economy is stable only if the government puts a sufficiently high weight on stabilizing public debt. If that does not hold the process of debt accumulation becomes explosive and, thus, the whole economy. This underlines the importance of the government sector as regards the stability of market economies.

Acknowledgements I thank two referees for valuable comments on an earlier version that helped to improve the paper.

[7]For $\beta > (<)\,\rho$ we must set $\phi < (>)\,0$ to get $b > 0$ on the BGP.

Appendix

Proof of Proposition 1

To prove this proposition we set $\dot{C}/C = \dot{B}/B$, which must hold on a BGP in the case of permanent public deficits, giving

$$\phi\, \psi\, c^{-\beta/(1-\beta+\gamma)} b^{-1} = \rho - \beta \qquad (A.1)$$

Substituting this relation in $\dot{b}/b$ gives,

$$\dot{b}/b = c - \rho - \psi c^{-\beta/(1-\beta+\gamma)}((1-\tau)(1-\alpha) + \rho\phi/(\rho-\beta))$$

From $(A.1)$ we know that $b > 0$ implies that ϕ and $\rho - \beta$ have the same sign so that $\phi/(\rho-\beta) > 0$ holds. With this, it is easily seen that the following relations hold,

$$\lim_{c\to 0}(\dot{b}/b) = -\infty, \quad \lim_{c\to\infty}(\dot{b}/b) = +\infty, \quad \partial(\dot{b}/b)/\partial c > 0.$$

This proves the existence of a unique $c^\star$ which solves $\dot{b}/b = 0$.

In case of a balanced budget or a slight deficit we have $b^\star = 0$ since public debt is constant while the capital stock grows over time or grows less than capital and, in addition, we also have $\phi = 0$ (see the proof of Lemma 7). Using this, the equation $\dot{c}/c$ can be written as

$$\dot{c}/c = \left(c - c^{-\alpha/((1-\alpha)+\gamma)}\psi(1-\tau)(1-(1-\alpha)) - \rho\right),$$

It is easily seen that the following relations hold,

$$\lim_{c\to 0}(\dot{c}/c) = -\infty, \quad \lim_{c\to\infty}(\dot{c}/c) = +\infty, \quad \partial(\dot{c}/c)/\partial c > 0.$$

This proves the existence of a unique $c^\star$ which solves $\dot{c}/c = 0$. □

Proof of Proposition 2

Setting $\dot{C}/C = \dot{B}/B$ implies

$$-\phi\,\psi\, c^{-\beta/(1-\beta+\gamma)} - \beta\, b = -\rho\, b$$

Substituting in $-\phi\,\psi\, c^{-\beta/(1-\beta+\gamma)} - \beta\, b$ by $-\rho\, b$ in (17) gives,

$$c - \rho - c^{-\alpha/((1-\alpha)+\gamma)}\,\psi\,(1-\tau)(1-(1-\alpha)) - \rho\, b = 0$$

From this we get by implicit differentiation:

$$\frac{dc}{db} = \frac{\rho}{1 + (\alpha/((1-\alpha)+\gamma))c^{-1-\alpha/((1-\alpha)+\gamma)}\psi\,(1-\tau)(1-(1-\alpha))} > 0$$

Since a higher c implies a lower balanced growth rate the proposition is proven. □

Proof of Lemma 7

To prove that lemma we see from (10) that setting $\phi = 0$ and $\beta = (1-\tau)$ gives $\dot{B} = 0$ (balanced budget scenario) and setting $\phi = 0$ and $\rho < \beta < (1-\tau)$ gives $0 < \dot{B}/B < \dot{C}/C = g$ (slight deficit). Both of these debt/deficit policies imply $b^{\star} = 0$ so that $\dot{b} = 0$ always holds. The balanced growth rate, then, is determined by the solution of $\dot{c} = 0$ with respect to c. Since $b^{\star} = 0$ and $\phi = 0$ hold both in the balanced budget scenario and for the slight deficit scenario, both scenarios imply the same $c^{\star}$ and, therefore, the same balanced growth rate. □

Proof of Proposition 3

To prove Proposition 3, we first note that the balanced budget rule (rule (i)) implies $b^{\star} = 0$ since public debt is constant while the capital stock grows over time. Further, $\phi = 0$ and $\beta = (1-\tau)r$ hold in this case.

The Jacobian of the dynamic system (17)–(18) is given by

$$J = \begin{bmatrix} a_{11} & -c(1-\tau)\,\alpha\,\psi\,c^{-\alpha/(1-\alpha+\gamma)} \\ 0 & -g \end{bmatrix},$$

with c and b evaluated at the rest point $\{c^{\star}, 0\}$ and a_{11} given by

$$a_{11} = c + (1-\tau)(1-\alpha)\,\psi\,(\alpha/(1-\alpha+\gamma))\,c^{-\alpha/(1-\alpha+\gamma)}.$$

The eigenvalues are $a_{11} > 0$, and $-g < 0$ so that the BGP is saddle point stable.

For slight deficits (rule (ii)) we also have $b^{\star} = 0$ and $\phi = 0$. The reaction coefficient now is $\rho < \beta < (1-\tau)r$ and the Jacobian J is obtained as,

$$J = \begin{bmatrix} a_{11} & -\beta\,c \\ 0 & \rho-\beta \end{bmatrix},$$

with c and b evaluated at the rest point $\{c^\star, 0\}$, a_{11} as above and where we used that $\dot{C}/C = \dot{K}/K$ holds on the BGP. Since $\rho < \beta$ holds for rule (ii) one eigenvalue is negative and one is positive implying saddle point stability, which proves the first part of the proposition.

In the case of permanent deficits (rule (iii)), the Jacobian matrix evaluated at the rest point of (17)–(18). The Jacobian is given by

$$J = \begin{bmatrix} a_{11} & -\beta\, c \\ a_{21} & \phi\, \psi\, c^{-\alpha/(1-\alpha+\gamma)} b^{-1} - \beta\, b \end{bmatrix},$$

with c and b evaluated at the rest point $\{c^\star, b^\star\}$ and a_{11} and a_{21} given by

$$a_{11} = c\left(1 + (\alpha/(1-\alpha+\gamma))\psi\, c^{-1-\alpha/(1-\alpha+\gamma)}(\phi + (1-\tau)(1-\alpha))\right)$$

$$a_{21} = b\left(1 + (\alpha/(1-\alpha+\gamma))\psi\, c^{-1-\alpha/(1-\alpha+\gamma)}(\phi(1+b^{-1}) + (1-\tau)(1-\alpha))\right)$$

The determinant of the Jacobian matrix can be computed as

$$\det J = (\rho - \beta)\left(c + (\alpha/(1-\alpha+\gamma))\psi\, c^{-\alpha/(1-\alpha+\gamma)}(1-\tau)(1-\alpha)\right) + \rho\,(\alpha/(1-\alpha+\gamma))\phi\, \psi\, c^{-\alpha/(1-\alpha+\gamma)}$$

From $\dot{C}/C = \dot{B}/B$, which must hold on a BGP, we have $\phi\, \psi\, c^{-\alpha/(1-\alpha+\gamma)} b^{-1} = \rho - \beta$. Using this we can rewrite the determinant as follows,

$$\det J = (\rho - \beta)\left(c + (\alpha/(1-\alpha+\gamma))\psi\, c^{-\alpha/(1-\alpha+\gamma)}(1-\tau)(1-\alpha)\right) + \rho\,(\alpha/(1-\alpha+\gamma))(\rho - \beta)\, b$$

For $\beta > \rho$ the determinant is negative since $b > 0$ holds.

For $\beta < \rho$ the determinant is positive. To show that the BGP is unstable we have to compute the trace of the Jacobian, $\operatorname{tr} J$, which is given by,

$$tr\, J = c - \beta\, b + (\alpha/(1-\alpha+\gamma))\psi\, c^{-\alpha/(1-\alpha+\gamma)}(\phi + (1-\tau)(1-\alpha)) + \phi\, \psi b^{-1} c^{-\alpha/(1-\alpha+\gamma)}.$$

To see that $\operatorname{tr} J$ is positive we first note that a positive value of b implies $\phi > 0$ for $\beta < \rho$. Further, from $\dot{c}/c = 0$ we get $c - \beta\, b = \rho + c^{-\alpha/(1-\alpha+\gamma)}(\phi + (1-\tau)(1-\alpha)) > 0$, so that the trace of the Jacobian is positive, too. Since the trace and the determinant are both positive, the BGP is unstable for $\beta < \rho$.

Finally, we note that on the BGP we have from (14) the relation $\rho = (1-\tau)r - g$. Thus, Proposition 3 is proven. □

Proof of Proposition 4

To prove that proposition we set $\dot{x}/x = 0$ and solve that equation with respect to c giving

$$c = \theta - \theta\,\frac{(1-\tau)L^d/L + \kappa(L-L^d)/L}{(1-\tau)\,(1-\gamma(1-\alpha))} + \beta\,b + (\phi-\tau)(L^d)^\alpha + (L^d)^\alpha - \delta + \kappa(L-L^d)\,x,$$

with $L^d = \alpha^{1/(1-\alpha)}x^{1/(\alpha-1)}$. Inserting that c in $\dot{c}$ leads to the following equation that we denote by f:

$$f = (1-\alpha)(1-\tau)\,\alpha^{\alpha/(1-\alpha)}x^{\alpha/(\alpha-1)} - \left(\frac{\theta(1-\tau)-\theta\kappa}{(1-\tau)(1-\gamma(1-\alpha))}\right)\alpha x^{1/(\alpha-1)}/L - (\rho+\delta) +$$
$$\theta(1-\kappa/((1-\tau)(1-\gamma(1-\alpha)))).$$

A solution $f = 0$ with respect to x gives a rest point of (31)–(33). As regards f we have

$$\lim_{x\to 0} f = -\infty, \quad \lim_{x\to\infty} f = -(\rho+\delta) + \theta\left(1 - \frac{\kappa}{(1-\tau)(1-\gamma(1-\alpha))}\right).$$

The first derivative of f is given by

$$\frac{\partial f}{\partial x} = -\left(\frac{\alpha}{1-\alpha}\right)x^{-1+\alpha/(\alpha-1)}\alpha^{\alpha/(1-\alpha)}(1-\alpha)(1-\tau) +$$
$$\left(\frac{1}{1-\alpha}\right)x^{-1+1/(\alpha-1)}\left(\frac{\theta(1-\tau)-\theta\kappa}{(1-\tau)(1-\gamma(1-\alpha))}\right)\alpha/L.$$

The second derivative of f is

$$\frac{\partial^2 f}{\partial x^2} = \frac{\alpha}{(1-\alpha)^2}\,x^{-2+\alpha/(\alpha-1)}\alpha^{\alpha/(1-\alpha)}(1-\alpha)(1-\tau) +$$
$$(-1)\,\frac{(2-\alpha)}{(1-\alpha)^2}\,x^{-2+1/(\alpha-1)}\left(\frac{\theta(1-\tau)-\theta\kappa}{(1-\tau)(1-\gamma(1-\alpha))}\right)\alpha/L.$$

Setting $\partial f/\partial x = 0$ gives

$$x = x_m = \frac{\alpha}{\alpha\,L\alpha^{\alpha/(1-\alpha)}(1-\alpha)(1-\tau)}\left(\frac{\theta(1-\tau)-\theta\kappa}{(1-\tau)(1-\gamma(1-\alpha))}\right).$$

Inserting x_m in $\partial^2 f/\partial x^2$ shows that the sign of the resulting expression is equivalent to

$$-\alpha\,\alpha^{\alpha/(1-\alpha)}(1-\alpha)(1-\tau) < 0.$$

This demonstrates that the function f reaches a maximum for $x = x_m$ and it has a unique turning point given by

$$x = x_w = x_m \cdot (1 + (1-\alpha)) .$$

Thus, the function f is concave-convex, starts at $-\infty$, reaches a maximum at $x = x_m$, has a turning point at $x = x_w$ and converges to $-(\rho+\delta)+\theta(1-\kappa/((1-\tau)(1-\gamma(1-\alpha))))$ for $x \to \infty$. This implies that for $-(\rho+\delta)+\theta(1-\kappa/((1-\tau)(1-\gamma(1-\alpha)))) > 0$ there exists a unique rest point of (31)–(33) and for $-(\rho+\delta)+\theta(1-\kappa/((1-\tau)(1-\gamma(1-\alpha)))) < 0$ there exist two rest points for Eqs. (31)–(33) or no rest point if f does not intersect the horizontal axis. □

Proof of Proposition 5

We know that a BGP is given for a value of x such that the function f in Proposition 4 equals zero. Looking at f it is immediately seen that this function does neither depend on the ratio of public debt to capital, b, nor on the parameters β and ϕ. □

Proof of Proposition 6

We compute the Jacobian matrix evaluated at the rest point of (31)–(33). For the balanced budget (rule (i)) and for the slight deficit (rule (ii)), the Jacobian is given by

$$J = \begin{bmatrix} c & c\left(\partial(\dot{C}/C)/\partial x - \partial(\dot{K}/K)/\partial x\right) & -\beta c \\ x & x\left(\partial(\dot{w}/w)/\partial x - \partial(\dot{K}/K)/\partial x\right) & -\beta x \\ 0 & 0 & a_{33} \end{bmatrix},$$

where c and x evaluated at the BGP and where we have used that $b = 0$ holds on the BGP with a balanced government budget or a slight deficit. In case of a balanced budget we have $a_{33} = -\dot{K}/K$ and for the slight deficit $a_{33} = \rho - \beta < 0$, since the slight deficit is obtained for $0 < \rho < \beta < (1-\tau)r$. The eigenvalues of that matrix are given by $-\dot{K}/K = -g < 0$ (balanced budget) and by $\rho - \beta < 0$ (slight deficit) and by the eigenvalues of the matrix J_1 which is

$$J_1 = \begin{bmatrix} c\, c\left(\partial(\dot{C}/C)/\partial x - \partial(\dot{K}/K)/\partial x\right) \\ x\, x\left(\partial(\dot{w}/w)/\partial x - \partial(\dot{K}/K)/\partial x\right) \end{bmatrix}$$

The determinant of that matrix is obtained as

$$\det J_1 = cx\left(\partial(\dot{w}/w)/\partial x - \partial(\dot{C}/C)/\partial x\right) = (-1)\, c\, x\, (\partial f/\partial x),$$

with f from the proof of Proposition 5. With a unique rest point of (31)–(33), f has a positive derivative at $f = 0$ implying that the determinant is negative so that J_1 has one negative and one positive eigenvalue.

In case of permanent deficits (rule (iii)) we have

$$J_2 = \begin{bmatrix} c\, c\left(\partial(\dot{C}/C)/\partial x - \partial(\dot{K}/K)/\partial x\right) & -\beta c \\ x\, x\left(\partial(\dot{w}/w)/\partial x - \partial(\dot{K}/K)/\partial x\right) & -\beta x \\ b\, b\left(\partial(\dot{B}/B)/\partial x - \partial(\dot{K}/K)/\partial x\right) & -\beta b + (\rho - \beta) \end{bmatrix},$$

The determinant is given by

$$\det J_2 = (-1)\, c\, x\, (\rho - \beta)\, (\partial f/\partial x)$$

and the trace, tr J_2, is

$$\text{tr}\, J_2 = c + x\left(\partial(\dot{w}/w)/\partial x - \partial(\dot{K}/K)/\partial x\right) + (-1)(\beta b - (\rho - \beta)) = C_1 - (\beta - \rho), \quad (A.1)$$

with C_1 containing terms that are independent of β and independent of c and b that are determined by β on the BGP.

Setting $\beta = 0$ we can explicitly compute the eigenvalues ev_i, $i = 1, 2, 3$ as

$$ev_1 = \rho,\ ev_{2,3} = (1/2)\left(\text{tr}\, J_1 \pm \sqrt{(\text{tr}\, J_1)^2 - 4 \det J_1}\right),$$

with J_1 as for the balanced budget that has one positive and one negative eigenvalues. The determinant of the Jacobian J_2 is negative in this case. This shows that two eigenvalues of J_2 are positive and one is negative. For reasons of continuity this also holds in an ϵ environment around $\beta = 0$.

If β is sufficiently large, i.e. at least larger than ρ, the determinant of the Jacobian J_2 is positive, since $(\partial f/\partial x) > 0$. Because of $\det J_2 = ev_1 \cdot ev_2 \cdot ev_3$ we know that in this case there is no negative eigenvalue or two negative eigenvalues (or eigenvalues with negative real parts in case of complex conjugate eigenvalues). Further, we know that we have for the trace, tr $J_2 = ev_1 + ev_2 + ev_3$. From (A.1) we see that tr J_2 monotonically declines with β so that it becomes negative for a sufficiently large β implying that there is at least one negative eigenvalue. But, because of $\det J_2 > 0$ there must be two negative eigenvalues in that case. □

References

Acemoglu, D. (1997). Technology, unemployment and efficiency. *European Economic Review, 41*, 525–533.

Aghion, P., & Howitt, P. (1994). Growth and unemployment. *Review of Economic Studies, 61*, 477–494.

Arico, F. (2003). Growth and unemployment: Towards a theoretical integration. *Journal of Economic Surveys, 17*, 419–455.

Benhabib, J., & Farmer, R. (1994). Indeterminacy and increasing returns. *Journal of Economic Theory, 63*, 19–41.

Bohn, H. (1998). The behaviour of U.S. public debt and deficits. *Quarterly Journal of Economics, 113*, 949–963.

Blanchard, O., & Katz, L.F. (1999). Wage dynamics: Reconciling theory and evidence. *American Economic Review, 89*, 69–74.

Blanchard, O., & Wolfers, J. (2000). The role of shocks and institutions in the rise of European unemployment: The aggregate evidence. *The Economic Journal, 110*, C1–C33.

Bräuninger, M. (2005). The budget deficit, public debt, and endogenous growth. *Journal of Public Economic Theory, 7*, 827–840.

Collard, F., & de la Croix, D. (2000). Gift exchange and the business cycle: The fair wage strikes back. *Review of Economic Dynamics, 3*, 166–193.

Fincke, B., & Greiner, A. (2011). Debt sustainability in selected euro area countries. Empirical evidence estimating time-varying parameters. *Studies in Nonlinear Dynamics & Econometrics, 15*(3), article 2.

Futagami, K., Iwaisako, T., & Ohdoi, R. (2008). Debt policy rule, productive government spending, and multiple growth paths. *Macroeconomic Dynamics, 12*, 445–462.

Greiner, A. (2008). Does it pay to have a balanced government budget? *Journal of Institutional and Theoretical Economics, 164*, 460–476.

Greiner, A. (2011). Economic growth, public debt and welfare: Comparing three budgetary rules. *German Economic Review, 12*, 205–22.

Greiner, A. (2013). Sustainable public debt and economic growth under wage rigidity. *Metroeconomica, 64*, 272–292.

Greiner, A., & Flaschel, P. (2010). Public debt and public investment in an endogenous growth model with real wage rigidities. *Scottish Journal of Political Economy, 57*, 68–84.

Greiner, A., & Fincke, B. (2015). *Public debt, sustainability and economic growth: Theory and empirics*. Heidelberg/New York: Springer.

Greiner, A., Köller, U., & Semmler, W. (2007). Debt sustainability in the European monetary union: Theory and empirical evidence for selected countries. *Oxford Economic Papers, 59*, 194–218.

Layard, P. R. J., Nickell, S. J., & Jackman, R. (1991). *Unemployment: Macroeconomic performance and the labor market*. Oxford: Oxford University Press.

Pissarides, C. A. (1990). *Equilibrium unemployment theory*. Oxford: Blackwell.

Raurich, X., Sala, H., & Sorolla, V. (2006). Unemployment, growth and fiscal policy: New insights on the Hysteresis hypothesis. *Macroeconomic Dynamics, 10*, 285–316.

Yakita, A. (2008). Sustainability of public debt, public capital formation, and endogenous growth in an overlapping generations framework. *Journal of Public Economics, 92*, 897–914.

Financial Intermediation and Directed Technical Change

Elmar Hillebrand

Abstract This paper develops an endogenous directed technical change growth model with financial intermediation. Technical change is driven by R&D investments of private agents in response to market incentives and can take different directions. Key feature is that innovators are capital constrained and need external funds to finance R&D effort. Financial intermediaries finance these ventures. The main theoretical result shows that credit interest rates—a "risk effect"—add to the determinants of directed technical change: Beside the price and the market size effect, the risk effect encourages innovations in those sectors, where the risk of innovation failure is lower. The degree of substitutability regulates the power of these different effects and determines how innovations respond to changes in relative factor supply, given that the risk effect is an additional determinant of directed technical change.

1 Introduction

There is little doubt that technical change is a key driving force of economic growth. Less clear, however, is the role of financial intermediation in the process of innovations and technical change: "pre-crisis" macroeconomics with regard to financial intermediation and economic growth offer deviating positions. Somewhat oversimplifying existing approaches, we can summarize these as follows. First, Some economists suggest that financial markets have at the most a minor relevance in the sense that the development in the financial sector follows"real"-sector activity (Robinson 1952) or that macroeconomic research overestimates the role of financial development when exploring the determinants of economic growth (Lucas 1988). Second, other economic researchers go one step further and proclaimed their view on financial matters with respect to economic growth by simply ignoring it (Chandavarkar 1992; Meier and Seers 1984; Stern 1989). Third and last, some researchers assign financial intermediation an active role that cannot be ignored when exploring

E. Hillebrand
EEFA Research Institute, Muenster, Germany
e-mail: e.hillebrand@eefa.de

B. Bökemeier, A. Greiner (eds.), *Inequality and Finance in Macrodynamics*,
Dynamic Modeling and Econometrics in Economics and Finance 23,
DOI 10.1007/978-3-319-54690-2_6

the mechanics of technical change and economic growth (Schumpeter 1912; Boyd and Prescott 1986; King and Levine 1993; Levine 1997), although the channels of this active influence are still a matter of scientific debate. The reasons for these different positions are (i) that in the standard general equilibrium framework with complete markets, financial intermediation is completely irrelevant. (ii) The financial sector became much more important for global economic activity over the last 50 years: according to the world bank, the ratio of total credit to global GDP increased from 48% in 1960 to 131% in 2015. (iii) The global financial and governmental debt crisis that started with the collapse of Lehman Brothers in September 2008 proved that financial market melt downs can be contagious to economic activity and can cause global recessions.

Empirical observations report that the share of research and development expenditures (R&D) on total investment expenditures in the U.S. increased from 8% to 20% between 1960 and 2011. While in the same time period, total business loans divided by total business investment in the U.S. fluctuated around 85%. Since firms raise capital to finance investments either by asking lenders for loans or by issuing bonds/stocks on financial markets, credit markets and firms' R&D effort are connected to some extent.

Moreover, interest rates on business loans differ across industries: Capital markets allocate capital to sectors with the highest expected return. But interest rates also reflect the risk inherent in an investment project that investors would have to bear. Since different investment projects compete for loans, projects linked with higher risk have to accept higher loan rates to compensate for the higher risk involved. Consequently, firms who are free to direct R&D effort to a certain industry, ceteris paribus choose the sector linked with the lowest risk of default, because the associated credit costs are the lowest. Together, this could indicate that financial intermediaries might affect the direction of firms' R&D effort and thus innovations and technical change in a systematic way.

To examine this impact of financial intermediation on the determinants of directed technology innovation, the present study develops a micro-founded general equilibrium model framework. The developed model combines elements from two different strands of literature and is therefore related to each of them. First, the study builds on the literature on endogenous directed technological change (Acemoglu 1998, 1999, 2002, 2007). The approach in its turn builds on the work of Schumpeter (1912), Romer (1990), Aghion and Howitt (1992), Grossman and Helpman (1993) and features some types of profit maximizing individuals who engage in research and development (R&D) in order to attain profits from inventing new commodities of greater quality. By extension, in the endogenous directed technical change approach, innovative firms target different sector- or factor specific technologies. The key feature is that expected profits from innovation determine the choice to which sector or factor the innovative firms direct their R&D effort in the first place. Consequently, the direction of innovations and thus technical change is now endogenous.

The literature on endogenous technical change cited above does not consider the sketched role of credit or financial intermediation in the process of R&D and innovations. So second, this study draws on the literature on financial intermediation and economic development (Schumpeter 1912; Boyd and Prescott 1986; King and Levine 1993; Levine 1997) which is more or less entirely based on the work of Schumpeter (1912) who states that financial intermediaries have to be considered in the process of economic growth, because they evaluate and finance entrepreneurs and thereby help to invent new products, i.e. they influence technical change. This literature does not address questions related to the direction and bias of technical change.

The present paper therefore combines the findings of these two different strands of research and introduces capital constraints and financial intermediaries into existing approaches of endogenous directed technical change. The paper is organized as follows. Section 2 introduces the model. The decentralized equilibrium solution and its properties are studied in Sect. 3. Section 4 studies the determinants of directed technical change given that costly innovations are financed via credit. Section 5 presents the properties of the considered balanced growth path equilibrium. Section 6 concludes, mathematical results and proofs can be found in the Appendix.

2 The Model

2.1 The Economy

The economy considered here is a single country or world economy which evolves infinitely in discrete time t, $t \in \{0, 1, 2, \ldots\}$. In each period two different generations of consumers populate the economy. Population is constant over time. Consumers are heterogeneous: Each generation consists of workers, entrepreneurs and innovators who differ in terms of their access to investment projects and consumption profiles. The other entities in the economy are: a sector of financial intermediation and a "real" side of goods production. The production contains three different stages: during the first stage, firms produce a variety of capital goods of different quality. Capital constrained innovators engage in R&D to increase the quality of those capital goods. Financial intermediaries supply business credit to finance innovators' R&D effort. Entrepreneurs produce capital goods of existing quality. The innovation activity, i.e. the rise in product quality is the engine of economic growth. The second stage contains two heterogeneous intermediate production sectors and during the last stage, firms produce one unique final good. The final good can be consumed, transferred into future periods and invested in capital goods production or in R&D. The final or consumption good serves as the numeraire: all prices, returns, payments and costs are measured in terms of the consumption good.

The economy features the following *ex-ante* information imperfection: Neither financial intermediaries nor individual innovators themselves know in advance, whether they are successful in the research lab. It is only public information that innovators are successful with a certain probability, so the investment projects feature idiosyncratic risk. This lack in information implies that financial intermediaries allocate some credit resources to innovators who will fail in the innovation process and thus default from credit liabilities. Financial intermediaries thus face idiosyncratic credit risk (see for instance Wong 1997).

2.2 *Production Sectors*

Goods production in the different stages is linked in a "downstream" manner.

2.2.1 Final Output

In the final stage, production combines inputs from two intermediate sectors to produce the final output of the economy. The final sector consists of a single representative firm which produces consumption good output in period t using the technology

$$Y_t = \left(\gamma Y_{m,t}^{\frac{\varepsilon-1}{\varepsilon}} + (1-\gamma) Y_{z,t}^{\frac{\varepsilon-1}{\varepsilon}}\right)^{\frac{\varepsilon}{\varepsilon-1}}, \tag{1}$$

where $\varepsilon \in [0, \infty[$ represents the constant elasticity of substitution between the inputs used in the production process, $\gamma \in]0, 1[$ is a distribution parameter and $Y_{m,t}$, $Y_{z,t}$ represent intermediate input quantities.

2.2.2 Intermediate Production

During the second production stage, firms produce these sector specific intermediate goods by combining two input factors: a sector-specific factor and a continuum variety of capital goods which complement this sector-specific factor. In both of these two stages, firms operate under perfect competition. The intermediate stage consists of the two different sectors, denoted by $j \in \{m, z\}$. Each sector consists of a single representative firm, and in each sector j firms produce intermediate good output using the technology

$$Y_{j,t} = \left(\int_0^1 q_t(\vartheta_j)^{1-\mu} x_t(\vartheta_j|q)^{\mu} d\vartheta_j\right) J_t^{1-\mu}, \tag{2}$$

where $\mu \in]0, 1[$ is a technology parameter, $J = \{M, Z\}$ represents a sector specific factor, constant in supply and $x(\vartheta_j|q)$ denotes the quantity input of a capital good type ϑ_j with quality q in sector j (at time t). This technology assumes the use of a continuum variety of different capital goods, whose mass is normalized to unity.

2.2.3 Capital Goods Production and Innovations

The undermost and first stage exhibits imperfect competition. Each capital good is supplied monopolistically either by an entrepreneur or by an innovator. Entrepreneurs hold patents to produce capital goods of existing quality. Innovators try to invent new generations of product with higher quality in some of the capital good lines. If innovation is successful, the innovators receive a patent on that innovation and they supply the newly invented good of higher quality to the intermediate firms. The entrepreneur on that product line makes zero profit. If innovation is not successful, the entrepreneur exclusively sells the capital good with existent quality level to the intermediate sector. Key feature here is, that innovators decide to which sector they direct their R&D effort on the basis of expected profits. This is the engine of endogenous directed technical change.

Innovators in sector j distribute across the different capital good lines ϑ_j. Each innovator observes the characteristics of her current state of the art product with quality q and spends h units of the final good for R&D to improve the quality of that commodity. With regard to the quality increase of an innovation, I make the following:

Assumption 1 *The quality of each new generation of J-complementary capital goods is exactly λ_j times the quality of the preceding product generation. Formally,*

$$q(n_{\vartheta_j}) \doteq \lambda_j q(n_{\vartheta_j} - 1) \quad \textit{for all } n_{\vartheta_j} = 0, 1, 2, \ldots \textit{ and } \vartheta_j \in [0, 1], \lambda_j > 1.$$

To save on notation, we write q instead of $q(n_{\vartheta_j})$ and simply mean the current quality level of capital good line ϑ_j. This step-by-step rise in the quality of a commodity is commonly referred to as innovation on a "quality ladder": whenever a firm is successful in the research lab and develops a blue print of a higher quality product, it "jumps" one step up the quality ladder. In order to ensure the possibility of permanent economic growth, every capital goods quality can be improved an unlimited number of times. To perform R&D, innovators borrow capital (measured in units of the final good) from financial intermediaries.[1] Accordingly, credit contracts between innovators and banks are signed on an individual level. Besides

[1]This lack of resources is in contrast to standard models of endogenous growth (Rivera-Batiz and Romer 1991; Romer 1990) and growth models of factor biased technical change (Acemoglu 2002), where each potential entrepreneur takes production costs from the jar of final goods. The assumption of capital-constrained innovators has been used in different strands of research. For an analysis regarding endogenous growth and finance see for instance King and Levine (1993). Moreover, Williamson (1986) assume external project finance to analyze optimal debt contracts under asymmetric information. As in Williamson (1986, 1987), Bernanke and Gertler (1989), the formulation presented here implies a fixed project size. In other words the decision, which sector the individual chooses to produce machines for, does not depend on the project size and the size of a project remains constant over time. A variable loan size is considered for instance in Gale and Hellwig (1985).

the final good, no other inputs are required in the R&D process.[2] The "size" of an innovation λ_j is exogenous and constant and differs with respect to the target sector $j = \{m, z\}$. Therefore λ_j is sector-specific and $\lambda_m \neq \lambda_z$ might hold. Independent of the intermediate sector $j = \{m, z\}$ a capital good is produced for, the unit costs are given ψ. Without loss of generality, I normalize $\psi \equiv 1$.

2.3 Heterogeneous Consumers

In each period a continuum of young consumers is born. Each consumer lives for two periods. Population is constant over time. The young generation is indexed by the superscript "y", members of the old generation are indexed with an "o". Consumers are heterogeneous and are either a worker, an entrepreneur or an innovator. So at each point in time, two different generations populate the economy and each generation consists of workers (w), entrepreneurs (e), and innovators (i), who differ in terms of their access to investment projects and consumption profiles. The subindex $\ell \in \{w, e, i\}$ identifies the different consumer types. Preferences over consumption are identical for all consumers and equal to:

Assumption 2 *Consumption preferences can be represented by an utility function $U : \mathbb{R}^2_+ \to \mathbb{R}$, which is defined as*

$$U(c^y_{\ell,t}, c^o_{\ell,t+1}) := \beta u(c^o_{\ell,t+1}), \tag{3}$$

β is the individual discount factor and the "instantaneous" utility function $u : \mathbb{R}_+ \to \mathbb{R}$, is logarithmic: $u(c) := log(c)$,

2.3.1 Workers

At each point in time a continuum of mass M of young workers is born. Each one is endowed with one unit of labor time. Workers supply their labor inelastically to the labor market when young and receive a "real" wage rate of $w_{M,t}$. At the beginning of their second period of life, when old, they retire. Workers only consume in the second period of life and, therefore, wish to transfer their current wealth into the next period. For this purpose, they supply their labor income to the deposit market and receive a deposit rate of r_{t+1} on these savings.

[2]This specification is referred to as the 'lab equipment' specification, first developed in Rivera-Batiz and Romer (1991).

2.3.2 Innovators

Next, a continuum of innovators is born whose mass is normalized to one. Innovators direct their investment to capital goods innovation for either intermediate sector on the basis of expected profits. Denote the share of innovators directing R&D investment to sector m by ι, then the share of innovators investing in R&D in sector z equals $1 - \iota$, where $\iota \in [0, 1]$ can vary in equilibrium.

Innovators group in an "alliance" to protect themselves from idiosyncratic risks in profits within one intermediate sector $j = \{m, z\}$. Since the investment projects contain idiosyncratic risk, we can apply a law of large number argument for a continuum of random variables, and conclude that the fraction of successful projects is equal to the ex-ante probability of successful innovation η_j. Aggregate average profits in sector j are given by $\eta_j \Pi_j^{(i)}$, so the income for innovators in sector $j \in \{m, z\}$ reads

$$\bar{\Pi}_{j,t} = \iota_j \eta_j \Pi_{j,t}. \tag{4}$$

2.3.3 Entrepreneurs

Finally, at the beginning of each period, a continuum of young entrepreneurs enters the economy. The mass of entrepreneurs holding patents for capital goods in each intermediate sectors $j \in \{m, z\}$ equals unity. This ensures identical masses of capital good lines and entrepreneurs in either sector. Each entrepreneur receives a patent to produce a capital good of current quality, so on each capital good line exactly one entrepreneur is present.

Entrepreneurs form an "entrepreneur association" to protect themselves from idiosyncratic risks in profits. Since realizations of individual innovation attempts are independent and thus uncorrelated, a fraction of the projects will be successful while the rest of the projects will be unsuccessful.[3] Entrepreneurs make profits in capital good lines without an innovator or where innovators' R&D is unsuccessful. Together this implies an income for the entrepreneur association of

$$\bar{\Pi}_t^e = (1 - \eta_m \iota)\Pi_{m,t}^e + (1 - (1 - \iota)\eta_z)\Pi_{z,t}^e. \tag{5}$$

[3] Using a law of large numbers type of reasoning for a continuum of random variables (see Uhlig 1996), the fraction of successful projects in sector $j \in \{m, z\}$ is equal to the ex-ante probability of successful innovation η_j and the fraction of unsuccessful projects is equal to the ex-ante probability of innovation failure $1-\eta_j$. Moreover, using the same type of reasoning, the ex-ante probability that an innovator tries to improve the quality of a certain capital good in sector j is equal to the frequency or mass of innovators in sector m, denoted by $\iota_m := \iota$ and sector z, denoted by $\iota_z := 1 - \iota$.

2.4 *Banking Sector*

Financial intermediation is described by a banking sector which consists of a single price taking representative bank.[4,5] This bank pools a large number of small amount private savings from the workers and provides capital resources for business investments to the innovators.

Successful innovators return credit plus interest to the bank. If not successful, the invested capital is lost and innovators default from credit liabilities. So the bank faces individual credit risk (See for instance Wong 1997). However, the returns of the individual innovators' investment projects are stochastically independent and consequently, for a large number, in fact a continuum of different projects, these idiosyncratic risks are completely diversified. Therefore, the randomness in the returns vanishes in aggregate terms and the bank completely diversifies individual credit risk.[6]

Since $\eta_m \neq \eta_z$ holds true in general, expected return from financial intermediation depends on the total size of capital resources devoted to either sector m or z: Denote total credit supply in sector $j = \{m, z\}$ at time t by $D_{j,t}$. The aggregate banking sectors' deposit demand reads D_t. Denote r_t the deposit interest rate, aggregate deposit cost are equal to $r_{t+1}D_t$. With credit interest rates $R_{j,t}, j = \{m, z\}$, the bank's return of funding then equals

$$\eta_m R_{m,t+1} D_{m,t} + \eta_z R_{z,t+1} D_{z,t}. \tag{6}$$

2.5 *Aggregation*

At the sectoral level, we need a quality index, describing the average development of capital goods quality. Then, one can derive R&D spending, capital goods investment and the sector specific outputs as functions of the average sectoral quality index, denoted by $Q_{m,t}$ in sector m and $Q_{z,t}$ in sector z respectively. The aggregate sectoral

---

[4]This study deals with "banks", because a bank represents the classic example of a "financial intermediary". Nevertheless, other institutions like credit unions, financial advisers or brokers, insurance companies or pension funds are also financial intermediaries who perform functions similar to banks in the context of the present study: here, a bank or financial intermediary (a) brings together borrowers and lenders and (b) performs certain functions mentioned above. In order to highlight the role of banks with respect to the intensity and direction of firms R&D effort, all financial transactions in the economy are entirely intermediated, no "private" borrowing or lending is possible (Diamond 1984; Williamson 1986).

[5]See also Boyd and Prescott (1986), King and Levine (1993), and Levine and Zervos (1998).

[6]Again, the underlying law of large numbers argument holds in principle only for averages of a countable sequence of random variables, where as the model here features a continuum of random variables. However, Uhlig (1996) showed how to obtain a law of large number for a continuum of uncorrelated random variables.

quality index for $t \geq 0$ is defined as[7]:

$$Q_{j,t} := \int_0^1 q_t(\vartheta_j)d\vartheta_j \quad \text{for} \quad j = \{m, z\}. \tag{7}$$

Index Q_j is a combination of the various $q(\vartheta_j)$'s and increases in the $q(\vartheta_j)$'s affect aggregate sectoral output to the extent that they raise $Q_{j,t}$. The index of aggregate quality in this economy is the average of qualities, since each intermediate sector uses a continuum of capital goods whose mass is normalized to unity.[8]

The dynamics of the average sector-specific quality improvements can now be derived as follows: for any industry ϑ_j in sector $j = \{m, z\}$, the probability of success is the same and is given by (34). One can show that the expected change in the average quality of capital goods in sector j between time t and $t + 1$ is given by

$$\mathbb{E}_t[Q_{j,t+1} - Q_{j,t}] = \iota_{j,t}\eta_{j,t}\Big(\lambda_j - 1\Big)Q_{j,t}. \tag{8}$$

Using the law of large numbers argument given in Uhlig (1996), we can conclude that the average sectoral capital goods quality develops deterministically over time—even though individual R&D projects exhibit stochastic returns, so that the expectations operator in the equation above can be dropped and the dynamic development of average sectoral capital goods quality $Q_{m,t}$ and $Q_{z,t}$ is given by

$$\frac{Q_{m,t+1} - Q_{m,t}}{Q_{m,t}} = \iota_t\eta_{m,t}\,(\lambda_m - 1) \quad \text{for all } t \geq 0, \;\; Q_{m,0} > 0 \text{ given}, \tag{9a}$$

$$\frac{Q_{z,t+1} - Q_{z,t}}{Q_{z,t}} = (1 - \iota_t)\eta_{z,t}\,(\lambda_z - 1) \quad \text{for all } t \geq 0, \;\; Q_{z,0} > 0 \text{ given}, \tag{9b}$$

where by Assumption 1, $\lambda_j > 1$ so $\iota_{j,t}\eta_{j,t}\left(\lambda_j - 1\right) > 0$ and $Q_{j,0}$ is some given initial quality level. According to Eqs. (9a) and (9b), the dynamic development of the average quality of the capital goods in each intermediate sector can be represented by a deterministic, linear first order difference equation. It is important to bear in mind that the average quality of sector-specific capital goods used in the production of the two intermediate goods Y_m and Y_z are different, allowing technical change to

[7]See for instance Barro and Sala-i Martin (2004) and Acemoglu (2009).

[8]Other aggregation types that reflect for instance a CES-aggregator type can be found in Barro and Sala-i Martin (2004, P.324 eq. (7.15)) or Acemoglu (2009). Note that the aggregator is linear homogeneous, which captures the plausible feature that if we increase all single productivities by a number $a > 0$ then the average sectoral productivity will increase by a.

be biased or directed to one intermediate sector. The average quality of capital goods Q_m and Q_z, determine aggregate productivity, while the quotient Q_z/Q_m determines the relative productivity of the factor Z.

Now, sectoral R&D expenditures can be computed from individual R&D expenditures in industries ϑ_j as

$$H_{j,t} = \int_0^1 h_t(\vartheta_j)d\vartheta_j \qquad j \in \{m, z\} \tag{10}$$

and aggregate R&D in period t is simply given by the sum of aggregate sectoral R&D:

$$H_t = H_{m,t} + H_{z,t}. \tag{11}$$

Similarly, spending on capital goods in sector j is

$$X_{j,t} = \int_0^1 x_t(\vartheta_j)d\vartheta_j \qquad j \in \{m, z\} \tag{12}$$

and aggregate capital goods spending is the sum over sectoral capital goods spending equal to

$$X_t = X_{m,t} + X_{z,t}. \tag{13}$$

Aggregate savings are given by total labor income of the workers

$$S_t = w_{M,t}M_t \tag{14}$$

and aggregate consumption in period t is given by consumption of old workers, old entrepreneurs and old innovators:

$$C_t = \sum_\ell C_{\ell,t}^{(o)}, \qquad \ell \in \{w, e, i\}. \tag{15}$$

3 The Decentralized Equilibrium

The following sections derive the competitive equilibrium of the economy where producers, consumers and banks behave optimally and markets clear.

3.1 Producer Behavior

3.1.1 Final Output

The single representative firm in final output production takes the distribution parameter $\gamma > 0$, prices of the final good p and of intermediate inputs p_m, p_z the production technology stated in (1) as given and solves the following optimization problem:

$$\max_{(Y_m, Y_z) \in \mathbb{R}^2_+} \left\{ p\mathcal{F}(Y_m, Y_z) - p_m Y_m - p_z Y_z \quad | \quad \mathcal{F}(Y_M, Y_Z) = (1) \right\} \tag{16}$$

A solution to (16) satisfies the following first order conditions which equate prices and marginal products of each production factor for all $t \geq 0$:

$$p_{m,t} = \gamma \left(\gamma Y_{m,t}^{\frac{\varepsilon-1}{\varepsilon}} + (1-\gamma) Y_{z,t}^{\frac{\varepsilon-1}{\varepsilon}} \right)^{\frac{1}{\varepsilon-1}} Y_{m,t}^{-\frac{1}{\varepsilon}} \tag{17a}$$

$$p_{z,t} = (1-\gamma) \left(\gamma Y_{m,t}^{\frac{\varepsilon-1}{\varepsilon}} + (1-\gamma) Y_{z,t}^{\frac{\varepsilon-1}{\varepsilon}} \right)^{\frac{1}{\varepsilon-1}} Y_{z,t}^{-\frac{1}{\varepsilon}}. \tag{17b}$$

The consumption good serves as the numeraire. Hence, the price of this consumption good is normalized to $p \equiv 1$ in all periods $t \geq 0$. The normalization of the final output price implies a normalization of intermediate goods prices. The following lemma describes this equivalence in price normalization of final output and intermediate goods prices for the case of N different intermediate goods.

Lemma 1 *Suppose, the production of final output uses $N \in \mathbb{N}$ different intermediate inputs Y_i, $i = 1, 2, \ldots, N$. Then the normalization of final output prices p is equivalent to setting the weighted sum of intermediate goods prices $\sum_{i=1}^{N} (\gamma_i^{\varepsilon} p_i^{1-\varepsilon})^{\frac{1}{1-\varepsilon}}$ equal to 1 in all periods. This weighted sum is referred to as the "ideal" price index or Fisher price index.*

In this study, the final goods sector combines $N = 2$ different intermediate goods. The ideal price index in this case is given by

$$p = \left(\gamma_m^{\varepsilon} p_m^{1-\varepsilon} + \gamma_z^{\varepsilon} p_z^{1-\varepsilon} \right)^{\frac{1}{1-\varepsilon}} = \left(\gamma^{\varepsilon} p_m^{1-\varepsilon} + (1-\gamma)^{\varepsilon} p_z^{1-\varepsilon} \right)^{\frac{1}{1-\varepsilon}} \equiv 1. \tag{18}$$

3.1.2 Intermediate Output

The firms in the two sectors m and z maximize profits taking the price of their product, p_m, p_z, the rental prices of the machines $p(\vartheta_m|q)$, $p(\vartheta_z|q)$ the quality of machines, $q(\vartheta_m)$, $q(\vartheta_z)$ and the input prices w_M, w_Z of the factors M and Z as given.

The decision problem of the representative firm in sector $j = \{m, z\}$ reads

$$\max_{(x(\vartheta_j|q),J)\in\mathbb{R}^2_+} \left\{ p_j Y_j - w_J J - \int_0^1 p(\vartheta_j|q)x(\vartheta_j|q)d\vartheta_j \mid Y_j = (2) \right\} \quad (19)$$

A solution to (19) satisfies the following first order optimality conditions which equate prices and marginal products of each production factor for all $t \geq 0$

$$(1-\mu)p_{j,t}J_t^{-\mu}\left(\int_0^1 q_t(\vartheta_j)^{1-\mu}x_t(\vartheta_j|q)^{\mu}d\vartheta_j\right) = w_{J,t}, \quad (20a)$$

$$\mu p_{j,t}J^{1-\mu}\left(q_t(\vartheta_j)^{1-\mu}x_t(\vartheta_j|q)^{\mu-1}\right) = p_t(\vartheta_j|q). \quad (20b)$$

Demand for capital goods of type ϑ_j of quality $q(\vartheta_j)$ in sector $j = \{m, z\}$ can be derived from (20b) as

$$x_t(\vartheta_m|q) = \left(\frac{\mu p_{m,t}}{p_t(\vartheta_m|q)}\right)^{\frac{1}{1-\mu}} q_t(\vartheta_m)M_t. \quad (21a)$$

$$x_t(\vartheta_z|q) = \left(\frac{\mu p_{z,t}}{p_t(\vartheta_z|q)}\right)^{\frac{1}{1-\mu}} q_t(\vartheta_z)Z_t. \quad (21b)$$

These equations imply that the quantity of used capital goods is increasing in the corresponding price of the intermediate good, p_m, p_z and in the firms' employment of the sector-specific factor $J = \{M, Z\}$. The capital goods use in sector j is decreasing in the price of the capital good $p(\vartheta_j|q)$. Intuitively, a greater price for the intermediate output increases the value of the marginal product of all factors, including that of capital goods. This encourages intermediate firms to purchase more capital goods. A greater level of factor employment J implies more use of the capital goods, hence capital goods demand rises. Moreover, because the demand curve for the capital goods is downward sloping, a higher cost implies lower demand.

3.1.3 Capital Goods Production

Innovators and entrepreneurs face a two-stage decision process within two consecutive periods of time. Consider an arbitrary period t. This period t is referred to as the "first" decision period and the subsequent period $t + 1$ is the "second" decision period. Innovators' first decision period is the research and planning period. In this stage, they choose optimal R&D expenditures. Entrepreneurs do not undertake any actions in the first decision period. In the second decision period, innovators then set the profit maximizing price of their capital good, given that they were successful in the first period. Entrepreneurs correspondingly set the price given that innovation on their product line was not successful or even did not take place at all. The decision problem is solved recursively. First we establish the optimal price for a capital goods producer to set in the second decision period, $t+1$. Afterwards, we derive optimal R&D expenditures and compute the probability of successful innovation, while we assume the following relationship between R&D expenditures and the probability of successful innovation in this study:

Assumption 3 *The individual probability of successful innovation $\eta_t(\vartheta_j)$, $j = \{m, z\}$ is a strictly concave function of R&D expenditures $h_t(\vartheta_j)$ and is defined as*

$$\eta_t(\vartheta_j) := h_t(\vartheta_j)^{\alpha}\phi_t(\vartheta_j), \qquad j = \{m, z\}, \tag{22}$$

where $\alpha \in]0, 1[$ and

$$\phi_t(\vartheta_j) := \frac{1}{\zeta} q_t(\vartheta_j)^{-\alpha} \tag{23}$$

captures the effects of the capital good's current quality ladder position.[9]

Profit Maximization

Taking capital goods demand from Eq. (21) as given, the "second" period decision problem of an innovator in period $t+1$ on capital goods line ϑ_j then reads:

$$\max_{p(\vartheta_j|q)\in\mathbb{R}_+} \left\{ \left(p_{t+1}(\vartheta_j|q) - 1\right) x_{t+1}(\vartheta_j|q) - R_{j,t+1}h_t(\vartheta_j) \;\middle|\; x_{t+1}(\vartheta_j|q) = (21)\right\}. \tag{24}$$

As the time index indicates, optimal R&D $h_t(\vartheta_j)$ is determined in the first decision stage and thus plays no role in the determination of the profit maximizing price. The first order optimality condition with respect to $p(\vartheta_j|q)$ gives the profit maximizing monopoly price of a capital good ϑ_j:

$$p_{t+1}(\vartheta_j|q) = \frac{1}{\mu}, \tag{25}$$

which is a constant markup over marginal cost and equal across industries ϑ_j. The resulting flow of monopoly profit for an innovator then can be computed as

$$\pi_{t+1}^{(i)}(\vartheta_j) = \bar{\pi}_{j,t+1} q_{t+1}(\vartheta_j) - R_{j,t+1} h_{t+1}(\vartheta_j) \tag{26}$$

where $\bar{\pi}_{j,t+1} := (1-\mu)\mu^{\frac{1+\mu}{1-\mu}} p_{j,t+1}^{\frac{1}{1-\mu}} J_{t+1}$.

[9] Since the mapping from R&D to success-probability is unbounded in general, the defined η's are not probabilities in a strict sense, i.e. they do not necessarily take values between zero and one. We account for this and set the parameters appropriately, so that the values of success-probabilities lie between zero and one.

An entrepreneur in a different industry $\vartheta_{j'}$ solves:

$$\max_{p(\vartheta_{j'}|q)\in\mathbb{R}_+} \left\{ \left(p_{t+1}(\vartheta_{j'}|q) - 1\right) x_{t+1}(\vartheta_{j'}|q) \;\mid\; x_{t+1}(\vartheta_{j'}|q) = (21)\right\} \tag{27}$$

The profit maximizing monopoly price of a capital good $\vartheta_{j'}$ is

$$p_{t+1}(\vartheta_{j'}|q) = \frac{1}{\mu} \tag{28}$$

and the monopoly profit for an entrepreneur e reads

$$\pi_{t+1}^{(e)}(\vartheta_{j'}) = \bar{\pi}_{j,t+1} q_{t+1}(\vartheta_{j'}). \tag{29}$$

Optimal R&D

In the first decision period, an innovator in capital goods line ϑ_j chooses R&D expenditures $h_t(\vartheta_j)$ to maximizes expected discounted profits from capital goods selling and takes the interest rates r_t, $R_{j,t+1}$, the price of the intermediate good $p_{j,t}$, the profit maximizing price determined in the second decision stage $p_{t+1}(\vartheta_j|q)$, quality level q_{t+1} factors J_{t+1} and the function ϕ_{ϑ_j} as given.

$$\max_{h_t(\vartheta_j)\in\mathbb{R}_+} \left\{ \frac{\eta_t(\vartheta_j)}{r_t} \left(\bar{\pi}_{j,t+1} q_{t+1}(\vartheta_j) - R_{j,t+1} h_t(\vartheta_j) \right) \;\mid\; \eta_t(\vartheta_j) = (22)\right\} \tag{30}$$

A solution to (30) satisfies the following first order conditions

$$\alpha \left(\bar{\pi}_{j,t+1} q_{t+1}(\vartheta_j) - R_{j,t+1} h_t(\vartheta_j) \right) = R_{j,t+1} h_t(\vartheta_j). \tag{31}$$

Solve for $h_t(\vartheta_j)$ gives

$$h_t(\vartheta_j) = \frac{\alpha}{1+\alpha} \frac{\bar{\pi}_{j,t+1}}{R_{j,t+1}} q_{t+1}(\vartheta_j). \tag{32}$$

Expected profits for an innovator on capital good line ϑ_j then can be computed as

$$\pi_{j,t+1}^{(i)}(\vartheta_j) = \frac{1}{1+\alpha} \bar{\pi}_{j,t+1} \eta(\vartheta_j) q_{t+1}(\vartheta_j). \tag{33}$$

To derive the sectoral probabilities of innovation success use definition 3, insert individual R&D-expenditures and the expression for ϕ_{ϑ_j} gives[10]

$$\eta_t(\vartheta_j) = \eta_{j,t} = \zeta^{-1}\left(\frac{\alpha}{1+\alpha}\frac{\bar{\pi}_{j,t+1}}{R_{j,t+1}}\right)^{\alpha}. \tag{34}$$

3.2 Consumer Behavior

Consumers utility is given in Assumption 2. People consume only in the second period of live and since utility is strictly increasing in consumption quantities all old consumers optimally spend their income entirely for consumption of the final good.

Denote aggregate consumption of old workers as $C^{(o)}_{w,t}$ and aggregate savings as S_t, then the aggregate budget constraint and optimal consumption of the old working population is given by

$$C^{(o)}_{w,t+1} = r_{t+1}S_t = r_{t+1}w_{M,t}M_t =: I^{(o)}_{w,t+1}. \tag{35}$$

Optimal consumption of the entrepreneurs is given by

$$C^{(o)}_{e,t} = \bar{\Pi}^e_t =: I^{(o)}_{e,t} \tag{36}$$

The aggregate income of the innovators is sector-specific since innovators within one intermediate sector form a "R&D-Syndicate". Therefore, the aggregate income of the syndicate depends also on the mass of innovators ι_j in sector j, since the more innovators try to improve the quality of capital goods in one sector, the greater is the income of the R&D-syndicate. This implies an optimal consumption profile for innovators in sector j equal to

$$C^{(o)}_{i,j,t} = \bar{\Pi}^i_{j,t} =: I^{(o)}_{i,j,t} \qquad j \in \{m, z\}. \tag{37}$$

So total consumption of all innovators is equal to

$$C^{(o)}_{i,t} = C^{(o)}_{i,m,t} + C^{(o)}_{i,z,t}. \tag{38}$$

[10]The function $\phi_t(\vartheta_j) := \frac{1}{\zeta}q_t(\vartheta_j)^{-\alpha}$ captures the effects of the current position on the quality ladder. The present study assumes that successful innovation becomes more difficult the more product generations have been previously invented. The function ϕ now states that this difficulty increases in proportion to the additional output that would be produced in case of successfully increasing the product quality from $q_t(\vartheta_j)$ to $q_{t+1}(\vartheta_j) = \lambda_j q_t(\vartheta_j)$ between t and $t+1$. The parameter ζ represents a cost of research measured in units of the final consumption good.

3.3 Banking Behavior

The competitive bank sector takes the interest rates r_t, $R_{m,t}$ and $R_{z,t}$ as given and chooses sectoral credit quantities to maximize expected profits from financial intermediation.

$$\max_{(D_{m,t},D_{z,t})\in\mathbb{R}^2_+} \left\{ \sum_{j\in\{m,z\}} \eta_j R_{j,t+1} D_{j,t} - r_{t+1} \sum_{j\in\{m,z\}} D_{j,t} \right\} \tag{39}$$

A solution to (39) satisfies the following first order optimality conditions:

$$R_{m,t+1} = \frac{r_{t+1}}{\eta_m} \quad \text{and} \quad R_{z,t+1} = \frac{r_{t+1}}{\eta_z}. \tag{40}$$

These conditions state the relationship between sectoral loan interest rates $R_{j,t}$, $j \in \{m, z\}$ and the deposit interest rate r_{t+1} and shows that loan rates contain the inverse of the success-probabilities η_m^{-1}, η_z^{-1} as risk premiums: the greater the chance of success in innovation, the lower is the corresponding rate of interest on business credit. Moreover, one can derive the following relationship between sectoral credit interest rates:

$$\eta_m R_{m,t+1} = \eta_z R_{z,t+1}. \tag{41}$$

This simply gives the condition that in equilibrium, the expected loan interest rates in the two intermediate sectors have to be equal.

Conditions (40) and (41) explain the relationship between loan rates paid by innovators and deposit rates paid to households and shows that sectoral equilibrium loan rates can differ from each other.

3.4 Market Clearing

3.4.1 Capital Markets

Total deposit supply is given by aggregate savings of young workers S_t. Market clearing on the deposit market requires that workers' aggregate savings are equal to the deposit demand of the representative aggregate bank:

$$S_t \stackrel{!}{=} D_t \qquad \forall t \geq 0. \tag{42}$$

Credit demand in each sector $j = \{m, z\}$ is given by aggregate sectoral R&D expenditures (Eq. (10)). Market clearing requires that in each sector, credit supply is equal to credit demand:

$$D_{j,t} \stackrel{!}{=} H_{j,t} \qquad j = \{m, z\}, \forall t \geq 0. \tag{43}$$

Moreover, the bank cannot supply more capital to the innovators than received from the workers, so we impose the condition that total deposit demand today determines total credit supply tomorrow:

$$D_t = \sum_{j=\{m,z\}} D_{j,t+1}, \qquad \forall t \geq 0. \tag{44}$$

Using (42) and (43), this implies

$$S_t = \sum_{j=\{m,z\}} H_{j,t+1}, \qquad \forall t \geq 0. \tag{45}$$

3.4.2 Factor Markets

Market clearing on the market for factor $J = \{M, Z\}$ then requires

$$J_t^S \stackrel{!}{=} J_t^D \qquad J \in \{M, Z\}, \forall t \geq 0. \tag{46}$$

3.4.3 Goods Markets

For each capital good ϑ_j, market clearing requires:

$$x_t^S(\vartheta_j|q) \stackrel{!}{=} x_t^D(\vartheta_j|q) \qquad \vartheta_j \in [0, 1], j \in \{m, z\}, \forall t \geq 0. \tag{47}$$

Market clearing on the markets for the intermediate goods requires

$$Y_{j,t}^S \stackrel{!}{=} Y_{j,t}^D \qquad j \in \{m, z\}. \tag{48}$$

Market clearing on the final goods market requires that final goods demand (consumption, investment in R&D and machine production) equals supply and gives the economy's resource constraint:

$$Y_t = X_t + H_t + C_t \qquad \forall t \geq 0, \tag{49}$$

where C_t is given in (15), X_t is given in (13) and H_t is given in (11).

Finally, we need a "market clearing" condition for the mass of innovators. For all $t \geq 0$:

$$\iota_{m,t} + \iota_{z,t} = 1. \tag{50}$$

3.5 Equilibrium

The next definition of an *equilibrium* of the present economy with financial intermediation is standard:

Definition 1 An equilibrium of the economy considered here is an allocation

$$\mathbf{A} = (Y_t, C_t, M_t, Z_t, (Y_{j,t}, X_{j,t}, D_{j,t}, H_{j,t}, \iota_{j,t}, (q_t(\vartheta_j), x_t(\vartheta_j|q))_{\vartheta_j \in [0,1]})_{j \in \{m,z\}}, D_{t+1})_{t \geq 0}$$

and a price system

$$\mathbf{P} = (r_t, w_{j,t}, R_{j,t}, p_{j,t}, p_t(\vartheta_j|q))_{t \geq 0}$$

such that:

(i) The allocation is consistent with the production technologies (1), (2), and the market clearing conditions/resource constraints (42), (43), (46), (47), (48), (49), and (50).
(ii) Banks behave optimally, i.e. Eqs. (40) and (41) hold for all $t \geq 0$.
(iii) Producers behave optimally, i.e. Eqs. (20), (25), (28), and (32) hold for all $t \geq 0$.
(iv) Consumers behave optimally with (profit) incomes determined by (4), (5), and (35) for all $t \geq 0$.
(v) Average capital goods qualities in sector $j \in \{m, z\}$ evolve according to (9), given initial quality levels $Q_{m,0} > 0$ and $Q_{z,0} > 0$.

It is useful to derive properties of some equilibrium variables that will be used in the following analysis. In a first step, we compute aggregate production of the two intermediate goods $Y_{m,t}$ and $Y_{z,t}$ as functions of the factors M_t, Z_t, output prices $p_{m,t}, p_{z,t}$ and aggregate sectoral quality indices $Q_{m,t}, Q_{z,t}$. Insert the intermediate firms equilibrium demand schedules for capital goods $x_t(\vartheta_m), x_t(\vartheta_z)$ stated in Eqs. (21a) and (21b) into the production functions of good Y_m and good Y_z given in (2). This yields for sector $j \in \{m, z\}$

$$Y_{j,t} = \frac{\mu_0}{\mu} p_{j,t}^{\frac{\mu}{1-\mu}} J_t Q_{j,t} \tag{51}$$

where $\mu_0 := (1 - \mu)\mu^{\frac{2\mu}{1-\mu}}$. Note that the integral parts in the production functions vanish due to the definition of the aggregate or average sectoral quality index given in (7). Taking the ratio of Eq. (51) for $j = m$ and $j = z$ gives

$$\frac{Y_{z,t}}{Y_{m,t}} = \left(\frac{p_{z,t}}{p_{m,t}}\right)^{\frac{\mu}{1-\mu}} \frac{Q_{z,t}}{Q_{m,t}} \frac{Z_t}{M_t}. \tag{52}$$

In a second step, we combine the optimality conditions of the final good sector stated in (17) with Eq. (52) and eliminate the term $\tilde{Y}_{j,t} := \frac{Y_{z,t}}{Y_{m,t}}$. This gives the expression for the relative price of the two goods $Y_{m,t}$ and $Y_{z,t}$ as a function of the relative factor supply and the relative average quality of the sector-specific capital goods:

$$\tilde{p}_t = \left(\tilde{\gamma}^{-\varepsilon} \tilde{Z}_t \tilde{Q}_t\right)^{-\frac{(1-\mu)}{\sigma}} \tag{53}$$

where $\tilde{\gamma} := \frac{1-\gamma}{\gamma}$, $\tilde{Z}_t := Z_t/M_t$, $\tilde{Q}_t := Q_{z,t}/Q_{m,t}$ and $\sigma := 1 + (1-\mu)(\varepsilon - 1)$.[11]

In a third step we derive the equilibrium credit and deposit interest rates and sectoral credit quantities. Note that the stated equations all contain various endogenous variables on their right hand side that will be determined next, so the treatment is incomplete at this stage. To get the equilibrium deposit interest rate r_t use (42) together with (44), (43), (10), and (40), we get

$$r_{t+1} = \frac{\alpha}{S_t}\left(\iota_t \mathbb{E}_t[\Pi_{m,t+1}] + (1-\iota_t)\mathbb{E}_t[\Pi_{z,t+1}]\right). \tag{55}$$

The equilibrium interest rate on deposits is decreasing in aggregate savings which is equivalent to deposit supply, and increasing in sectoral expected profits from innovation. Insert (55) into (40) to get the credit interest rates for sector j:

$$R_{j,t+1} = \frac{\alpha}{\eta_j S_t}\left(\mathbb{E}_t[\Pi_{m,t+1}] + \mathbb{E}_t[\Pi_{z,t+1}]\right), \qquad j \in \{m, z\}. \tag{56}$$

[11]As in the standard directed technical change model of Acemoglu (2002), σ is the elasticity of substitution between the two factors M and Z, since one can show that

$$\sigma = -\left(\frac{dlog(\frac{w_Z}{w_M})}{dlog(\frac{Z}{M})}\right)^{-1}. \tag{54}$$

It will turn out in brief that the elasticity of substitution has a crucial role on the direction of technical change. If the two intermediate goods Y_m and Y_z are gross substitutes ($\varepsilon > 1 \Leftrightarrow \sigma > 1$), then the two factors M and Z are gross substitutes. If Y_m and Y_z are gross complements ($\varepsilon < 1 \Leftrightarrow \sigma > 1$) then M and Z are gross complements.

Take the ratio of (56) to compute the relative credit interest rate $\tilde{R}_t$:

$$\tilde{R}_t = \frac{R_{z,t}}{R_{m,t}} = \frac{\eta_{m,t}}{\eta_{z,t}} = \tilde{\eta}_t^{-1}, \tag{57}$$

so the relative credit interest rate is inversely related to the ratio of success-probabilities in innovation effort.

To get equilibrium credit in sector j, use the market clearing conditions in the capital markets (42) and (43) together with the condition that total deposit demand in t determines total credit supply in $t+1$ (44). Then, credit in sector m reads

$$D_{m,t-1} = \frac{S_{t-2}}{1 + \tilde{R}_t \tilde{\pi}_t \tilde{Q}_t \tilde{\iota}_t}, \tag{58}$$

which can be solved for

$$D_{m,t-1} = \frac{\iota_{t-1} R_{z,t} \bar{\pi}_{m,t} Q_{m,t} S_{t-2}}{\iota_{t-1} R_{z,t} \bar{\pi}_{m,t} Q_{m,t} + (1 - \iota_t) R_{m,t} \bar{\pi}_{z,t} Q_{z,t}}. \tag{59}$$

Similar, credit demand in sector z is given by

$$D_{z,t-1} = \frac{S_{t-2} \tilde{R}_t \tilde{\pi}_t \tilde{Q}_t \tilde{\iota}_t}{1 + \tilde{R}_t \tilde{\pi}_t \tilde{Q}_t \tilde{\iota}_t}, \tag{60}$$

which is equivalent to

$$D_{z,t-1} = \frac{(1 - \iota_{t-1}) R_{m,t} \bar{\pi}_{z,t} Q_{z,t}}{\iota_{t-1} R_{z,t} \bar{\pi}_{m,t} Q_{m,t} + (1 - \iota_{t-1}) R_{m,t} \bar{\pi}_{z,t} Q_{z,t}} S_{t-2}. \tag{61}$$

The expressions in (59) and (61) reveal that credit in sector $j \in \{m, z\}$ is increasing in aggregate savings and shares of innovators ι_j. Credit in sector m is lower for greater $R_{m,t}$, greater (lower) for greater $R_{z,t}$, increasing in $Q_{m,t}$ and decreasing in $Q_{z,t}$.

The ratio of relative credit in sector z is then equal to:

$$\tilde{D}_t = \tilde{\iota}_t \tilde{p}_{t+1}^{\frac{1}{1-\mu}} \tilde{Z}_{t+1} \tilde{Q}_{t+1} \tilde{R}_{t+1}. \tag{62}$$

It will turn out next that the right hand side equals relative expected profits from innovation. This is intuitively clear since sectoral credit is equal to sectoral R&D expenditures and the latter are constant share of expected profits.

4 Determinants of Directed Technical Change

Next, we derive the determinants of directed innovations on a balanced growth path and then analyze how these determinants response to changes in relative factor supply. After presenting the economics of the direction of technical change, I show the existence and uniqueness of the BGP-equilibrium.

4.1 The Demand for Innovations

In the following we analyze how the production side of the economy determines the return to different types of innovation—the demand for innovation. The next section then discusses the other side of this equation, the cost of different innovations or the supply side of innovations.

Average profits from capital goods selling for old innovators in sector m read

$$\Pi_{m,t} = (1-\mu)\mu^{\frac{1+\mu}{1-\mu}} p_{m,t}^{\frac{1}{1-\mu}} M_t Q_{m,t} \tag{63}$$

and in sector z

$$\Pi_{m,t} = (1-\mu)\mu^{\frac{1+\mu}{1-\mu}} p_{z,t}^{\frac{1}{1-\mu}} Z_t Q_{z,t}. \tag{64}$$

The net present discounted value of expected profits for young innovators in sector j, denoted by $V_{j,t}, j \in \{m, z\}$, are then given by

$$V_{j,t} = \frac{\mathbb{E}_t[\Pi_{j,t+1}]}{r_t}. \tag{65}$$

one can show that the V's are equal to

$$V_{m,t} = \mu_0 p_{m,t+1}^{\frac{1}{1-\mu}} M_{t+1} Q_{m,t+1} R_{m,t+1}^{-1} \qquad V_{z,t} = \mu_0 p_{z,t+1}^{\frac{1}{1-\mu}} Z_{t+1} Q_{z,t+1} R_{z,t+1}^{-1}. \tag{66}$$

where $\mu_0 := (1-\mu)\mu^{\frac{1+\mu}{1-\mu}}$. So the *larger $V_{z,t}$ in relation to $V_{m,t}$, the greater is the reward to develop Z-augmenting capital goods* (see Acemoglu 2002, p.789). The two profit equations given in (66) suggest that capital goods producers' profits increase with greater output prices, with greater factor use and with greater productivity in the intermediate sectors and decrease with larger interest rates on business credit. The latter effect simply follows from the relationship between credit interest rates and success-probabilities given in (40).

To show and discuss these determinants in more detail, take the ratio of $V_{z,t}$ and $V_{m,t}$. This gives

$$\frac{V_{z,t}}{V_{m,t}} = \underbrace{\left(\frac{p_{z,t+1}}{p_{m,t+1}}\right)^{\frac{1}{1-\mu}}}_{\substack{\text{price}\\ \text{effect}}} \times \underbrace{\frac{Z_{t+1}}{M_{t+1}}}_{\substack{\text{market size}\\ \text{effect}}} \times \underbrace{\left(\frac{R_{z,t+1}}{R_{m,t+1}}\right)^{-1}}_{\substack{\text{risk}\\ \text{effect}}} \times \underbrace{\left(\frac{Q_{z,t+1}}{Q_{m,t+1}}\right)}_{\substack{\text{productivity}\\ \text{effect}}}. \tag{67}$$

The conclusions drawn from Eq. (67) represent the central results of this paper. The next two theorems comprehend these main findings.

Theorem 7 *The direction of technical change—whether technical change will favour relatively scarce or abundant factors—is determined by four different market forces: the* price effect, *the* market size effect, *the* productivity effect *and the* risk effect.

- The *price effect*: Since $\mu \in]0,1[$, the relative profitability of inventing new Z-augmenting capital goods is increasing in the relative price $p_{z,t+1}/p_{m,t+1}$. Therefore, the higher this relative price is, the greater is the return on developing new Z-complementary technologies (Acemoglu 2002, p.789).

  Naturally, relatively scarce factors are relatively more expensive. Thus, the price effect directs technical changes to technologies or sectors that complement scarce input factors and thus command higher commodity prices.
- The *market size effect*: The relative profitability of inventing Z-complementary technologies increases in the relative supply of the factors Z_{t+1}/M_{t+1}. Therefore, the larger this relative factor supply, the greater is the return on the development of new Z-complementary technologies (Acemoglu 2002, p.789).

  In this study, the market for a technology is determined by the factors that use this technology. An increase in the supply of a factor leads to a larger market for capital goods that complement this factor. The market size effect encourages innovations in sectors that use the more abundant factor. Hence, this effect works in the opposite direction compared to the price effect.
- The *risk effect*: The risk effect results from the fact that $V_{z,t}/V_{m,t}$ decreases in $R_{z,t+1}/R_{m,t+1}$. The lower the relative probability of successful innovation in sector z, η_z/η_m, the higher is the corresponding relative loan rate R_z/R_m and the risk of default on external credit reflected in the interest rate on loans. Thus, the risk effect directs innovations to sectors with a higher probability of research success. This illustrates the significance of the assumption, how the current η_j's relate to the number of previous innovations in any industry ϑ_j and sector m, z with respect to the direction of technical change.
- The *productivity effect*: The ratio $V_{z,t}/V_{m,t}$ increases in $Q_{z,t+1}/Q_{m,t+1}$ and thus the productivity effect encourages innovations in sectors with a higher productivity.

The result established in theorem 7 resembles the effects stated amongst others in Acemoglu (2002) plus the risk effect. Altogether the net effect of these partly counteracting forces determines the direction of technical change. The following theorem establishes the second fundamental result:

Theorem 8 *If innovators are capital constrained and R&D is financed by loans, then credit interest rates influence the direction of technical change.*

First, credit interest rates contain a risk premium given by the inverse of the probability of successful innovation. The risk premia and the loan rates relate inversely to another: the lower the probability of successful innovation in any one sector, the higher is the risk of default on credit obligations by innovators who try to invent new capital goods in that sector. Consequently, banks compensate for the higher risk and demand increasing risk premia and thus charge innovators higher interest rates on credit. A relatively higher interest rate—in, say, sector z—ceteris paribus implies relatively increasing credit costs for innovators whose plan is to invent new Z-complementary capital goods. To keep profits constant, innovators would have to cut R&D investment. This would result in a lower probability of success in sector z. The expected profitability of developing new Z-complementary capital goods would decline and innovators would direct R&D effort to the relatively less expensive sector: The presence of banks adds an additional component to the determinants of directed technical change through the relative interest rate charged on loans $R_{z,t}/R_{m,t}$, which is equivalent to the inverse of the probabilities of successful innovation $(\eta_{z,t}/\eta_{m,t})^{-1}$.

Second, R&D expenditures are proportional to (expected) profits. Credit costs determine equilibrium profits and thus the amount of resources devoted to R&D in equilibrium. This in turn influences indirectly the probability of successful innovation and thus the *rate* of technical change in both sectors.[12]

[12]As a side note, the literature on induced innovation (e.g. Hicks 1932; Habakkuk 1962; Kennedy 1964; Dranakis and Phelps 1965; Samuelson 1965) states that relative factor prices influence the type of technological progress. In particular, the literature argues that innovations are directed at "more expensive" factors. In the discussion here, I confine myself to the role of (output)-prices, market size effects et cetera. However, we can show the similarity between the present approach and a formulation that considers factor costs as determinants to develop new technologies rather than output prices: Combine the capital goods demand stated in Eqs. (21a) and (21b) with the first order conditions of the intermediate firms with respect to the factors M and Z given in (20a). Using these expressions we can rewrite (63), (64) as

$$\Pi_m = \mu w_M M \quad \text{and} \quad \Pi_z = \mu w_Z Z. \tag{68}$$

Then one can express the relative profitability of developing Z-complementary capital goods in terms of factor costs w_M, w_Z and market sizes M and Z and the relative profitability of developing new Z-complementary capital goods then reads

$$\frac{\Pi_z}{\Pi_m} = \frac{w_Z}{w_M}\frac{Z}{M}. \tag{69}$$

For purposes of a compact notation, I employ the following notation to all relative equilibrium variables from now on: Let a_m and a_z denote arbitrary sector-specific variables for sector m and sector Z respectively. Then we define the *relative variable with respect to sector* Z as $\tilde{a} := \frac{a_z}{a_m}$. Then relative expected profits of innovation in sector z, given in Eq. (67), equal

$$\tilde{V}_t = \tilde{p}_{t+1}^{\frac{1}{1-\mu}} \tilde{Z}_{t+1} \tilde{Q}_{t+1} \tilde{R}_{t+1}^{-1}. \tag{70}$$

To get final expressions of relative profits and relative productivity, we have to eliminate endogenous prices $p_{m,t}$, $p_{z,t}$ and also loan rates $R_{m,t}$ and $R_{z,t}$.

Using (34), insert $\bar{\pi}_j := (1-\mu)\mu^{\frac{1+\mu}{1-\mu}} p_j^{\frac{1}{1-\mu}} J$ for $j \in \{m, z\}$, use (56) and take the ratio gives

$$\tilde{\eta}_t = \left(\frac{\bar{\pi}_{z,t+1}}{\bar{\pi}_{m,t+1}}\right)^{\frac{\alpha}{1-\alpha}} = \left(\tilde{p}_{t+1}^{\frac{1}{1-\mu}} \tilde{Z}_{t+1}\right)^{\frac{\alpha}{1-\alpha}}. \tag{71}$$

Note that the factors M_t and Z_t are constant in supply, so as long as the relative price $\tilde{p}_t$ is constant, $\tilde{\eta}_t$ is also constant. Since relative credit interest rates relate inversely to the ratio of success-probabilities, relative credit interest rates read

$$\tilde{R}_t = \left(\tilde{p}_{t+1}^{\frac{1}{1-\mu}} \tilde{Z}_{t+1}\right)^{-\frac{\alpha}{1-\alpha}}. \tag{72}$$

To highlight the role of the risk effect in determining the direction of technical change, we eliminate the relative price $\tilde{p}_t$, as given in (53), from (72) in a first step. The risk effect then reads

$$\tilde{R}_t = \left(\tilde{\gamma}^{\varepsilon} \tilde{Z}_{t+1}^{\sigma-1} \tilde{Q}_{t+1}^{-1}\right)^{-\frac{\alpha}{(1-\alpha)\sigma}}, \tag{73}$$

where $\sigma := 1 + (1-\mu)(\varepsilon - 1)$ is the elasticity of substitution between the two factors $M$and Z.[13] Inspection of (73) reveals that the response of the risk effect to

Equation (69) indicates a higher incentive to innovate for factors that are more expensive. This result shows the equivalence of the approach presented here (and for instance by Acemoglu 2002) that considers output prices, and the approach in the induced innovation literature cited above, which concentrates on factor input prices.

[13]For the purpose of this study it is important to distinguish the cases where $\varepsilon \lesseqgtr 1$ and thus $\sigma \lesseqgtr 1$. For the relevant parameter values of ε and σ, the exponent is always positive, i.e. $1/(1-\alpha)\sigma > 0$, since $\alpha \in]0, 1[$ and $\sigma > 0$ by assumption.

an increase in the relative supply of factors $\tilde{Z} := Z/M$ depends on the size of σ (since $1-\alpha > 0$). In fact (73) implies the following proposition:

Proposition 1 *If factors M and Z are gross complements,* $\varepsilon < 1 \Leftrightarrow \sigma < 1$, *the risk effect increases with an increase in the relative supply of factor* $\tilde{Z}$.

If factors M and Z are gross substitutes, $\varepsilon > 1 \Leftrightarrow \sigma > 1$, *the risk effect decreases with an increase in the relative supply of factor* $\tilde{Z}$.

After elimination of $\tilde{p}_t$, as given in (53), relative expected profits from innovation in sector z, given in Eq. (67), read

$$\tilde{V}_t = \left(\tilde{\gamma}^{\varepsilon}\tilde{Z}^{\sigma-1}\tilde{Q}^{\sigma-1}\tilde{R}^{-\sigma}\right)^{\frac{1}{\sigma}}. \tag{74}$$

Ignore the risk effect in Eq. (74) for the moment (and suppose that $\tilde{R} = 1$). The parameter σ has a crucial impact on the direction of technical changes. If $\sigma > 1$, M and Z are gross substitutes and an *increase* in the relative factor supply $\tilde{Z} := Z/M$ (either because the supply of Z increases or equivalently, the supply of M decreases) will *increase* the relative profitability of inventing Z-complementary capital goods $\tilde{V}_t$. On the one hand, if $\tilde{Z}$ increases, factor Z becomes relatively more abundant. This translates into a larger market for the capital goods that complement Z, which *increases* $\tilde{V}$ (this is referred to as the "market-size effect"). On the other hand, if $\tilde{Z}$ increases, factor M becomes relatively scarce and thus relatively more expensive. This translates into higher prices of goods that use M-complementary capital goods in the production process and thereby increase the profit from inventing those M-complementary capital goods. Thus $\tilde{V}$ *decreases* (this represents the "price-effect"). Consequently, the price effect and the market-size effect work in opposite directions. If $\sigma > 1$, $\tilde{V}$, as given in (74), *increases with an increase* in $\tilde{Z}$ and we can conclude that the market size effect dominates the price effect. If $\sigma < 1$, $\tilde{V}$ *decreases with an increase* in $\tilde{Z}$ and the price effect dominates the market-size effect. Together this implies that the parameter σ regulates whether the price effect dominates the market size effect.

Next consider the response of relative expected profits to a change in the relative factor supply $\tilde{Z}$, given that the risk effect is an additional determinant in the direction of technical changes (and compare this case to the hypothetical situation above, where we treated the risk effect $\tilde{R}$ as equal to 1). For that, insert $\tilde{R}$ from Eq. (73) into (74):

$$\tilde{V}_t = \left(\gamma^{\varepsilon}\tilde{Z}_{t+1}^{\sigma-1}\tilde{Q}_{t+1}^{-\varphi}\right)^{\frac{1}{\sigma(1-\alpha)}}, \qquad \varphi := 1-\sigma(1-\alpha). \tag{75}$$

Inspection of (75) shows first that relative expected profits basically respond to an increase in the relative factor supply $\tilde{Z}$ in the manner described above: If $\sigma > 1$, $\tilde{V}_t$, as given in (75), *increases with an increase* in $\tilde{Z}$ and the market size effect dominates the price effect. If $\sigma < 1$, $\tilde{V}_t$ *decreases with an increase* in $\tilde{Z}$ and the price effect dominates the market-size effect. So the result derived by Acemoglu (2002) that the elasticity of substitution plays a crucial role in determining the direction of technical change does not change if we consider capital constrained firms and financial intermediation.

Additionally, comparing (74) (and still assume $\tilde{R} = 1$ there) with (75) reveals that the response of relative profits to an increase in $\tilde{Z}$ changes, if we additionally account for the risk effect in the determinants of technical change:

Proposition 2 *Relative expected profits $\tilde{V}_t$ including the risk effect respond stronger to an increase in relative factor supply $\tilde{Z}$. This result holds independent of the size of the elasticity of substitution.*

Taken together, Propositions 1 and 2 imply that the elasticity of substitution between factors regulates the way in which innovators' incentive to invent Z-complementary capital goods responds if the relative factor supply changes in the first place and additionally, this response changes if the risk effect of private sector lending enters the determinants of directed technical change.[14]

4.2 The Supply of Innovations

So far, the equilibrium analysis has studied how the production side of the economy determines the return on different types of innovations. In the next step, we consider the other side of this equation and analyze the costs of different innovations. As stated in the model section, only the final good is being used to generate innovations. This approach implies that future relative costs of innovations are unaffected by the composition of current R&D and therefore remain constant.

To have innovators who are willing to invest in R&D in both sectors, expected profits from improving the productivity of the capital goods in both sectors have to be equal:

$$V_{m,t} = V_{z,t}. \tag{77}$$

The condition states that in terms of expectation, it is equally profitable to invest in generating innovations on M- and Z-complementary capital goods. Hence the

[14] To see another important aspect of the elasticity of substitution, consider the relative price $\tilde{w}$ of the (relative) factors Z and M, $\tilde{Z}$. The relative factor reward can be computed as

$$\tilde{w} = \tilde{\gamma}^{\frac{\varepsilon}{\sigma}} \tilde{Z}^{-\frac{1}{\sigma}} \tilde{Q}^{\frac{\sigma-1}{\sigma}}. \tag{76}$$

This relative factor price is decreasing in the relative supply $\tilde{Z}$, since $\sigma > 0$. This constitutes the usual substitution effect: the more abundant factor is substituted for the less abundant one, and has a lower marginal product (see Acemoglu 2002, p. 790). For a derivation of the expression given in (76), see Appendix "Proof of Lemma 1".

average productivity of the sectoral capital goods $Q_{m,t}$ and $Q_{z,t}$ can both grow along the balanced growth path. Now resolve the condition above as $\tilde{V}_t$, insert Eqs. (73), (74) and resolve for $\tilde{Q}$ gives

$$\tilde{Q} = \left(\tilde{\gamma}^{\varepsilon}\tilde{Z}^{\sigma-1}\right)^{\frac{1}{\varphi}}, \tag{78}$$

where $\varphi := 1 - \sigma(1-\alpha)$. The notable feature is that with the direction of technical change endogenized, the relative productivity of technology $\tilde{Q}$ is determined by the relative factor supply $\tilde{Z}$ and the elasticity of substitution σ. Moreover, the expression for $\tilde{Q}$ implies the following proposition:

Proposition 3 *If the two factors are gross substitutes and* $\frac{1}{1-\alpha} > \sigma > 1$, *an increase in* $\tilde{Z}$ *will raise* $\tilde{Q}$. *Hence, Eq. (78) implies that if these two conditions hold, the physical productivity of the relatively more abundant factor tends to be higher. If the two factors are gross complements,* $\sigma < 1$, *an increase in* $\tilde{Z}$ *will lower* $\tilde{Q}$. *Thus, Eq. (78) implies that the physical productivity of the relatively more scarce factor tends to be higher if* $\sigma < 1$.

Since $\sigma > 1 \Leftrightarrow \varepsilon > 1$, a higher level of the relative physical productivity $\tilde{Q}$ corresponds to Z-biased technical change, because if $\varepsilon > 1$, the relative marginal product of Z, $\tilde{w}$ as given in Eq. (76), is increasing in $\tilde{Q}$. Thus Z-augmenting technical change is also Z-biased and technological advances will be endogenously biased in favor of the more abundant factor: with gross substitutability an increase in $\tilde{Q}$ is relatively biased toward Z.

In case of gross complementarity ($\sigma < 1$), the condition $\sigma < 1 < \frac{1}{1-\alpha}$ is automatically fulfilled and the result stated in the proposition above holds generally. However, since $\sigma < 1 \Leftrightarrow \varepsilon < 1$, the decrease in the relative physical productivity translates into higher value of the relative marginal product $\tilde{w}$, as given in (76). Therefore, even if $\sigma < 1$, technological advances will again be endogenously biased in favor of the more abundant factor: with gross complementarity, a decrease in $\tilde{Q}$ is relatively biased toward Z. Taken together this gives us the

Proposition 4 *Consider the directed technical change model with capital constraints and financial intermediation. Then, if* $\sigma < \frac{1}{1-\alpha}$ *an increase in the relative abundance of the factor Z, denoted by* $\tilde{Z} := Z/M$, *always induces technological change relatively biased toward factor Z.*

According to this proposition, the induced technological change increases the relative marginal product of the factor becoming more abundant. This holds, as long as $\sigma \neq 1$. If instead $\sigma = 1$, the elasticity of substitution between factors is equal to one, and technological change is not biased towards any one of the factors. Since marginal factor productivities are equal to factor prices, the relative equilibrium bias therefore describes the impact of technology on relative factor prices at given factor proportions. Note that this equilibrium bias is not a result of the banks' financial intermediation. So in other words, the implementation of capital constrained firms and a bank sector does not alter the findings of the standard directed technical change model introduced by Acemoglu (2002), but adds a component to the determinants of technology bias.[15]

5 Properties of the Balanced Growth Path Equilibrium

To solve for the balanced growth path equilibrium, we need equations (a) relating the growth rate of sectoral output to sectoral R&D expenditures and (b) relating the rate of aggregate growth to sectoral output growth rates. The growth rates of sectoral or average capital goods quality were given in Eqs. (9a) and (9b), sectoral outputs as functions of average sectoral capital goods qualities were derived in (51). The latter two equations reveal that intermediate outputs are constant over time, as long as factors M and Z and prices p_m and p_z are constant over time. In this case, the relationship between sectoral output and R&D expenditures over time equals:

$$\frac{Y_{m,t+1} - Y_{m,t}}{Y_{m,t}} = \frac{Q_{m,t+1} - Q_{m,t}}{Q_{m,t}} \tag{79}$$

and

$$\frac{Y_{z,t+1} - Y_{z,t}}{Y_{z,t}} = \frac{Q_{z,t+1} - Q_{z,t}}{Q_{z,t}}. \tag{80}$$

Before we get final expressions for the equilibrium deposit interest rate and also rates of aggregate output growth, we establish a result in the following Lemma that helps to simplify the rest of the analysis.

Lemma 2 *If innovators are indifferent between intermediate sectors m and z, i.e.*

$$V_{m,t} = V_{z,t} \qquad t \geq 0, \tag{81}$$

the growth rates of average sectoral capital goods qualities are identical:

$$g_{m,t} := \frac{Q_{m,t+1}}{Q_{m,t}} - 1 = \frac{Q_{z,t+1}}{Q_{z,t}} - 1 =: g_{z,t} \qquad t \geq 0. \tag{82}$$

[15] The result is confined to the case considered here, where technologies are factor augmenting. So the equilibrium bias crucially depends on the assumptions about the set of possible technologies. If we include non-factor-augmenting technologies into the set of possibilities, examples can be constructed, where an increase in the abundance of a factor induces technology to be biased against this factor and the stated equilibrium bias no longer remains valid. For a detailed analysis of the necessary and sufficient conditions of (relative and absolute) equilibrium bias of technology development, see Acemoglu (2007).

The condition (81) states that in expected terms, it is equally profitable to invest in generating innovations on M-and Z-complementary capital goods. In this case, innovators are indifferent and R&D is directed to both sectors. Hence, the average productivity of the sectoral capital goods $Q_{m,t}$ and $Q_{z,t}$ can both grow. In fact, the lemma proves that in this case, the growth rates of $Q_{m,t}$, denoted by $g_{m,t}$, and $Q_{z,t}$, denoted by $g_{z,t}$ are identical and we can define

$$g_t := g_{m,t} = g_{z,t}. \tag{83}$$

After elimination of endogenous prices etc., the equilibrium deposit interest rate given in (55) reads:

$$r_{t+1} = \alpha_0 \bar{W}_{t+1}^{\frac{\alpha}{\sigma-1}} \left(\frac{\mu(1+g_t)^2}{\zeta \left(\gamma^{\varepsilon} M_{t+1}^{\sigma-1}\right)^{\frac{\alpha}{\varphi}}} \right)^{1-\alpha}, \tag{84}$$

where still $\alpha_0 := \alpha/(1+\alpha)$, $\bar{W}_t := \left(\gamma^{\varepsilon} M_t^{\sigma-1}\right)^{\frac{\alpha}{\varphi}} + \left((1-\gamma)^{\varepsilon} Z_t^{\sigma-1}\right)^{\frac{\alpha}{\varphi}}$, $\varphi := 1-\sigma(1-\alpha)$. We use this deposit interest rate to get final expressions for the success-probabilities and equilibrium credit interest rates. The probability of successful innovation in sector m can be computed as

$$\eta_{m,t} = \left(\frac{\bar{W}_{t+1}^{\frac{1}{\sigma-1}}}{\zeta^{\frac{1-\alpha}{\alpha}} \mu(1+g_t)^2 \left(\gamma^{\varepsilon} M_{t+1}^{\sigma-1}\right)^{\frac{1-\alpha}{\varphi}}} \right)^{\alpha} \tag{85}$$

and in sector z

$$\eta_{z,t} = \left(\frac{\left(\gamma^{\varepsilon} M_{t+1}^{\sigma-1}\right)^{\frac{\alpha}{\varphi}} \bar{W}_{t+1}^{\frac{1}{\sigma-1}}}{\zeta^{\frac{1-\alpha}{\alpha}} \mu(1+g_t)^2 \left((1-\gamma)^{\varepsilon} Z_{t+1}^{\sigma-1}\right)^{\frac{1}{\varphi}}} \right)^{\alpha}. \tag{86}$$

Insert prices, success-probabilities and aggregate savings into (56) this leads directly to the following credit interest rates in sector m:

$$R_{m,t+1} = \frac{\alpha(\mu(1+g_t)^2}{(1+\alpha)} \tag{87}$$

and in sector z, innovators pay an interest rate on credit equal to

$$R_{z,t+1} = \frac{\alpha\mu(1+g_t)^2}{(1+\alpha)} \left(\tilde{\gamma}^{\varepsilon} \tilde{Z}^{\sigma-1}\right)^{\frac{\alpha}{\varphi}}. \tag{88}$$

To see that the growth rate of sectoral output $g_t := g_{m,t} = g_{z,t}$ is also the rate of final output, consider the production function for final output given in (1) for period $t+1$ and use the relationship between the dynamic development of average capital goods quality and intermediate output in sector $j = \{m, z\}$ as given in (79) and (80) final output in period $t+1$ reads

$$Y_{t+1} = \left(\gamma \left((1+g_{m,t}) Y_{m,t} \right)^{\frac{\varepsilon-1}{\varepsilon}} + \left((1-\gamma)(1+g_{z,t}) Y_{z,t} \right)^{\frac{\varepsilon-1}{\varepsilon}} \right)^{\frac{\varepsilon}{\varepsilon-1}}. \tag{89}$$

Since $g_{m,t} = g_{z,t} = g_t$ we get

$$Y_{t+1} = (1+g_t) \left(\gamma Y_{m,t}^{\frac{\varepsilon-1}{\varepsilon}} + (1-\gamma) Y_{z,t}^{\frac{\varepsilon-1}{\varepsilon}} \right)^{\frac{\varepsilon}{\varepsilon-1}} = (1+g_t) Y_t. \tag{90}$$

Furthermore, on the balanced growth path, output and consumption growth must be equal.[16]

The equilibrium mass of innovators in sector $j = \{m, z\}$ is given by:

$$\iota_{m,t} = \frac{\eta_{z,t} \bar{\lambda}_z}{\eta_{m,t} \bar{\lambda}_m + \eta_{z,t} \bar{\lambda}_z}, \tag{91}$$

$$\iota_{z,t} = \frac{\eta_{m,t} \bar{\lambda}_m}{\eta_{m,t} \bar{\lambda}_m + \eta_{z,t} \bar{\lambda}_z}, \tag{92}$$

where $\bar{\lambda}_j := \lambda_j - 1$ and λ_j represents the exogenous and constant sector-specific rise of quality in case of innovation. After elimination of the $\eta_{j,t}$'s the mass of innovators in sector m reads:

$$\iota_{m,t} := \iota_t = \frac{\bar{\lambda}_z}{\bar{\lambda}_z + \left(\tilde{\gamma}^{\varepsilon} \tilde{Z}^{\sigma-1} \right)^{\frac{\alpha}{\varphi}} \bar{\lambda}_m}, \tag{93}$$

[16] Just take the consumers' income profiles for two consecutive periods, use the dynamic development of capital goods quality together with the fact that consumers spend their income entirely, it follows that:

$$\frac{C^{(o)}_{W,t+1}}{C^{(o)}_{W,t}} = \frac{C^{(o)}_{e,t+1}}{C^{(o)}_{e,t}} = \frac{C^{(o)}_{i,t+1}}{C^{(o)}_{i,t}} = 1 + g_t.$$

and in sector z the mass of innovators equals

$$\iota_{z,t} := 1 - \iota_t = \frac{\bar{\lambda}_m}{\bar{\lambda}_m + \left(\tilde{\gamma}^{\varepsilon}\tilde{Z}^{\sigma-1}\right)^{\frac{-\alpha}{\varphi}} \bar{\lambda}_z}. \tag{94}$$

Finally, the equilibrium growth rate of the economy is given by an implicit expression:

$$g_t(1+g_t)^{2\alpha} = \frac{\bar{\lambda}_z\bar{\lambda}_m}{\bar{\lambda}_z + \left(\tilde{\gamma}^{\varepsilon}\tilde{Z}_{t+1}^{\sigma-1}\right)^{\frac{\alpha}{\varphi}} \bar{\lambda}_m} \left(\frac{\left(\gamma^{\varepsilon}M_{t+1}^{\sigma-1}\right)^{\frac{\alpha}{\varphi}} \bar{W}_{t+1}^{\frac{1}{\sigma-1}}}{\zeta^{\frac{1-\alpha}{\alpha}}\mu} \right)^{\alpha}. \tag{95}$$

Note that the term on the right hand side is constant over time, as long as the factors M and Z are constant. Define this term as

$$\bar{\Lambda} := \frac{\bar{\lambda}_z\bar{\lambda}_m}{\bar{\lambda}_z + \left(\tilde{\gamma}^{\varepsilon}\tilde{Z}_{t+1}^{\sigma-1}\right)^{\frac{\alpha}{\varphi}} \bar{\lambda}_m} \left(\frac{\left(\gamma^{\varepsilon}M_{t+1}^{\sigma-1}\right)^{\frac{\alpha}{\varphi}} \bar{W}_{t+1}^{\frac{1}{\sigma-1}}}{\zeta^{\frac{1-\alpha}{\alpha}}\mu} \right)^{\alpha}. \tag{96}$$

So in equilibrium, the growth rate has to be such that the equality condition in (95) holds. This is equivalent to the problem of finding the root(s) of the function

$$F(g_t) := g_t(1+g_t)^{2\alpha} - \bar{\Lambda}. \tag{97}$$

The economy's growth rate is unique if the function $F(\cdot)$ has only one root. The following Lemma proves the conditions under which F has one zero point. We use the results to characterize the balanced growth path equilibrium afterwards.

Lemma 3 *Let the function F given in (97) be defined on the interval $[g_u, g^o]$, with $0 < g_u < g^o$. Then for appropriate values of g_u, g^o there exist one g_t^* such that $F(g_t^*) = 0$.*

Together, these derivations prove the following

Proposition 5 *Consider the directed technological change model described above. Then, there exists a unique balanced growth path equilibrium in which the relative technologies are given by*

$$\tilde{Q}^* = \left(\tilde{\gamma}^{\varepsilon}\tilde{Z}^{\sigma-1}\right)^{\frac{1}{\varphi}}, \quad \textit{with } \varphi := 1 - \sigma(1-\alpha). \tag{98}$$

The interest rate of the economy is given by (84), and consumption, sectoral and aggregate output grow at the rate stated in (95).[17]

[17] See also (Acemoglu 2009, p. 509).

6 Conclusions

This paper developed a macroeconomic model to analyze the general relationship between financial intermediation and the direction of technical change. Existing approaches allow innovations on technologies to be directed towards different factors or sectors: firms can invest resources to develop new sector or factor specific technologies. Accordingly, the relative profitability of generating innovations for these specific technologies shape the direction of technical change. This work complements these approaches and shows that if innovating firms lack the necessary capital resources to finance their investments, four market forces determine the relative profitability of the different types of technology: (i) the *price effect*, which encourages innovations on technologies that use more expensive factors; (ii) the *market size effect*, which directs innovations towards technologies that use the more abundant factor; (iii) the *productivity effect*, which increases the incentive to develop technologies for sectors with a higher productivity; and last but not least (iv) the *risk effect*. This *risk effect* creates incentives to develop technologies complementing sectors where the risk of innovation failure is lower and originates from banks' external funding of innovators' investment projects.

Furthermore, the elasticity of substitution between the factors has a crucial role in determining the relative power of the price and the market size effect. If the two intermediate goods and, therefore, the two factors are gross substitutes, the market size effect is relatively more powerful than the price effect. Therefore technical changes are directed towards the factor with the larger market. In contrast, if the two intermediate goods and therefore also the two factors are gross complements, the price effect is relatively more powerful and technical changes are directed towards the relatively more scarce factor. This role of the elasticity of substitution remains unaltered if the risk effect is an additional component in the determinants of directed technical change. The presence of capital constraints and banks, however, influences the response of relative profits of technology invention to an increase in relative factor supply: If the factors are gross substitutes, relative profits including the risk effect respond more strongly to an increase in the relative factor supply compared to relative profits without the risk effect and if the factors are gross complements, relative profits including the risk effect responds less strongly. So the presence of capital constraints and banks lead to a stronger response of relative profits to factor supply changes.

Finally, the analysis also yields the relative equilibrium bias hypothesis in Proposition 4: irrespective of the elasticity of substitution, an increase in the relative supply of a factor always induces technological change that is biased in favor of that factor. This implies that the market size effect is always relatively more powerful than the price effect.

Appendix: Mathematical Appendix

Proof of Lemma 1

(i) Take the relative demand for the intermediate goods stated in Eq. (17) for any two goods i, j, solve for Y_i and insert the result into the total production costs:

$$\sum_{i=1}^{N} p_i Y_i = \sum_{i=1}^{N} p_i \left(\frac{\gamma_j}{\gamma_i}\right)^{-\varepsilon} \left(\frac{p_i}{p_j}\right)^{-\varepsilon} Y_j = \gamma_j^{-\varepsilon} p_j^{\varepsilon} Y_j \sum_{i=1}^{N} \gamma_i^{\varepsilon} p_i^{1-\varepsilon} = \left(\frac{p_j}{\gamma_j}\right)^{\varepsilon} Y_j \sum_{i=1}^{N} \gamma_i^{\varepsilon} p_i^{1-\varepsilon}. \tag{99a}$$

(ii) Since the production of the final output uses N intermediate inputs, Eq. (1) changes to

$$Y = \left(\sum_{i=1}^{N} \gamma_i Y_i^{\frac{\varepsilon-1}{\varepsilon}}\right)^{\frac{\varepsilon}{\varepsilon-1}}.$$

Use this expression and the relative factor demand from (17) for any two goods i, j to manipulate the firms return PY:

$$\begin{aligned} PY &= P\left(\sum_{i=1}^{N} \gamma_i Y_i^{\frac{\varepsilon-1}{\varepsilon}}\right)^{\frac{\varepsilon}{\varepsilon-1}} = P\left(\sum_{i=1}^{N} \gamma_i \left(\left(\frac{\gamma_j\, p_i}{\gamma_i\, p_j}\right)^{-\varepsilon} Y_j\right)^{\frac{\varepsilon-1}{\varepsilon}}\right)^{\frac{\varepsilon}{\varepsilon-1}} \\ &= P\left(\sum_{i=1}^{N} \gamma_i \left(\frac{\gamma_j\, p_i}{\gamma_i\, p_j}\right)^{1-\varepsilon} Y_j^{\frac{\varepsilon-1}{\varepsilon}}\right)^{\frac{\varepsilon}{\varepsilon-1}} = P\left(\left(\frac{p_j}{\gamma_j}\right)^{\varepsilon-1} Y_j^{\frac{\varepsilon-1}{\varepsilon}} \sum_{i=1}^{N} \gamma_i^{\varepsilon} p_i^{1-\varepsilon}\right)^{\frac{\varepsilon}{\varepsilon-1}} \\ &= P\left(\frac{p_j}{\gamma_j}^{\varepsilon} Y_j \left(\sum_{i=1}^{N} \gamma_i^{\varepsilon} p_i^{1-\varepsilon}\right)^{\frac{\varepsilon}{\varepsilon-1}}\right). \end{aligned} \tag{99b}$$

Now combine the right hand sides of Eqs. (99a) and (99b) to get

$$P\left(\sum_{i=1}^{N} \gamma_i^{\varepsilon} p_i^{1-\varepsilon}\right)^{\frac{\varepsilon}{\varepsilon-1}} = \sum_{i=1}^{N} p_i^{1-\varepsilon} \gamma_i^{\varepsilon}.$$

Solve for P gives the final result and proves the corollary stated in the text:

$$P = \left(\sum_{i=1}^{N} p_i^{1-\varepsilon} \gamma_i^{\varepsilon}\right)^{\frac{1}{1-\varepsilon}} \equiv 1. \tag{99c}$$

Derivation of $\tilde{\omega}$ (Eq. (76))

The first order optimality conditions for the factors Z and Z and in intermediate goods production are given by

$$(1-\mu)p_m M^{-\mu}\left(\int_0^1 q_{\vartheta_m}^{1-\mu} x_{\vartheta_m}^{\mu} d\vartheta_m\right) = w_M, \tag{100a}$$

$$(1-\mu)p_z Z^{-\mu}\left(\int_0^1 q_{\vartheta_z}^{1-\mu} x_{\vartheta_z}^{\mu} d\vartheta_z\right) = w_Z. \tag{100b}$$

The capital goods demand functions of type ϑ_j of quality q in sector $j = \{m, z\}$ are equal to

$$x_{\vartheta_m} = \left(\mu^2 p_m\right)^{\frac{1}{1-\mu}} q_{\vartheta_m} M. \tag{101a}$$

$$x_{\vartheta_z} = \left(\mu^2 p_z\right)^{\frac{1}{1-\mu}} q_{\vartheta_z} Z. \tag{101b}$$

Inserting these capital goods demand functions into the corresponding Eqs. (100a), (100b) respectively and take the ratio yields:

$$\tilde{w} := \frac{w_z}{w_m} = \left(\frac{p_z}{p_m}\right)^{\frac{1}{1-\mu}} \left(\frac{Q_z}{Q_m}\right) = \tilde{p}^{\frac{1}{1-\mu}} \tilde{Q}. \tag{102}$$

Insert $\tilde{p}$ from (53) gives (76) in the text:

$$\tilde{w} = \left(\tilde{\gamma}^{\varepsilon} \tilde{Z}^{-1} \tilde{Q}^{\sigma-1}\right)^{\frac{1}{\sigma}}. \tag{103}$$

Proof of Lemma 2

Suppose condition (81) holds and

$$V_{m,t} = V_{z,t} \qquad t \geq 0. \tag{104}$$

This is equivalent to

$$\tilde{V}_t = 1 \qquad t \geq 0. \tag{105}$$

To prove the statement in the Lemma, we have to show that this implies

$$g_{m,t} := \frac{Q_{m,t+1}}{Q_{m,t}} - 1 = \frac{Q_{z,t+1}}{Q_{z,t}} - 1 =: g_{z,t} \qquad t \geq 0. \tag{106}$$

The relative productivity of sectoral average capital goods quality was given in (78):

$$\tilde{Q}_t = \left(\tilde{\gamma}^{-\varepsilon}\tilde{Z}^{\sigma-1}\right)^{\frac{1}{1-\sigma(1-\alpha)}} \qquad \forall t \geq 0. \tag{107}$$

Solve for $Q_{z,t}$:

$$Q_{z,t} = \left(\tilde{\gamma}^{-\varepsilon}\tilde{Z}^{\sigma-1}\right)^{\frac{1}{1-\sigma(1-\alpha)}} Q_{m,t} \quad \forall t \geq 0. \tag{108}$$

The growth rates $g_{m,t}$ and $g_{z,t}$ are equal to

$$\frac{Q_{m,t+1}}{Q_{m,t}} = 1 + \iota\eta_m\left(\lambda_m - 1\right) \qquad \text{for all } t \geq 0 \tag{109}$$

and

$$\frac{Q_{z,t+1}}{Q_{z,t}} = 1 + (1-\iota)\eta_z\left(\lambda_z - 1\right) \qquad \text{for all } t \geq 0. \tag{110}$$

Since (107) holds for all $t \geq 0$, we have

$$Q_{z,t+1} = \left(\tilde{\gamma}^{-\varepsilon}\tilde{Z}^{\sigma-1}\right)^{\frac{1}{1-\sigma(1-\alpha)}} Q_{m,t+1}. \tag{111}$$

Now insert (111) into the enumerator and (108) into the denominator on the left-hand side of (110) yields:

$$1 + g_{z,t} = \frac{Q_{z,t+1}}{Q_{z,t}} = \frac{\left(\tilde{\gamma}^{-\varepsilon}\tilde{Z}^{\sigma-1}\right)^{\frac{1}{1-\sigma(1-\alpha)}} Q_{m,t+1}}{\left(\tilde{\gamma}^{-\varepsilon}\tilde{Z}^{\sigma-1}\right)^{\frac{1}{1-\sigma(1-\alpha)}} Q_{m,t}} = \frac{Q_{m,t+1}}{Q_{m,t}} = 1 + g_{m,t}. \tag{112}$$

Hence $g_{m,t} = g_{z,t}$ and we can conclude that if the productivity of capital goods in both sectors grows, the growth rates are identical. This proves the statement in the lemma.

Proof of Lemma 3

The function F is defined on the interval $[g_u, g^o]$ with $0 < g_u < g^o$ and according to (97) given by

$$F(g_t) := (g_t(1 + g_t)^{2\alpha} - \bar{\Lambda}. \tag{113}$$

First, note that $F(\cdot)$ is continuous on its domain and strictly increasing in its argument g_t. Then, for a given parameter value $\bar{\Lambda} > 0$, we can choose a g_u such that $g_u < \bar{\Lambda}$. Then $F(g_u) < 0$. Since $\bar{\Lambda}$ is constant, we can find a g^o with $g^o > \bar{\Lambda}$. Then $F(g^o) > 0$. Together this implies $F(g_o)F(g^u) < 0$ and we can conclude using the intermediate value theorem that g_t^* in $]g_u, g^o[$ exist with $F(g_t^*) = 0$. This already shows that g_t^* is a root of F. Uniqueness of g_t^* follows, since F is strictly increasing in g_t. Therefore, the growth rate of the balanced growth equilibrium is unique.

References

Acemoglu, D. (1998). Why do new technologies complement skills? directed technical change and wage inequality. *Quarterly Journal of Economics, 113*, 1055–1090.

Acemoglu, D. (1999). Labor and capital augmenting technical change. NBER Working Paper 69 (7544).

Acemoglu, D. (2002). Directed technical change. *Review of Economic Studies, 69*, 781–809.

Acemoglu, D. (2007). Equilibrium bias of technology. *Econometrica, 75*(5), 1371–1409.

Acemoglu, D. (2009). *Introduction to modern economic growth*. Princeton: Princeton University Press.

Aghion, P., & Howitt, P. (1992). A model of growth through creative destruction. *Econometrica, 110*, 323–351.

Barro, R. J., & Sala-i Martin, X. (2004). *Economic growth*. Cambridge: The MIT Press.

Bernanke, B., & Gertler, M. (1989). Agency costs, net worth and business fluctuations. *American Economic Review, 79*, 14–31.

Boyd, J. H., & Prescott, E. (1986). Financial intermediary-coalitions. *Journal of Economic Theory, 38*(2), 211–232.

Chandavarkar, A. (1992). Of finance and development: Neglected and unsettled questions. *World Development, 20*(1), 133–142.

Diamond, D. (1984). Financial intermediation and delegated monitoring. *The Review of Economic Studies, 51*(3), 393–414.

Dranakis, E., & Phelps, E. (1965). A model of induced invention, growth and distribution. *Economic Journal, 76*, 823–840.

Gale, D., & Hellwig, M. (1985). Incentive-compatible debt contracts: The one-period problem. *Review of Economic Studies, 52*(4), 647–664.

Grossman, G., & Helpman, E. (1993). *Innovation and growth in the global economy*. Cambridge: The MIT Press.

Habakkuk, J. (1962). *American and British technology in the nineteenth century: Searching for labor saving inventions*. Cambridge: Cambridge University Press.

Hicks, J. (1932). *The theory of wages*. London: Macmillan.

Kennedy, C. (1964). Induced bias in innovation and the theory of distribution. *Economic Journal, 74*, 541–547M.

King, R.G., & Levine, R. (1993). Finance, entrepreneurship and growth: Theory and evidence. *Journal of Monetary Economics, 32*(3), 513–542.

Levine, R. (1997). Financial development and economic growth: views and agenda. *Journal of Economic Literature, 35*, 688–726.

Levine, R., & Zervos, S. (1998). Stock markets, banks, and economic growth. *American Economic Review, 88*(3), 537–58.

Lucas, R. J. (1988). On the mechanics of economic development. *Journal of Monetary Economics, 22*, 3–42.

Meier, G., & Seers, D. (1984). *Pioneers in development.* New York: Oxford University Press for the World Bank.

Rivera-Batiz, L., & Romer, P. (1991). Economic integration and endogenous growth. *The Quarterly Journal of Economics, 2*, 531–555.

Robinson, J. (1952). The generalization of the general theory. *The rate of interest and other essays* (pp. 67–142), London: Macmillan.

Romer, P. (1990). Endogenous technological change. *Journal of Political Economy, 102*, 71–102.

Samuelson, P. (1965). A theory of induced innovation along Kennedy-Weisaecker lines. *Review of Economics and Statistics, 47*, 343-356.

Schumpeter, J. (1912). *Theorie der wirtschaftlichen Entwicklung*. Berlin: Duncker & Humblot.

Stern, N. (1989). The economics of development: A survey. *Economic Journal, 99*(397), 597–685.

Uhlig, H. (1996). A law of large numbers for large economies. *Economic Theory, 8*(1), 41–50.

Williamson, S. D. (1986). Costly monitoring, financial intermediation and equilibrium credit rationing. *Journal of Monetary Economics, 18*, 159–179.

Williamson, S. D. (1987). Costly monitoring, loan contracts and equilibrium credit rationing. *Quarterly Journal of Economics, 102*, 135–145.

Wong, K. P. (1997). On the determinants of bank interest margins under credit and interest rate risk. *Journal of Banking and Finance, 21*(2), 251–271.

King, R. G., & Levine, R. (1993). Finance, entrepreneurship, and growth: Theory and evidence. Journal of Monetary Economics, 32(3), 513–542.
Levine, R. (1997). Financial development and economic growth: Views and agenda. Journal of Economic Literature, 35, 688–726.
[illegible]

Sustainability of Public Debt in an AK Model with Complex Tax System

Atsumasa Kondo

Abstract This paper theoretically investigates the role of the tax system in sustaining the public debt. The paper explicitly derives the critical level of the public debt-to-GDP ratio that is compatible with a balanced growth path. If the ratio exceeds this critical level at time 0, then it diverges to $+\infty$ as time passes. Analyzing a situation where the government marginally increases the consumption tax rate, the paper reveals the extent to which the government can then cut the income tax rate while maintaining the sustainability of public debt. Tax rates that are compatible with the balanced growth are also derived as a function of the initial level of debt-to-GDP ratio.

1 Introduction

Along with many other developed countries, Japan faces high levels of accumulated public debt. The Organization for Economic Cooperation and Development (OECD) predicted that Japan's public debt-to-gross domestic product (GDP) ratio, which is monotonically increasing, would exceed 2.30 in 2016 (OECD "Economic Outlook 98", November 2015).[1] To avoid fiscal bankruptcy, the Japanese government has established an overriding objective: Japan's primary balance should be in surplus until 2020. To achieve this objective, the government has discussed implementing fiscal reforms (including increasing the consumption tax rate) to secure its financial resources.

The purpose of this paper is to theoretically study the role of the tax system in sustaining public debt. For that aim, this paper explicitly derives a critical level of public debt-to-GDP ratio that is compatible with a balanced growth path (BGP). If the ratio exceeds this critical level at time 0, then it diverges to $+\infty$ as time passes,

[1] http://www.paris-oecd.diplo.de/contentblob/4658486/Daten/6022232/Economic_Outlook.pdf.

A. Kondo
Shiga University, 1-1-1, Banba, Hikone, Shiga 522-0069, Japan
e-mail: a-kondo@biwako.shiga-u.ac.jp

B. Bökemeier, A. Greiner (eds.), *Inequality and Finance in Macrodynamics*, Dynamic Modeling and Econometrics in Economics and Finance 23,
DOI 10.1007/978-3-319-54690-2_7

which means that the public debt cannot be sustained. If the public debt-to-GDP ratio is at the critical level at time 0, then it remains constant over time; this can be interpreted as the public debt being sustainable. This paper highlights how the critical level of the public debt-to-GDP ratio depends on the tax system. The tax system includes the labor income and interest income tax rates, and the taxation on the consumption behavior of households.

The analysis reveals the "marginal rate of substitution of the tax rates". That is, if the government increases the consumption tax rate at the marginal level, then to what extent can it cut the income tax rate while maintaining a sustainable public debt. The analysis may be of interest to policy makers in those countries with high levels of accumulated public debt. Further, the approach used explicitly derives a tax rate that is compatible with the sustainability of public debt as a function of the initial level of debt-to-GDP ratio, given other parameters. For example, if the public debt-to-GDP ratio increases from 2.30 to 2.40, then it is possible to calculate the extent to which the government must increase the consumption tax rate to avoid fiscal bankruptcy.

Many studies examine the sustainability of public debt. Bräuninger (2005), Yakita (2008) and Arai (2011) theoretically study the relationship between a deficit ratio and economic growth rates, by using overlapping generations models. These studies find that there is a critical level of public debt-to-GDP ratio for sustainable public debt. If the public debt is greater than the critical level at the initial time point, then it is no longer sustainable. Greiner (2011) shows the existence of a critical level of public debt that is sustainable in a case where the primary surplus-to-GDP ratio is independent of the public debt-to-GDP ratio. This critical level depends on the difference between the real interest rate and the economic growth rate. Although Kondo (2007, 2012) and Kondo and Kitaura (2009) explicitly derive the critical level of public debt, they assume a lump-sum tax. Hence, they do not consider the interaction between different taxations for the sustainability of public debt. Kondo's (2016) model accounts for income and consumption taxes, and derives the critical level of the public debt-to-GDP ratio. However, that paper looks at properties of consumption tax in comparison with other taxes. Hence, the tax rates for labor income and interest income are supposed to be same. The present paper identifies the tax rates for the labor income and the interest income, to reflect the complex tax system of the real world.

Many empirical studies focus on the sustainability of public debt. Greiner et al. (2007) and Fincke and Greiner (2011, 2012) test whether governments are properly subject to Bohn's (1998) rule for fiscal sustainability; i.e., governments should increase the primary surplus-to-GDP ratio to maintain a healthy fiscal status if public debt-to-GDP ratio increases. Greiner et al. (2007) and Fincke and Greiner (2012) primarily focus on European countries, while Fincke and Greiner (2011) focuses on the United States, Germany and Japan. Bökemeier (2015) investigates the relationship between public deficits and economic growth rates for central and eastern European countries.

Many studies consider the tax system from a macroeconomics perspective. Schmitt-Grohe and Uribe (1997) demonstrate that under the balanced-budget rule, economic fluctuations driven from expectation are more likely to arise than under a case in which public deficit is counted in. This is because under the balanced-budget rule, it is impossible to make fine adjustments of the public expenditure in response to economic fluctuations. They assume that public expenditures are exclusively financed from income taxation. Giannitsarou (2007) shows that if a government fully finances its expenditure by consumption taxation, then such fluctuations will disappear. While Schmitt-Grohe and Uribe (1997) and Giannitsarou (2007) assume constant tax rates, Nourry et al. (2013) and Greiner and Bondarev (2015) include a state-dependent consumption tax, in which the consumption tax rate increases in line with the consumption level of consumers. However, those studies focus on the local indeterminacy of the equilibrium paths near steady states, and do not highlight the sustainability of public debt. Nourry et al. (2013) study the case of balanced budgets, and Greiner and Bondarev (2015) postulate Bohn's (1998) rule, for the public debt to be sustainable.

The rest of this paper is organized as follows. Section 2 builds a model on which the subsequent analysis is based. Section 3 analyzes an equilibrium path, with a particular focus on a BGP. Section 4 presents the main results. Section 5 briefly concludes the paper.

2 Model

This section sets up a basic model. Think of an economy with an infinite time horizon, in which economic activities are conducted at each time point $t \in [0, \infty)$.[2] The economy consists of households, firms and a government.

2.1 *Household*

The economy contains many identical households. A representative household maximizes a discounted integral of utilities from now to the future. The instantaneous utility function is represented by a utility function $u(C) = \left(C^{1-\sigma} - 1\right) / (1-\sigma)$, where $C\ (\geq 0)$ stands for consumption. The parameter $\sigma > 0\ (\sigma \neq 1)$ represents the inverse of the elasticity of consumption, which is independent from time. It is well-known that the logarithmic utility function is potentially included as the case of $\sigma = 1$. The household supplies the labor service L^S to a representative firm. The labor service is assumed to be inelastically supplied and thus, it becomes constant over time. The wage rate is denoted by $w(t)$. The household purchases consumption

[2] The symbol $[0, \infty)$ stands for the set $\{x \in \mathbb{R} : 0 \leq x < \infty\}$, where $\mathbb{R}$ is the set of real numbers.

goods, and partly saves its income for a future consumption at each time point. The total real asset is denoted by

$$W(t) = B(t) + K(t) \tag{1}$$

where $B(t)\ (\in \mathbb{R})$ and $K(t)\ (\geq 0)$ are the public debt and the physical capital, respectively. Their initial levels $B(0)$ and $K(0)$ are given for the household. The saving bears real interest, which is denoted by $r(t)$. The consumption behavior and the income of the household are levied by the government. The consumption tax rate is $\tau_C > 0$, and the labor income and interest income tax rates are denoted by $\tau_L \in (0, 1)$ and $\tau_W \in (0, 1)$, respectively. These tax rates are supposed to be constants. The feasible path of consumption and saving stream is subject to budget constraints and the no Ponzi-game (NPG) condition. The household's behavior can be summarized as the following maximizing problem:

$$\max_{C(t),\, W(t)} \int_0^{\infty} e^{-\rho t} \frac{C(t)^{1-\sigma} - 1}{1-\sigma} dt \tag{2}$$

subject to

$$(1 + \tau_C)\, C(t) + \dot{W}(t) = (1 - \tau_L)\, w(t)\, L^S + (1 - \tau_W)\, r(t)\, W(t) \tag{3}$$

$$\lim_{t\to\infty} e^{-(1-\tau_W)\int_0^t r(s)ds}\, W(t) = 0 \tag{4}$$

where $\rho > 0$ stands for the subjective discount rate. The time derivative is represented by the dot symbol: $\dot{(\ )} = d/dt$. In what follows, the time index is often omitted if no ambiguity arises.

2.2 Firms

The representative firm produces output Y, by using labor L^D and physical capital K. The production process is influenced by the economy-wide capital level $\overline{K}$. Although each firm takes $\overline{K}$ as given, the condition $\overline{K} = K$ is required in an equilibrium. The production function regarding the firm is given by

$$Y = AK^{\alpha} \left(\overline{K}L^D\right)^{1-\alpha}, \tag{5}$$

where $\alpha \in (0, 1)$ and $A\ (> 0)$ are constant parameters that represent a capital share and a technology level, respectively. The representative firm maximizes its static profit at each time point. Its behavior is summarized as the following maximizing problem:

$$\max_{K,\, L^D \geq 0} AK^{\alpha}\left(\overline{K}L^D\right)^{1-\alpha} - rK - wL^D \tag{6}$$

$$\overline{K}\,(\geq 0) : \text{given}$$

2.3 Government

The government obtains tax revenues from taxation on consumption, labor income, and interest income. It also obtains revenues from issuing public bonds B^G. The government's budget constraint is given by

$$\dot{B^G} = rB^G - (T - G) \tag{7}$$

where

$$T = \tau_C C + \tau_L wL + \tau_W rW \tag{8}$$

is the total tax revenue and G is the public spending. Equation (7) can be regarded as an accounting identity. For simplicity, this paper assumes that the public spending is wastefully used. I also assume that the government uses a constant fraction of GDP Y for the public spending, i.e.,

$$G = gY \tag{9}$$

where $g = [0, 1)$ is a constant. The initial level of the public debt $B^G(0)\,(= B(0))$ is given for the government.

In addition to (7), the government must be subject to the NPG condition:

$$\lim_{t \to \infty} e^{-(1-\tau_W)\int_0^t r(s)ds} B^G(t) = 0. \tag{10}$$

The condition (10) means that the present value of the public debt must converge to 0 in the remote future. This condition implies that the government plans to raise revenues by levying tax from now to the future to just compensate for their expenditure plan $\{G(t)\}$, and to repay the accumulated debt at the initial time point $B^G(0)$ (for this point, see Greiner 2013, 2015).

2.4 Market Clearing Conditions

In an equilibrium, markets for the good, labor and the public bond are simultaneously cleared. That is, the following equations must hold:

$$Y = C + \dot{K} + gY, \quad L^S = L^D, \quad B^G = B, \tag{11}$$

for any $t \geq 0$. The equations in (11) are those for the good, the labor service and the public bonds, respectively. Henceforth, I simply denote the equilibrium value of labor and public debt by L and B, respectively. Further, as mentioned in Sect. 2.2, an average economy-wide level of capital is required to be equal to the capital level of the representative firm in an equilibrium:

$$\overline{K} = K. \tag{12}$$

As a consequence, it holds from (5) that

$$Y = AKL^{1-\alpha}. \tag{13}$$

3 Equilibrium

This section derives the equilibrium path, focusing on a BGP. Based on the discussions in this section, Sect. 4 investigates the initial level of public debt-to-capital (or debt-to-GDP) ratio that is compatible with the BGP.

3.1 Equilibrium Path

In an equilibrium, the following four conditions are satisfied:

(i) given the time paths of prices $\{w(t), r(t)\}$, the tax rates (τ_C, τ_L, τ_W) and the initial level of assets $(B(0), K(0))$, the households maximize (2) subject to (3) and (4),
(ii) given the time paths of prices $\{w(t), r(t)\}$ and an average level of economy-wide capital $\{\overline{K}(t)\}$, the firms maximize (6) for each t,
(iii) given $\{r(t)\}$ and the initial level of public debt $B^G(0)$, the government determines the policy variables $((\tau_C, \tau_L, \tau_W, g), \{B^G(t)\})$ subject to (7) and (10),
(iv) all of the markets are simultaneously cleared (11), and an average level of economy-wide capital $\overline{K}$ coincides with the physical capital K for each t.

As first order conditions of the maximizing problem of the firm (6), it holds in the equilibrium that

$$r = \alpha A L^{1-\alpha}, \tag{14}$$

$$w = (1-\alpha) A K L^{-\alpha}. \tag{15}$$

Equation (14) is obtained from the condition $\overline{K} = K$.

Next, the consumption path in an equilibrium is investigated. The (present-value) Hamiltonian associating with the maximizing problem of the representative household (2)–(4) is defined by

$$\mathcal{H} = e^{-\rho t}\frac{C^{1-\sigma}-1}{1-\sigma} + \lambda\left[(1-\tau_L) w L^s + (1-\tau_W) r W - (1+\tau_C) C\right],$$

where λ is called a co-state variable. As conditions $\partial\mathcal{H}/\partial C = 0$ and $\partial\mathcal{H}/\partial W = -\dot{\lambda}$, the following holds:

$$\frac{\dot{C}}{C} = \frac{(1-\tau_W)\alpha A L^{1-\alpha} - \rho}{\sigma}. \tag{16}$$

Thus, it holds that

$$C(t) = C(0) e^{\theta t}, \tag{17}$$

where θ is defined as

$$\theta \equiv \frac{(1-\tau_W)\alpha A L^{1-\alpha} - \rho}{\sigma}. \tag{18}$$

The variable θ stands for the economic growth rate on the BGP. Note that among the tax system (τ_C, τ_L, τ_W), only the tax rate for the interest income τ_W influences the economic growth rate on the BGP. As seen in (4), the interest tax rate τ_W affects the discount rate for the future activities, and hence, affects the long-term growth rate.[3]

The dynamics of capital stock is derived from the first equation of the market clearing conditions (11):

$$\dot{K} = (1-g) A K L^{1-\alpha} - C.$$

[3]This point was also pointed out by Kondo (2016) although he deals with the case of $\tau_L = \tau_W$.

It holds from (17) that

$$\dot{K} = (1 - g) AKL^{1-\alpha} - C(0) e^{\theta t}.$$

Solving this differential equation, I obtain

$$K = e^{(1-g)AL^{1-\alpha}t} \left[K(0) - C(0) \frac{1}{(1-g) AL^{1-\alpha} - \theta} \right] + \frac{C(0)}{(1-g) AL^{1-\alpha} - \theta} e^{\theta t}. \tag{19}$$

It is clear from (19) that the capital path is on the BGP if and only if the initial level of consumption is chosen at

$$C(0) = \left[(1-g) AL^{1-\alpha} - \theta \right] K(0), \tag{20}$$

given $K(0) > 0$.

Here, the following assumption is made.

Assumption 1 $0 \leq \theta < (1 - \tau_W) \alpha AL^{1-\alpha} < (1-g) AL^{1-\alpha}$, where θ is defined in (18).

This assumption implies that $\theta < (1-g) AL^{1-\alpha}$, which guarantees that the economic variables (C, K) are positive on a BGP (see (20)). The first inequality $0 \leq \theta$ means that the economic growth rate θ is non-negative on the BGP, which is satisfied if ρ is sufficiently near 0. The second inequality $\theta < (1 - \tau_W) \alpha AL^{1-\alpha}$ $(= (1 - \tau_W) r)$ guarantees that the NPG conditions (4) and (10) are satisfied on the BGP (see the BGP conditions (21) and (27)).

3.2 *Balanced Growth Path*

The conditions of the BGP are obtained from (17), (19) and (13) as

$$\begin{aligned} C(t) &= C(0) e^{\theta t}, \\ K(t) &= K(0) e^{\theta t}, \\ Y(t) &= AK(0) L^{1-\alpha} e^{\theta t}, \end{aligned} \tag{21}$$

with the consumption level (20) given $K(0) > 0$.

Using these conditions (21), I study the dynamic path of the public debt. Substituting (8) and (9) into (7) yields

$$\dot{B} = rB - (\tau_C C + \tau_L wL + \tau_W rW - gY).$$

It holds from (1), (13), (14) and (15) that

$$\begin{aligned}\dot{B} &= (1-\tau_W)\, rB - \left(\tau_C C + \tau_L wL + \tau_W rK - gAKL^{1-\alpha}\right) \\ &= (1-\tau_W)\,\alpha AL^{1-\alpha}B \\ &\quad - \left[\tau_C C + \tau_L (1-\alpha)\, AKL^{1-\alpha} + \tau_W \alpha AL^{1-\alpha}K - gAKL^{1-\alpha}\right].\end{aligned}$$

Thus,

$$\dot{B} = (1-\tau_W)\,\alpha AL^{1-\alpha}B - \left[\tau_C C + \left\{\tau_L (1-\alpha) + \tau_W \alpha - g\right\} AKL^{1-\alpha}\right]. \tag{22}$$

Substituting the BGP conditions with respect to $C(t)$ and $K(t)$ shown in (21) into (22) yields

$$\begin{aligned}\dot{B} &= (1-\tau_W)\,\alpha AL^{1-\alpha}B \\ &\quad - \left[\tau_C C(0) + \left\{(1-\alpha)\,\tau_L + \alpha\tau_W - g\right\} AL^{1-\alpha}K(0)\right] e^{\theta t}.\end{aligned}$$

It follows from (20) that the following holds:

$$\dot{B} = (1-\tau_W)\,\alpha AL^{1-\alpha}B - HK(0)\, e^{\theta t}, \tag{23}$$

where H is a variable defined by

$$H \equiv \tau_C \left[(1-g)\, AL^{1-\alpha} - \theta\right] + \left[(1-\alpha)\,\tau_L + \alpha\tau_W - g\right] AL^{1-\alpha}. \tag{24}$$

From (23), I obtain

$$\begin{aligned}B &= e^{(1-\tau_W)\alpha AL^{1-\alpha}t}\left[B(0) + \frac{H}{\theta - (1-\tau_W)\,\alpha AL^{1-\alpha}} K(0)\right] \\ &\quad - \frac{H}{\theta - (1-\tau_W)\,\alpha AL^{1-\alpha}} K(0)\, e^{\theta t}.\end{aligned}$$

From (18), it holds that

$$\begin{aligned}B &= e^{(1-\tau_W)\alpha AL^{1-\alpha}t}\left[B(0) - \frac{\sigma H}{\rho - (1-\tau_W)\,\alpha AL^{1-\alpha}(1-\sigma)} K(0)\right] \\ &\quad + \frac{\sigma H}{\rho - (1-\tau_W)\,\alpha AL^{1-\alpha}(1-\sigma)} K(0)\, e^{\theta t}.\end{aligned} \tag{25}$$

Thus, if and only if

$$B(0) = \frac{H}{\rho - (1-\tau_W)\,\alpha AL^{1-\alpha}(1-\sigma)} \sigma K(0), \tag{26}$$

then the public debt is on the BGP:

$$B(t) = B(0)\, e^{\theta t}. \tag{27}$$

The set of Eqs. (21) and (27) represents the BGP in the present economy.

Here, the following assumptions are made.

Assumption 2 $2(1-\alpha)\,\tau_L + \alpha\tau_W - g > 0$.

Assumption 3 σ is sufficiently near 1.

Assumption 2 is met if the public spending-to-GDP ratio g is sufficiently small in relation to the income tax rates (τ_L, τ_W). Together with Assumption 1, Assumption 2 guarantees that the variable H is positive (see (24)). Assumptions 1–3 imply that $B(0)$ is positive (see (26)); i.e., the government is a borrower. This is a reasonable assumption given the high level of accumulated public debt in many advanced countries in the real world. Section 4 offers a numerical example that satisfies Assumptions 1–3.

The public debt-to-GDP ratio on the BGP is of great interest. It can be derived from (21), (27) and (26) as

$$\begin{aligned} \frac{B}{Y} &= \frac{\sigma H}{AL^{1-\alpha}\left[\rho - (1-\tau_W)\,\alpha A L^{1-\alpha}(1-\sigma)\right]} \\ &= \frac{\tau_C\left[(1-g)\,AL^{1-\alpha} - \theta\right] + \left[(1-\alpha)\,\tau_L + \alpha\tau_W - g\right]AL^{1-\alpha}}{AL^{1-\alpha}\left[\rho - (1-\tau_W)\,\alpha A L^{1-\alpha}(1-\sigma)\right]}\sigma \end{aligned} \tag{28}$$

The following proposition is established.

Proposition 4 (Balanced Growth Path) *In the present economy, the BGP is given by* (21) *and* (27)*, with the initial level of the consumption* (20) *and the public debt* (26) *given* $K(0) > 0$. *The economic growth rate* θ *and the variable* H *are shown in* (18) *and* (24)*, respectively. The public debt-to-GDP ratio on the BGP is given by* (28).

4 Sustainability of Public Debt

This section analyzes the initial level of public debt-to-GDP ratio that is compatible with the BGP. The section further examines how the level depends on the tax system (τ_C, τ_L, τ_W). A numerical example is given to illustrate the analyses.

4.1 Critical Level of Public Debt-to-GDP Ratio

It is clear from (25) that if $B(0) > \frac{H}{\rho-(1-\tau_W)\alpha AL^{1-\alpha}(1-\sigma)}\sigma K(0)$, then the public debt-to-capital ratio diverges to $+\infty$ as time passes. Further, from (21) and (25), the public debt-to-GDP ratio also diverges to $+\infty$, which contradicts the NPG condition (10). Thus, the initial level of public debt must be

$$B(0) \leq \frac{H}{\rho - (1-\tau_W)\alpha AL^{1-\alpha}(1-\sigma)}\sigma K(0).$$

Using $Y(0) = AL^{1-\alpha}K(0)$, the critical level of public debt-to-GDP ratio that is compatible with the BGP is given by

$$\frac{B(0)}{Y(0)} = \frac{H}{AL^{1-\alpha}\left[\rho - (1-\tau_W)\alpha AL^{1-\alpha}(1-\sigma)\right]}\sigma.$$

Substituting the definition of H (24) into the right hand side of the above inequality, I get a public debt-to-GDP ratio that is compatible with the BGP as follows:

$$\frac{B(0)}{Y(0)} = \frac{\begin{array}{c}\tau_C\left[(1-g)AL^{1-\alpha} - \left\{(1-\tau_W)\alpha AL^{1-\alpha} - \rho\right\}/\sigma\right] \\ + \left\{(1-\alpha)\tau_L + \alpha\tau_W - g\right\}AL^{1-\alpha}\end{array}}{AL^{1-\alpha}\left[\rho - (1-\tau_W)\alpha AL^{1-\alpha}(1-\sigma)\right]}\sigma \equiv \varphi. \tag{29}$$

Under Assumptions 1–3, it holds that $\varphi > 0$, where the variable φ is defined in (29). Equation (29) explicitly shows how the tax system (τ_C, τ_L, τ_W) effects the public debt sustainability.

I sum up this result as a theorem.

Theorem 9 (Critical Level of Public Debt-to-GDP Ratio) *The initial level of public debt-to-GDP ratio that is compatible with the BGP is given by* (29).

The relationship in (29) is complicated; hence, I describe the case in which the utility function of the household is logarithmic. Substituting $\sigma = 1$ into (29) yields

$$\varphi = \frac{\begin{array}{c}\tau_C\left[(1-g)AL^{1-\alpha} - \left\{(1-\tau_W)\alpha AL^{1-\alpha} - \rho\right\}\right] \\ + \left\{(1-\alpha)\tau_L + \alpha\tau_W - g\right\}AL^{1-\alpha}\end{array}}{AL^{1-\alpha}\rho}. \tag{30}$$

This result is summarized as the following corollary.

Corollary 2 (Critical Level of Public Debt-to-GDP Ratio for the Log-Utility Case) *Assume that the utility function of the representative household is logarithmic. Then, the initial level of public debt-to-GDP ratio that is compatible with the BGP is given by* (30).

4.2 Marginal Rates of Substitution of the Tax Rates

The main purpose of this paper is to investigate how the critical level of the sustainable public debt-to-GDP ratio φ depends on the tax system (τ_C, τ_L, τ_W). To do this, I derive the marginal effects of the tax rates on φ. Some calculations from (29) yield

$$\frac{\partial\varphi}{\partial\tau_C} = \frac{(1-g)\,AL^{1-\alpha}\sigma - (1-\tau_W)\,\alpha AL^{1-\alpha} + \rho}{AL^{1-\alpha}\left[\rho - (1-\tau_W)\,\alpha AL^{1-\alpha}(1-\sigma)\right]},$$

$$\frac{\partial\varphi}{\partial\tau_L} = \frac{(1-\alpha)\,\sigma}{\rho - (1-\tau_W)\,\alpha AL^{1-\alpha}(1-\sigma)},$$

$$\frac{\partial\varphi}{\partial\tau_W} = \frac{\begin{array}{c}(\tau_C+\sigma)\left[\rho-(1-\tau_W)\,\alpha AL^{1-\alpha}(1-\sigma)\right] \\ -(1-\sigma)\,AL^{1-\alpha}\left[\begin{array}{c}\tau_C\left\{(1-g)\,\sigma-(1-\tau_W)\,\alpha+\frac{\rho}{AL^{1-\alpha}}\right\} \\ +\left\{(1-\alpha)\,\tau_L+\alpha\tau_W-g\right\}\sigma\end{array}\right]\end{array}}{\left[\rho-(1-\tau_W)\,\alpha AL^{1-\alpha}(1-\sigma)\right]^2}\alpha.$$

It is easily ascertained that under Assumptions 1–3, the sign of $\partial\varphi/\partial\tau_i$ is positive for any $i = C, L, W$.

By using the implicit function theorem, I can derive the marginal rate of policy substitutions.

$$-\frac{d\tau_L}{d\tau_C} = \frac{\partial\varphi/\partial\tau_C}{\partial\varphi/\partial\tau_L} = \frac{(1-g)\,AL^{1-\alpha}\sigma - (1-\tau_W)\,\alpha AL^{1-\alpha} + \rho}{(1-\alpha)\,AL^{1-\alpha}\sigma}, \tag{31}$$

$$\begin{aligned}-\frac{d\tau_W}{d\tau_C} &= \frac{\partial\varphi/\partial\tau_C}{\partial\varphi/\partial\tau_W} \\ &= \frac{(1-g)\,AL^{1-\alpha}\sigma-(1-\tau_W)\,\alpha AL^{1-\alpha}+\rho}{\dfrac{\begin{array}{c}(\tau_C+\sigma)\left[\rho-(1-\tau_W)\,\alpha AL^{1-\alpha}(1-\sigma)\right] \\ -(1-\sigma)\,AL^{1-\alpha}\left[\begin{array}{c}\tau_C\left\{\begin{array}{c}(1-g)\,\sigma \\ -(1-\tau_W)\,\alpha+\frac{\rho}{AL^{1-\alpha}}\end{array}\right\} \\ +\left\{(1-\alpha)\,\tau_L+\alpha\tau_W-g\right\}\sigma\end{array}\right]\end{array}}{\rho-(1-\tau_W)\,\alpha AL^{1-\alpha}(1-\sigma)}\alpha AL^{1-\alpha}},\end{aligned} \tag{32}$$

$$-\frac{d\tau_W}{d\tau_L} = \frac{\partial\varphi/\partial\tau_L}{\partial\varphi/\partial\tau_W} \tag{33}$$

$$
= \frac{(1-\alpha)\,\sigma}{(\tau_C+\sigma)\left[\rho-(1-\tau_W)\,\alpha AL^{1-\alpha}\,(1-\sigma)\right] - \dfrac{(1-\sigma)\,AL^{1-\alpha}\left[\begin{array}{l}\tau_C\left\{\begin{array}{c}(1-g)\,\sigma\\ -(1-\tau_W)\,\alpha+\frac{\rho}{AL^{1-\alpha}}\end{array}\right\}\\ +\{(1-\alpha)\,\tau_L+\alpha\tau_W-g\}\,\sigma\end{array}\right]}{\rho-(1-\tau_W)\,\alpha AL^{1-\alpha}\,(1-\sigma)}\,\alpha}.
$$

How are results (31)–(33) interpreted from an economic standpoint? First, see (31). Think of a situation in which the economy is on the BGP. Then, the debt-to-GDP ratio remains constant over time at the level shown in (29). Suppose that the government plans to increase the consumption tax rate τ_C at the marginal level. In such a situation,

Q. To what extent can the government cut the labor tax rate $\tau_L$while sustaining the public debt-to-GDP ratio at the constant level (29)?
A. The answer is given by (31).

If the government changes τ_C and τ_L, then the behavior of the households is influenced, and thus the equilibrium path is affected. That effect is captured by the conditions of BGP (20), (21), (26) and (27). Accordingly, the critical level of the public debt-to-GDP ratio is affected by the policy change (see (29)). The Eq. (31) explicitly demonstrates that the government can cut the labor tax rate by the level shown in (31). The relationships (32) and (33) can be interpreted in much the same way.

The following theorem is established:

Theorem 10 (Marginal Rates of Substitution of the Tax Rates) *Assume that the economy is on the BGP, and that the government marginally increases the consumption tax rate τ_C. Then, the government can cut the labor tax rate τ_L by the level shown in* (31), *or cut the interest tax rate τ_W by the level shown in* (32) *while maintaining the initial level of the public debt-to-GDP ratio at a constant level* (29). *In much the same way, if the government marginally increases the labor tax rate, then it can cut the interest tax rate by the level shown in* (33) *while maintaining the debt-to-GDP ratio at a constant level* (29).

The relationships in (31)–(33) are complicated; hence I describe the case in which the utility functions of the households are logarithmic. Substituting $\sigma = 1$ into (31)–(33) yields

$$
-\frac{d\tau_L}{d\tau_C} = \frac{\partial\varphi/\partial\tau_C}{\partial\varphi/\partial\tau_L} = \frac{(1-g)\,AL^{1-\alpha}-(1-\tau_W)\,\alpha AL^{1-\alpha}+\rho}{(1-\alpha)\,AL^{1-\alpha}}, \tag{34}
$$

$$
-\frac{d\tau_W}{d\tau_C} = \frac{\partial\varphi/\partial\tau_C}{\partial\varphi/\partial\tau_W} = \frac{(1-g)\,AL^{1-\alpha}-(1-\tau_W)\,\alpha AL^{1-\alpha}+\rho}{(\tau_C+1)\,\alpha AL^{1-\alpha}}, \tag{35}
$$

$$
-\frac{d\tau_W}{d\tau_L} = \frac{\partial\varphi/\partial\tau_L}{\partial\varphi/\partial\tau_W} = \frac{1-\alpha}{(\tau_C+1)\,\alpha}. \tag{36}
$$

These results are summarized as the following corollary.

Corollary 3 (Marginal Rates of Substitution of the Tax Rates for the Log-Utility Case) *Let the utility function of the representative household be logarithmic. Assume that the economy is on the BGP, and that the government marginally increases the consumption tax rate* τ_C*. The government can cut the labor tax rate* τ_L *by the level shown in* (34) *or cut the interest tax rate* τ_W *by the level shown in* (35) *while maintaining the initial level of the public debt-to-GDP ratio* φ *at a constant level* (30)*. In much the same way, if the government marginally increases the labor tax rate* τ_L*, then it can cut the interest tax rate* τ_W *by the level shown in* (36) *while maintaining the debt-to-GDP ratio* φ *at a constant level* (30).

4.3 Required Tax Rates for the Sustainability of the Public Debt

This section derives the tax rates that are compatible with the sustainable public debt, given the initial level of the public debt-to-GDP ratio $B(0)/Y(0)$. In other words, the Eq. (29) is solved for each tax rate, e.g., τ_C, given $B(0)/Y(0)$ and other tax rates, e.g., (τ_L, τ_W).

Solving the Eq. (29) yields

$$\tau_C = \frac{\begin{array}{c} AL^{1-\alpha}\left[\rho - (1-\tau_W)\,\alpha AL^{1-\alpha}(1-\sigma)\right]\frac{B(0)}{Y(0)} \\ -\left[(1-\alpha)\,\tau_L + \alpha\tau_W - g\right]AL^{1-\alpha}\sigma \end{array}}{(1-g)\,AL^{1-\alpha}\sigma - \left[(1-\tau_W)\,\alpha AL^{1-\alpha} - \rho\right]}. \tag{37}$$

In much the same way, it holds that

$$\tau_L = \frac{\begin{array}{c} AL^{1-\alpha}\left[\rho - (1-\tau_W)\,\alpha AL^{1-\alpha}(1-\sigma)\right]\frac{B(0)}{Y(0)} \\ -\tau_C\left[(1-g)\,AL^{1-\alpha}\sigma - (1-\tau_W)\,\alpha AL^{1-\alpha} + \rho\right] \\ -(\alpha\tau_W - g)\,AL^{1-\alpha}\sigma \end{array}}{(1-\alpha)\,AL^{1-\alpha}\sigma}, \tag{38}$$

$$\tau_W = \frac{\begin{array}{c} \tau_C\left[(1-g)\,\sigma - \alpha + \rho\frac{1}{AL^{1-\alpha}}\right] + \left[(1-\alpha)\,\tau_L - g\right]\sigma \\ -\left[\rho - \alpha AL^{1-\alpha}(1-\sigma)\right]\frac{B(0)}{Y(0)} \end{array}}{\alpha\left[AL^{1-\alpha}(1-\sigma)\frac{B(0)}{Y(0)} - \tau_C - \sigma\right]}. \tag{39}$$

Note that under Assumptions 1–3,

$$\frac{\partial\tau_i}{\partial\left(B(0)/Y(0)\right)} > 0,$$

for any $i = C, L, W$. This means that when the initial level of public debt-to-GDP ratio is high, the tax rate τ_i must be set at a high level given other parameters, e.g., $\tau_j\,(j \neq i)$.

The following theorem is established.

Theorem 11 (Required Levels of Tax Rates) *The tax rates for consumption, labor income and interest income that are required for the economy to be on the BGP are presented by* (37), (38) *and* (39), *respectively.*

For simplicity, consider the log-utility case. Then, substituting $\sigma = 1$ into (37)–(39) I obtain that

$$\tau_C = \frac{AL^{1-\alpha}\rho\frac{B(0)}{Y(0)} - [(1-\alpha)\,\tau_L + \alpha\tau_W - g]\,AL^{1-\alpha}}{(1-g)\,AL^{1-\alpha} - [(1-\tau_W)\,\alpha AL^{1-\alpha} - \rho]}, \tag{40}$$

$$\tau_L = \frac{\begin{array}{c}AL^{1-\alpha}\rho\frac{B(0)}{Y(0)} - \tau_C\left[\{(1-g)-(1-\tau_W)\,\alpha\}\,AL^{1-\alpha} + \rho\right] \\ -\,(\alpha\tau_W - g)\,AL^{1-\alpha}\end{array}}{(1-\alpha)\,AL^{1-\alpha}}, \tag{41}$$

$$\tau_W = \frac{\rho\frac{B(0)}{Y(0)} - \tau_C\left[(1-g) - \alpha + \rho\frac{1}{AL^{1-\alpha}}\right] - (1-\alpha)\,\tau_L + g}{\alpha\,(\tau_C + 1)}. \tag{42}$$

These results are summarized as the following corollary.

Corollary 4 (Required Levels of Tax Rates for the Log-Utility Case) *Assume that the utility function of the representative household is logarithmic. Then, the tax rates for consumption, labor income and interest income that are required for the economy to be on the BGP are presented by* (40), (41) *and* (42), *respectively.*

4.4 *Numerical Example*

This section offers a numerical example, which illustrates the analyses presented above. I specify the parameters, except for (τ_C, τ_L, τ_W) and $B(0)/Y(0)$, as $A = L = \sigma = 1$, $\alpha = 0.25$, $\rho = 0.1$, $g = 0.15$. In what follows, I often specify $\tau_L = \tau_W (= \tau_C) = 0.2$, and it is easily ascertained that this parameter constellation satisfies Assumptions 1–3.

From (30), it holds that

$$\varphi = 7\tau_C + \frac{15}{2}\tau_L + \frac{5}{2}\tau_W + \frac{5}{2}\tau_C\tau_W - \frac{3}{2}, \tag{43}$$

which shows how the sustainability of public debt depends on the tax system (τ_C, τ_L, τ_W). If $\tau_C = \tau_L = \tau_W = 0.2$, then it is obtained from (43) that $\varphi = 2$.

That is, if the tax rates are all set at 20%, then 200% of the public debt-to-GDP ratio is allowed for on-going economic growth.

The marginal rates of substitution of the tax rates are calculated from (34)–(36) as

$$-\frac{d\tau_L}{d\tau_C} = \frac{1}{3}\tau_W + \frac{14}{15}, \tag{44}$$

$$-\frac{d\tau_W}{d\tau_C} = \frac{14 + 5\tau_W}{5(\tau_C + 1)}, \tag{45}$$

$$-\frac{d\tau_W}{d\tau_L} = \frac{3}{\tau_C + 1}. \tag{46}$$

See (44). Since $\tau_W \in [0, 1)$, it holds that $-d\tau_L/d\tau_C \in \left[\frac{14}{15}, \frac{19}{15}\right)$. Especially, if $\tau_W = 0.2$, then $-d\tau_L/d\tau_C = 1$. In other words, assume that the tax rate on interest income is 20%, and that it is fixed. In such a situation, if a government increases the consumption tax rate by 1%, then it can reduce the tax rate for labor income by 1% while maintaining a sustainable debt policy.

The required tax rates for the sustainability of public debt are derived from (40)–(42) as

$$\tau_C = \frac{2\frac{B(0)}{Y(0)} - 15\tau_L - 5\tau_W + 3}{5\tau_W + 14}, \tag{47}$$

$$\tau_L = \frac{2}{15}\frac{B(0)}{Y(0)} - \frac{14}{15}\tau_C - \frac{1}{3}\tau_W - \frac{1}{3}\tau_C\tau_W + \frac{1}{5}, \tag{48}$$

$$\tau_W = \frac{2\frac{B(0)}{Y(0)} - 14\tau_C - 15\tau_L + 3}{5(\tau_C + 1)}. \tag{49}$$

Assume that $\tau_L = \tau_W = 0.2$, and see (47). Then, the relationship between the initial level of the public debt-to-GDP ratio and the consumption tax rate that is required for the sustainability of public debt is given by

$$\tau_C = \frac{2}{15}\frac{B(0)}{Y(0)} - \frac{1}{15}.$$

When $B(0)/Y(0) = 2.30$, the required consumption tax rate is $\tau_C = 0.24$. If $B(0)/Y(0) = 2.40$, then the required consumption tax rate is $\tau_C \doteqdot 0.253$. Thus, if the public debt-to-GDP ratio increases from 2.30 to 2.40, then the consumption tax rate must be increased by approximately 1.3% to maintain fiscal sustainability.

5 Conclusion

This paper explicitly derives the critical level of public debt that is compatible with a balanced economic growth. The model considers a complex tax system—a combination of tax rates for consumption, labor income and interest income. Analyzing the critical level reveals the marginal rates of substitution of tax rates. That is, by how much a government can reduce one tax rate when another one is slightly increased while maintaining a sustainable debt policy. Further, the required tax rates to sustain a given level of public debt-to-GDP ratio are computed.

Although the results obtained in this paper are relevant to economists and to policy makers in countries that are suffering from a highly accumulated public debt, the model used is somewhat simple. Some extensions should be implemented in possible future research. First, this paper assumes that the population is constant over time. Given the declining birth rates observed in many advanced countries, it is desirable to take a population decline into account. Second, progressive income tax rates should be included. Finally, it is important to include the productive expenditure of the government.

Acknowledgements This research is financially supported by Ryosui Gakujutsu Foundation of Shiga University. The author would like to thank an anonymous referee for his/her helpful comments and advice.

References

Arai, R. (2011). Productive government expenditure and fiscal sustainability. *FinanzArchiv/Public Finance Analysis, 67*(4), 327–351.

Bohn, H. (1998). The Behavior of U. S. Public Debt and Deficits. *Quarterly Journal of Economics, 113*, 949–963.

Bökemeier, B. (2015). Economic growth and the public deficit in EU member states in Central and Eastern Europe. *Romanian Journal of Fiscal Policy, 6*(1), 47–53.

Bräuninger, M. (2005). The budget deficit, public debt, and endogenous growth. *Journal of Public Economic Theory, 7*(5), 827–840.

Fincke, B., & Greiner, A., (2011). Do large industrialized economies pursue sustainable debt policies? A comparative study for Japan, Germany and the United States. *Japan and the World Economy, 23*(3), 202–213.

Fincke, B., & Greiner, A. (2012). How to assess debt sustainability? some theory and empirical evidence for selected Euro area countries. *Applied Economics, 44*(28), 3717–3724.

Giannitsarou, C. (2007). Balanced budget rules and aggregate instability: The role of consumption taxes. *The Economic Journal, 117*(523), 1423–1435.

Greiner, A. (2011). Sustainability of public debt: Some theoretical considerations. *Economics Bulletin, 31*(4), 3311–3319.

Greiner, A. (2013). Sustainable public debt and economic growth under wage rigidity. *Metroeconomica, 64*(2), 272–292.

Greiner, A. (2015). Public debt, productive public spending and endogenous growth. *Japanese Economic Review, 66*(4), 520–535.

Greiner, A., & Bondarev, A. (2015). Public debt and aggregate stability with endogenous growth and a state-dependent consumption tax. *Economics Bulletin, 35*(4), 2601–2609.

Greiner, A., Köler, U., & Semmler, W. (2007). Debt sustainability in the European Monetary Union: Theory and empirical evidence for selected countries. *Oxford Economic Papers, 59*, 194–218.

Kondo, A. (2007). On the sustainability of government borrowing in a dynamic general equilibrium. *Pacific Economic Review, 12*(5), 565–576.

Kondo, A. (2012). A note on public-debt sustainability in an economy with declining fertility. *FinanzArchiv/Public Finance Analysis, 68*(2), 153–164.

Kondo, A. (2016). Interconnection of fiscal policies on sustainability of public debt. Center of Risk Research, Shiga University, Discussion Paper A-20.

Kondo, A., & Kitaura, K. (2009). Does Deflation impinge on a government's fiscal standing?. *Pacific Economic Review, 14*(5), 651–656.

Nourry, C., Seegmuller, T., & Venditti, A. (2013). Aggregate instability under balanced-budget consumption taxes: A re-examination. *Journal of Economic Theory, 148*(5), 1977–2006.

Schmitt-Grohe, S., & Uribe, M. (1997). Balanced-budget rules, distortionary taxes, and aggregate instability. *Journal of Political Economy, 105*(5), 976–1000.

Yakita, A. (2008). Sustainability of public debt, public capital formation, and endogenous growth in an overlapping generations setting. *Journal of Public Economics, 92*, 879–914.

Demographic Change and the Rates of Return to Risky Capital and Safe Debt

Wolfgang Kuhle

Abstract This paper studies how the upcoming demographic transition will affect the returns to risky capital and safe government debt. Using a neoclassical two-generations-overlapping model, we show that the entrance of smaller cohorts into the labor market will lower both interest rates. The risky rate, however, will react more sensitive than the risk-free rate. Consequently, the risk premium declines when an economy transitions from high fertility to low fertility.

1 Introduction

Neoclassical models of the Solow (1956) and Swan (1956) type predict a positive relation between the growth rate for population and the rate of return to capital. In more refined models, involving uncertainty and more than one asset, several authors have recently examined the consequences of the demographic transition with regard to the rates of return to risky capital and safe debt.[1] However, while these studies tend to agree that the overall level of interest will fall when the baby boom generation retires, it remains an open question whether the demographic transition will affect both rates of return in the same manner. In particular, Brooks (2002) and Geanakoplos et al. (2004) project that the risky rate will deteriorate more severely than the return to safe government bonds. That is, the equity premium would fall during the demographic transition. At the same time, Brooks (2004) and Börsch-Supan et al. (2007) project an increase in the equity premium for that period.

Against the background of this ambiguous evidence we develop a tractable equilibrium model, which allows us to derive the analytical conditions under which equity premia rise and fall during a demographic transition. Put differently, the

[1] See Poterba (2001) for a survey.

W. Kuhle
Max Planck Institute for Research on Collective Goods, Kurt-Schumacher-Str. 10, 53113 Bonn, Germany
e-mail: kuhle@coll.mpg.de

B. Bökemeier, A. Greiner (eds.), *Inequality and Finance in Macrodynamics*, Dynamic Modeling and Econometrics in Economics and Finance 23,
DOI 10.1007/978-3-319-54690-2_8

purpose of the current paper is to complement these previous studies, which were based exclusively on simulations, by developing a tractable model which allows to obtain analytic results on the relation between the growth rate for population and the two key interest rates.

The paper is organized as follows. In Sect. 2, we discuss a baseline case where households work in the first period only, aggregate technology shocks are log-normal, and utility is of the Epstein and Zin (1989) type. In this setting, we show that (1) both rates of return increase with the growth rate for population, (2) the risky rate to capital reacts more sensitive than the risk-free rate, i.e., the equity premium increases with the growth rate of population. In Sect. 3, we show that the results from Sect. 2 carry over to a framework with a general concave utility function and shocks which are no longer log-normal. Moreover, we discuss a model where households work in both periods of life. In this setting it turns out that there is a "human capital effect" to agents' portfolio choice problem, which counteracts the positive relation between population growth and equity premium. Section 4 questions whether agents are indeed able to anticipate the general equilibrium changes associated with a decade long demographic transition correctly. If agents are unable to predict how demographic change affects the marginal product of capital in the future, then equity premia are very high (low) in periods where large (small) cohorts enter the labor market.[2] Finally, Sect. 5 concludes.

2 The Model

The economy is inhabited by overlapping generations who live for two periods; one period of work is followed by one period of retirement. During the first period of life, each individual supplies one unit of labor inelastically and population evolves according to:

$$N_{t+1} = (1 + n_{t+1})N_t, \tag{1}$$

where N_t is the size of the cohort born at time t and $1 + n_{t+1}$ is the number of children raised by each member of cohort t.

Output is produced according to a continuous, concave, constant returns to scale, aggregate production function $F(K_t, N_t)$. This production process is subject to an aggregate technology shock z_t, which follows a log-normal distribution. For simplicity we assume that this shock is on average neutral. Per capita output y_t is

[2]Similarly, even if agents can anticipate how given changes in labor supply change interest rates, they may not be able to anticipate these changes in labor supply correctly. That is, agents born in the first half of the twentieth century may have been surprised by the drastic increase in female labor force participation.

therefore given by:

$$y_t = z_t f(k_t); \quad f'() > 0, \quad f''() < 0, \quad E[z_t] = 1, \quad \forall t. \tag{2}$$

Once the respective realization of the shock z_t is known, each firm will rent capital and hire labor up to the point where the respective marginal products are equal to the market prices:

$$R_t = z_t \frac{\partial F(K_t, N_t)}{\partial K_t} = z_t f'(k_t), \tag{3}$$

$$w_t = z_t \frac{\partial F(K_t, N_t)}{\partial N_t} = z_t (f(k_t) - f'(k_t) k_t). \tag{4}$$

2.1 Government Debt

In our model, the government is the only entity that can supply safe debt. The budget constraint of the government is given by:

$$B_t + N_t \tau_t = r_t B_{t-1}, \tag{5}$$

where B_{t-1} is the amount of outstanding and B_t the amount of newly issued debt in period t. Lump-sum taxes are denoted by τ_t. The rate of (gross) interest on government debt which was issued at time $t-1$ is denoted by r_t. This rate of interest earned on government debt is deterministic, i.e., at time t the government issues debt with a guaranteed rate of return r_{t+1}. Risk-averse individuals will therefore be willing to hold safe debt even if its rate of return is below the expected risky rate. As in Bohn (1998) and Smetters (2006), we assume that the government holds the debt to GDP ratio constant over time. This assumption is indeed consistent with the Maastricht criterion on government debt for countries in the eurozone. If policy is characterized by such a constant debt output ratio ρ, we have[3]:

$$\frac{B_t}{Y_t} = \rho \quad \forall t. \tag{6}$$

Solving (5) for per capita taxes τ, using (6) and (1), yields:

$$\tau_t = \left(\frac{1}{(1+n_t)} r_t y_{t-1} - y_t \right) \rho. \tag{7}$$

[3]There are obviously many different debt policies perceivable. However, the following results will be valid for all perceivable debt policies provided that taxes τ_t and the amount of debt B_t which is issued at time t do not depend on variables that are not yet realized in period t, e.g., the future capital intensity k_{t+1}.

2.2 Households

The representative household lives for two periods and supplies labor inelastically in the first period only. Towards the end of the first period, the household faces a consumption/saving and a portfolio allocation decision. Preferences over current and future consumption, $c_{t,1}$ and $c_{t+1,2}$, are described by a simplified Epstein and Zin (1989) utility function:

$$U_t = \ln(c_{t,1}) + \frac{\beta}{1-\phi}\ln(E_t[(c_{t+1,2})^{1-\phi}]); \quad 0 < \phi, \quad 0 < \beta < 1. \tag{8}$$

The utility function in (8) exhibits an elasticity of intertemporal substitution of unity. Hence, the individual savings/consumption decision is independent of the interest rate, since income and substitution effects cancel and precautionary savings neither dampen nor amplify private thrift. This assumption is reasonable as long as the (ambiguous) influence of changes in the rate of interest on savings is not too large. The advantage of this specification can be seen in the coefficient of relative risk aversion ϕ with respect to second-period consumption, which allows us to study the entire scope of the portfolio choice problem.

Recalling the taxes levied by the government τ_t, the value of wealth owned by the consumer when young can be written as[4]:

$$\Omega_t \equiv w_t - \tau_t. \tag{9}$$

For given values of lifetime wealth Ω, the individual chooses to hold assets amounting to:

$$a_t \equiv b_t + h_t, \tag{10}$$

where b_t and h_t are the amounts of riskless bonds and risky capital, respectively. Denoting the portfolio share of risky assets by $\gamma_t \equiv \frac{h_t}{a_t}$ and the share of risk-free assets by $(1-\gamma_t) \equiv \frac{b_t}{a_t}$ yields, according to (8) and (9), the following household problem:

$$\max_{a,\gamma} U_t = \ln(\Omega_t - a_t) + \beta\ln(a_t) + \frac{\beta}{1-\phi}\ln\left(E_t\left[(\gamma_t R_{t+1} + (1-\gamma_t)r_{t+1})^{1-\phi}\right]\right), \tag{11}$$

with the corresponding first-order condition for the optimal portfolio size:

$$a_t = \frac{\beta}{1+\beta}\Omega_t, \tag{12}$$

[4]The individual receives his wage w_t after the realization of z_t is known. Note also that taxes are known once z_t is known.

where the propensity to save out of wealth is $\frac{\beta}{1+\beta}$. The portfolio choice is characterized by the familiar implicit condition for γ_t:

$$E_t\Big(\left[\gamma_t R_{t+1} + (1-\gamma_t) r_{t+1}\right]^{-\phi} (R_{t+1} - r_{t+1})\Big) = 0. \tag{13}$$

Using a second-order Taylor series approximation, Campbell and Viceira (2002) show that the corresponding optimal portfolio share can be approximated as:

$$\gamma_t(R_{t+1}, r_{t+1}; \phi) = \frac{E_t[\ln(R_{t+1})] - \ln(r_{t+1}) + \frac{1}{2}\sigma_t^2}{\phi\sigma_t^2} = \ln\left(\frac{E_t[R_{t+1}]}{r_{t+1}}\right)\frac{1}{\phi\sigma_t^2}, \tag{14}$$

where $\sigma_t^2 = Var[\ln(R_{t+1})] = Var[\ln(z_{t+1})]$ and $\ln(E_t[R_{t+1}]) = E_t[\ln(R_{t+1})] + \frac{1}{2}\sigma_t^2$.[5] Once the investment opportunities are changing, the individual will adjust his portfolio according to:

$$\gamma_r \equiv \frac{\partial\gamma_t}{\partial r_{t+1}} = -\frac{1}{r_{t+1}}\frac{1}{\phi\sigma_t^2} < 0, \quad \gamma_{f'} \equiv \frac{\partial\gamma_t}{\partial(f'(k_{t+1}))} = \frac{1}{E_t[R_{t+1}]}\frac{1}{\phi\sigma_t^2} > 0. \tag{15}$$

The decisive property of the portfolio adjustment behavior in (15) is:

$$\left.\frac{d(f'(k_{t+1}))}{dr_{t+1}}\right|_{d\gamma_t=0} = -\frac{\gamma_r}{\gamma_{f'}} = \frac{f'(k_{t+1})}{r_{t+1}} > 1, \tag{16}$$

where (16) indicates that, for positive expected equity premia, the share devoted to the risky asset reacts more sensitive with respect to the risk-free rate than the risky rate. That is, an increase in both rates of return, which leaves the equity premium (19) unchanged, results in a lower portfolio share in the risky asset.

2.2.1 Equilibrium

Having completed the partial analysis of the firm, the government, and the household, we can now turn towards the equilibrium conditions for the bond and equity markets. Capital market clearing requires:

$$(1 + n_{t+1})k_{t+1} = \gamma_t \frac{\beta}{1+\beta}\Omega_t. \tag{17}$$

The bond market equilibrium condition reads:

$$y_t\rho = (1-\gamma_t)\frac{\beta}{1+\beta}\Omega_t. \tag{18}$$

[5]The rate of return $R_{t+1} = z_{t+1}f'(k_{t+1})$ inherits its log-normal distribution from the technology shock z_{t+1}. Thus, $\ln(R_{t+1})$ follows a normal distribution.

Taken together, Eqs. (17), (18), (14), and (3) define the time path of the capital intensity k, the safe rate of return r, the optimal portfolio share γ, and the risky rate R. Finally, the resulting expected equity premium $E_t[\Pi_{t+1}]$ is given by:

$$E_t[\Pi_{t+1}] = E_t[R_{t+1}] - r_{t+1} = f'(k_{t+1}) - r_{t+1}. \tag{19}$$

2.3 Baby Boom and Equity Premium

We can now consider the consequences of the entrance of a large/small cohort into the labor market. Taking the current state of the economy $(k_{t-1}, k_t, z_{t-1}, z_t)$ as given, we differentiate (17) and (18) with respect to dn_{+1}, $d\gamma_t$, dk_{t+1}. Hence, after recalling (3) and (14), we have:

$$\frac{dk_{t+1}}{dn_{t+1}} = -\frac{k_{t+1}}{(1+n_{t+1})} < 0, \quad \therefore \quad \frac{d(f'(k_{t+1}))}{dn_{t+1}} = f''(k_{t+1})\frac{dk_{t+1}}{dn_{t+1}} > 0 \tag{20}$$

and

$$\frac{d\gamma_t}{dn_{t+1}} = 0, \quad \therefore \quad \gamma_{f'}\frac{d(f'(k_{t+1}))}{dn_{t+1}} + \gamma_r\frac{dr_{t+1}}{dn_{t+1}} = 0. \tag{21}$$

The expressions in (20) indicate that a change in the growth rate of population does not change government taxes (7). Hence the value of life-cycle income Ω_t, out of which individuals save a constant fraction, remains unchanged. Thus, an increase in the relative size of the next cohort lowers the capital intensity and increases the expected future return on risky investments. The expressions in (21) follow from the bond market equilibrium condition. They indicate that, for $\frac{d(f'(k_{t+1}))}{dn_{t+1}} > 0$, the government has to offer a higher riskless rate $\frac{dr_{t+1}}{dn_{t+1}} > 0$ to sell a given amount of debt. Using Eq. (19), and the chain rule, the expected change in the equity premium is:

$$\frac{d(E_t[\Pi_{t+1}])}{dn_{t+1}} = \left(\frac{d(f'(k_{t+1}))}{dr_{t+1}} - 1\right)\frac{dr_{t+1}}{dn_{t+1}} >_{|(16),(21)} 0. \tag{22}$$

Equation (22), has two components. First, $\frac{dr_{t+1}}{dn_{t+1}} > 0$ shows that, as we have seen in our discussion of Eqs. (20) and (21), increased fertility increases the overall level of interest. That is, the entrance of large cohorts into the labor market reduces the capital labor ratio and increases the return to physical capital, and the government has to offer a higher safe rate to sell its debt. Second, $\frac{d(f'(k_{t+1}))}{dr_{t+1}} - 1$ reflects that the equity premium increases in the risky rate, and decreases in the safe rate, creating a net effect which is at first sight ambiguous. However, due to the higher sensitivity of the portfolio shares with respect to the risk-free rate, (16), the government does

Fig. 1 Demographic change and portfolio adjustment

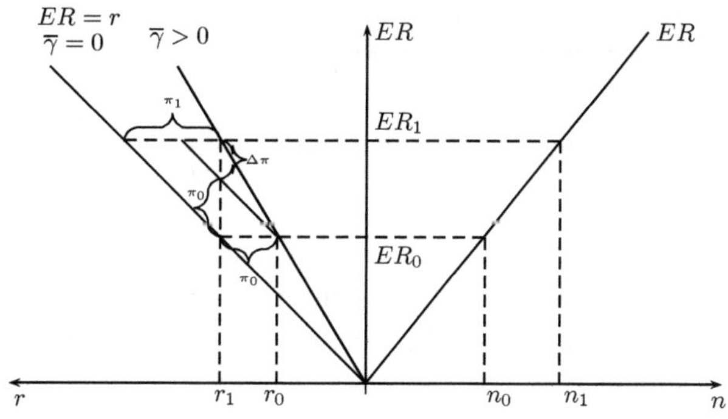

not need to raise its risk-free rate one for one with the expected risky rate to sell its debt. Accordingly, the equity premium increases with fertility.

The right-hand side of Fig. 1 illustrates how an increase in the fertility rate from n_0 to n_1 increases the expected rate of return to risky physical capital. That is, Eq. (20) indicates that increases in fertility reduce the capital-labor ratio, which, in turn, increases the (expected) marginal product of capital. The left-hand side of Fig. 1 illustrates how this increase in expected returns to capital affects the agents' portfolio choice. The two lines $\bar{\gamma} = 0$ and $\bar{\gamma} > 0$ indicate those pairs of safe and expected risky returns for which agents invest a constant share $\bar{\gamma}$ into the risky asset. The first line $\bar{\gamma} = 0$ is the 45° line along which the equity premium is zero, and agents do not invest into the risky asset. The second line illustrates the equilibrium level $\bar{\gamma} > 0$, in which agents hold risky capital and the debt that the government sells. In particular, we can identify the equilibrium r_0, ER_0, which obtains for low fertility, and the equilibrium r_1, ER_1 that obtains if fertility is high. Drawing horizontal lines from the $\bar{\gamma} > 0$ line to the 45° line, $\bar{\gamma} = 0$, illustrates the respective equity premia π_0 and π_1. Finally, due to the asymmetric portfolio adjustment effect, discussed in (16), we know that the slope of the $\bar{\gamma} > 0$ line must be larger than one, and thus the equity premium increases with fertility, i.e., $\Delta\pi > 0$.

3 Extensions

In this section, we will briefly discuss the robustness of our foregoing results. In particular, we analyze the role of a potential human capital effect. Moreover, we consider a more general portfolio choice setting, which is no longer based on CRRA preferences and log-normal shocks.

3.1 The Effect of Human Capital

If households' second-period labor endowment is given by θ, the demographic transition will not only affect the rates of return. It also changes the present value of the labor endowment through the induced factor-price changes. Consequently, the household problem now reads:

$$\max_{a,\gamma} U_t = \ln(\Omega_t - a_t) + \beta \ln(a_t) + \frac{\beta}{1-\phi} \ln\left(E_t\left[(\gamma_t R_{t+1} + (1-\gamma_t) r_{t+1})^{1-\phi}\right]\right), \tag{23}$$

where life-cycle wealth is given by $\Omega_t \equiv w_t - \tau_t + \theta \frac{w_{t+1}}{R_{t+1}}$.[6] Solving the household problem in the same manner as before, we obtain our two market-clearing conditions:

$$(1 + n_{t+1})k_{t+1} = \gamma_t \frac{\beta}{1+\beta}\Omega_t - \theta \frac{w_{t+1}}{R_{t+1}} \tag{24}$$

$$y_t \rho = (1-\gamma_t)\frac{\beta}{1+\beta}\Omega_t. \tag{25}$$

While the bond market clearing condition remains unaltered, the capital market clearing condition has to reflect that households only buy capital in excess of what they already hold as human capital. Moreover, Ω_t is now a function of the future capital intensity. To trace out the comparative statics of our system, it is useful to note that (24), (25), and (14) are separable. Beginning with (24) and (25), we determine how changes in the growth rate for population affect the risky rate and the portfolio share. Subsequently we differentiate (14) to determine the change in the safe rate. Taken together, we obtain:

$$\frac{dk_{t+1}}{dn_{t+1}} = \frac{k_{t+1}}{\frac{1}{1+\beta}\theta \frac{f(k_{t+1})f''(k_{t+1})}{f'^2(k_{t+1})} - (1+n_{t+1})} < 0, \tag{26}$$

$$\therefore \quad \frac{df'(k_{t+1})}{dn_{t+1}} = f''(k_{t+1})\frac{dk_{t+1}}{dn_{t+1}} > 0,$$

$$\frac{d\gamma_t}{dn_{t+1}} = \frac{(1-\gamma_t)\theta \frac{-f(k_{t+1})f''(k_{t+1})}{f'^2(k_{t+1})}}{\Omega_t}\frac{dk_{t+1}}{dn_{t+1}} < 0, \tag{27}$$

$$\therefore \quad \gamma_{f'}\frac{d(f'(k_{t+1}))}{dn_{t+1}} + \gamma_r \frac{dr_{t+1}}{dn_{t+1}} < 0.$$

[6] Given our specification of the production sector, second-period wage income and capital are perfectly correlated (perfect substitutes), i.e., $corr(w_{t+1}, R_{t+1}) = \frac{cov(w_{t+1}, R_{t+1})}{\sigma_w \sigma_R} = 1$.

The relations in (26) and (27) reveal that an increase in the population growth rate lowers the capital intensity and increases both rates of return.[7] The decrease in the share of the risky asset in the portfolio reflects that the present value of life-cycle wealth declines as higher birth rates reduce future wages. Hence, as the supply of government debt is unchanged, the share of the safe asset must be larger in the new equilibrium. That is, as the demand for safe bonds shrinks due to the decrease in Ω_t, the government has to offer a higher safe interest rate to sell its debt. Differentiation of (19) and (14) yields the change in the equity premium:

$$\frac{dE_t[\Pi_{t+1}]}{dn_{t+1}} = \left(\frac{d(f'(k_{t+1}))}{dr_{t+1}} - 1\right)\frac{dr_{t+1}}{dn_{t+1}} = \left(\frac{\frac{d\gamma_t}{dn_{t+1}}\frac{dn_{t+1}}{dr_{t+1}}}{\gamma_{f'}} - \frac{\gamma_r}{\gamma_{f'}} - 1\right) \gtreqless 0, \quad (28)$$

Equation (28) features two countervailing effects. First, the pure portfolio adjustment effect, which, as we have observed earlier, will once again increase the equity premium, i.e., $-\frac{\gamma_r}{\gamma_{f'}} - 1 > 0$. Second, due to the human capital effect, the portfolio share of the risky asset decreases $\frac{\frac{d\gamma_t}{dn_{t+1}}\frac{dn_{t+1}}{dr_{t+1}}}{\gamma_{f'}} < 0$. Taken together, the resulting change in the risk premium is ambiguous. In particular, (27) shows that the human capital effect rises with the labor endowment θ. This may explain why Börsch-Supan et al. (2007) and Brooks (2004) find that falling birth rates increase the equity premium in their large-scale OLG models, where households hold lots of unrealized human capital. Accordingly, the simulations of Brooks (2002) and Geanakoplos et al. (2004) yield the opposite result for smaller three- and four-generation-overlapping models.

Figure 2 illustrates the relation between the human capital effect and the equity premium: The left-hand side of Fig. 2 graphs a reduction $\Delta\pi_h$ in the equity premium, which obtains if the human capital effect is large, i.e., if the government has to offer a much-increased safe rate r_1, to increase the share of agents' wealth invested in the safe asset from $(1 - \bar{\gamma}_0)$ to $(1 - \bar{\gamma}_1)$. The right-hand side of Fig. 2 depicts a case where the human capital effect is small, and the equity premium increases if a large cohort enters the labor market due to the dominant portfolio adjustment effect.

3.2 *Portfolio Choice*

In the previous section, attention was confined to an economy where the TFP shock was log-normal and utility of the CRRA variety. In this section, we show that the asymmetry in the portfolio adjustment, which was driving our earlier results,

[7]From (26), we have $\frac{df'(k_{t+1})}{dn_{t+1}} > 0$; in turn, we find that $\frac{d\gamma}{dn_{t+1}} < 0$, together with $\gamma_r < 0$ and $\gamma_{f'} > 0$, implies that $\frac{dr_{t+1}}{dn_{t+1}} > 0$.

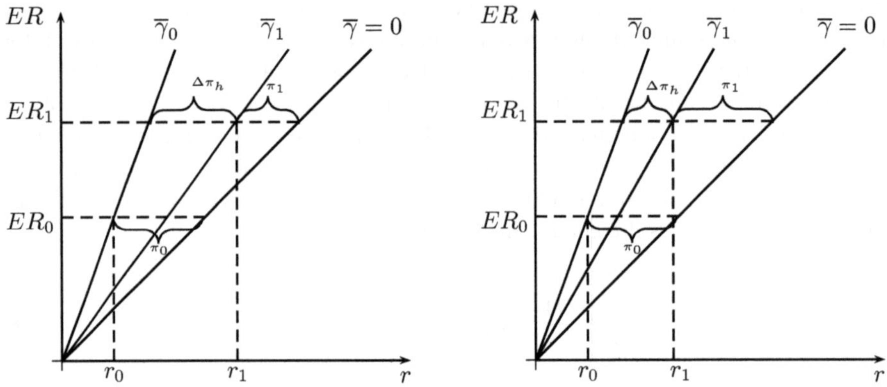

Fig. 2 The human capital effect and portfolio adjustment

carries over to a more general setting where utility is only assumed to be concave. Moreover, TFP shocks are no longer assumed to be log-normal. In this case, the two-period two-asset problem is given by[8]:

$$\max_{\gamma_t} U(\Omega_t - s_t) + \beta E_t[U((r_{t+1} + \gamma_t(R_{t+1} - r_{t+1}))s_t)]. \tag{30}$$

[8]Note that there are two ways to think about the savings decision:

$$\max_{\gamma_t, s_t} U(\Omega_t - s_t) + \beta E_t[U((r_{t+1} + \gamma_t(R_{t+1} - r_{t+1}))s_t)].$$

Expanding (30) at the point $\gamma = 0$ and $s = \bar{s}$, we have:

$$\begin{aligned} & U(\Omega_t - \bar{s}_t) + \beta U(\bar{s}_t r_{t+1}) \\ & + \beta U'(\bar{s}_t r_{t+1}) E[R_{t+1} - r_{t+1}]\bar{s}_t\gamma_t + \frac{1}{2}\beta U''(\bar{s}_t r_{t+1})\Big(E[R_{t+1} - r_{t+1}]\bar{s}_t\gamma_t\Big)^2 \\ & + \Big(-U'(\Omega_t - \bar{s}_t) + \beta U'(\bar{s}_t r_{t+1}) r_{t+1}\Big)(s_t - \bar{s}_t) \\ & + \frac{1}{2}\Big(U''(\Omega - \bar{s}_t) + \beta U''(\bar{s}_t r_{t+1}) r_{t+1}^2\Big)(s_t - \bar{s}_t)^2 \\ & + \gamma_t E[R_{t+1} - r_{t+1}]\beta U'(\bar{s}_t r_{t+1})\Big(\frac{U''(\bar{s}_t r_{t+1})}{U'(\bar{s}_t r_{t+1})} r_{t+1}\bar{s}_t + 1\Big)(s_t - \bar{s}_t). \end{aligned} \tag{29}$$

The last term in (29) indicates the interaction between the size and the composition of the portfolio (cross derivative). There are now two ways to think of our approximation in the main text: (i) The household chooses savings according to the usual Euler equation and chooses the portfolio shares according to the Taylor approximation (31). Put differently, the household chooses savings according to a precise rule. The portfolio shares, however, rely on an approximation. (ii) The household chooses both s and γ according to (29). In this (less appealing) case, there would be an additional component in the FOC for γ of ambiguous sign.

For simplicity, we normalize the time preference rate $\beta = 1$. A second-order Taylor-series expansion of the objective function at the point where $\gamma = 0$ yields the portfolio problem:

$$\max_{\gamma_t, s_t} U(s_t r_{t+1}) + U'(s_t r_{t+1}) E_t[\gamma_t s_t (R_{t+1} - r_{t+1})] \\ + \frac{1}{2} U''(s_t r_{t+1}) E_t[(\gamma_t s_t (R_{t+1} - r_{t+1}))^2]. \tag{31}$$

The corresponding optimal portfolio share is thus given by:

$$\gamma_t^* = -\frac{U'(s_t r_{t+1})}{U''(s_t r_{t+1})} \frac{E[R_{t+1} - r_{t+1}]}{s_t(\sigma_R^2 + E_t[(R_{t+1} - r_{t+1})]^2)} \\ =_{|CRRA} \frac{1}{\phi} \frac{E[R_{t+1} - r_{t+1}]}{\sigma_R^2 + E_t[(R_{t+1} - r_{t+1})]^2}. \tag{32}$$

If we assume $\gamma_{f'} > 0$, we find that $\gamma_r < 0$.[9] Moreover, the portfolio adjustment is once again asymmetric[10]:

$$\gamma_r + \gamma_{f'} = -\frac{2E_t[\Pi_{t+1}]f'(k_{t+1})\sigma_z^2}{\phi(\sigma_R^2 + E_t[(R_{t+1} - r_{t+1})]^2)^2} < 0 \quad \therefore -\frac{\gamma_r}{\gamma_{f'}} > 1. \tag{33}$$

For small equity premia we may follow Campbell and Viceira (2002, 2001) and regard $E_t[(R_{t+1} - r_{t+1})]^2$ as very small. This simplifies the approximate portfolio share in (32) such that we have:

$$\gamma_t^* = -\frac{U'(s_t r_{t+1})}{U''(s_t r_{t+1})} \frac{E[R_{t+1} - r_{t+1}]}{s_t \sigma_R^2}. \tag{34}$$

Compared to the special case studied by Campbell and Viceira (2002, 2001) the formula in (34) has the advantage that we neither require log-normal shocks nor CRRA utility. Moreover, it is interesting to note that the numerator now reads $E[R_{t+1} - r_{t+1}]$ rather than $log(E[R_{t+1}]) - log(r_{t+1})$.[11] For the portfolio adjustment,

[9] The respective partial derivatives are $\gamma_{f'} = \frac{\sigma_R^2 - E[\Pi]^2 - E[\Pi]2f'\sigma_z^2}{C}$ and $\gamma_r = \frac{-\sigma_R^2 + E[\Pi]^2}{C}$ where C is a positive constant. Hence, if $\gamma_{f'} > 0$, then $\gamma_r < 0$.

[10] For large equity premia, the term $E_t[(R_{t+1} - r_{t+1})]^2$ in the denominator grows very large compared to $E_t[(R_{t+1} - r_{t+1})]$ in the numerator. Increases in $f'(k)$ may now in principle decrease the portfolio share in the risky asset.

[11] Apparently, for small rates of return, there is not much difference between the two formulas. For small x, we have the first-order Taylor-series $log(1 + x) = 0 + 1 \cdot x = x$. Thus, $log(E[R_{t+1}]) - log(r_{t+1}) \approx E[R_{t+1} - r_{t+1}]$. Moreover, in the denominator, $var(zf') \approx var(log(z))$ if the variance of z is small and the (net) rate of return close to zero, i.e., $f' \approx 1$. With regard to our analysis of the equity premium, however, the expression in (34) has the disadvantage that the relative risk aversion is now a function of savings and changes during the transition. In this case, it is not

with CRRA utility, we now obtain:

$$\gamma_{f'} \leqq |\gamma_r|. \quad (35)$$

Where the inequality holds if shocks are multiplicative as specified in (2). Otherwise, if shocks are additive, the adjustment is symmetric. In reality one can expect some sort of multiplicative (TFP) component and thus our comparative statics with $\gamma_{f'} < |\gamma_r|$ shall point in the right direction.

4 Discussion

We have considered the consequences of the demographic transition with regard to the two key interest rates. In our log-linear example at the outset, the asymmetric portfolio adjustment behavior was relating the equity premium positively to the growth rate of population. A higher level of interest was associated with a higher risk premium and vice versa.

In a more general setting with a human-capital effect, this result was dampened. An increase in the growth rate for population lowers the present value of second-period human capital and increases savings in the risky asset. Moreover, the present value of life-cycle wealth falls. This effect lowers the demand for safe second-period consumption and forces the government to increase the safe rate to sell its debt. If this effect is sufficiently large, the initial conclusion may be reversed. Second, we relaxed our assumptions on the stochastic processes and the utility function to conclude that the asymmetry in the portfolio adjustment is robust.

The different settings discussed so far have in common that they implicitly assume a high level of rationality: Agents have to anticipate the equilibrium changes associated with a decade-long demographic transition correctly to choose their portfolios as described above. If households do not understand that faster population growth dilutes the capital stock, which increases capital's marginal product, and thus the risky rate, they do not demand a higher safe return to buy safe bonds. Figure 3 depicts such a case: Agents do not anticipate that increased fertility will eventually increase future capital returns from ER_0 to ER_1. They are therefore still willing to buy the government debt at rate r_0 resulting in an interest pair A_1, which is not located on the $\bar{\gamma}$ line that summarizes the $E[R], r$ pairs that are consistent with rational expectations. This departure from rational expectations means that the *objective* expected equity premium increases buy $\Delta\pi = ER_1 - ER_0 > 0$. That is, the equity premium would increase by more than it did in our baseline model of Sect. 2.3, where agents anticipate the relation between fertility and capital returns properly such that, as in Fig. 1, the equilibrium is always located on the $\bar{\gamma}$ line.

possible to derive appealing conditions for the relation between the risk premium and the growth rate for population.

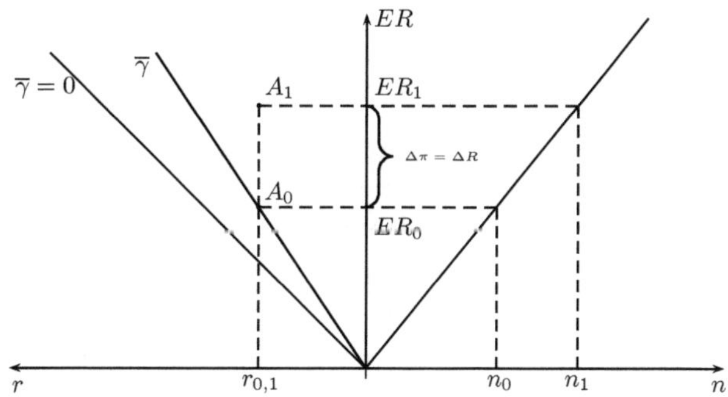

Fig. 3 Myopic adjustment

5 Conclusion

Motivated by the conflicting results of previous simulation-based studies, we have discussed the link between the entrance of smaller cohorts into the labor market and the equity premium in a simple two-generations-overlapping economy. In this framework, the entrance of a large (small) cohort into the labor market leads to a higher (lower) expected equity premium. While both rates of return rise (fall), the risky rate will rise (fall) by more than the riskless rate.

In the present setting, the positive link between the growth rate for population and the equity premium is indirect. It operates through the capital widening effect which increases the expected risky rate and forces the government to offer a higher riskless rate to sell its debt. The increase in the equity premium is solely due to the asymmetric portfolio adjustment behavior, i.e., the fact that agents react less to a 1% change in the risky rate as compared to a 1% change in the safe rate of return.

Myopia on the side of the households will amplify these effects. In the case of fully myopic households, there is a strong positive relation between the equity premium and population growth. This observation seems to be of interest in the context of the equity premium puzzle: If the economic impact of large cohorts born between 1950 and 1970, and the increase in female labor force participation, were not correctly anticipated by investors, then it is not surprising that equity premia were very high in the last 50 years. In turn, our findings suggest that equity premia should be unusually low in the coming decades, when large cohorts retire.

Acknowledgements I thank the anonymous referee, Axel Boersch-Supan, Brian Cooper, Nataliya Demchenko, Alexander Ludwig, Edgar Vogel, Matthias Weiss, and seminar participants in Mannheim for comments.

References

Bohn, H. (1998). Risk sharing in a stochastic overlapping generations economy. In *University of California at Santa Barbara, Economics Working Paper Series* (Vol. 1076, pp. 1–44).

Brooks, R. (2002). Asset-market effects of the baby boom and social-security reform. *American Economic Review, 92*(2), 402–406.

Brooks, R. (2004). The equity premium and the baby boom. http://repec.org/esNAWM04/up.25430.1048539824.pdf.
Börsch-Supan, A., Ludwig, A., & Sommer, M. (2007). Aging and asset prices. In *MEA Discussion Paper* (Vol. 129, pp. 1–75).
Campbell, J. Y., & Viceira, L. M. (2001). Appendix to strategic asset allocation. http://kuznets.fas.harvard.edu/~campbell/papers.html.
Campbell, J. Y., & Viceira, L. M. (2002). *Strategic asset allocation*. Oxford: Oxford University Press.
Epstein, L. G., & Zin, S. E. (1989). Substitution, risk aversion, and the temporal behavior of consumption and asset returns: A theoretical framework. *Econometrica, 57*(4), 937–969.
Geanakoplos, J., Magill, M., & Quinzii, M. (2004). Demography and the long-run predictability of the stock market. In *Cowels Foundation Discussion Paper* (Vol. 1380R, pp. 1–53).
Poterba, J. M. (2001). Demographic structure and asset returns. *Review of Economics and Statistics, 83*(4), 565–584.
Smetters, K. (2006). Risk sharing across generations without publicly owned equities. *Journal of Monetary Economics, 53*(7), 1493–1508.
Solow, R. M. (1956). A contribution to the theory of economic growth. *Quarterly Journal of Economics, 70*(1), 65–94.
Swan, T. W. (1956). Economic growth and capital accumulation. *The Economic Record, 32*, 334–361.

Financing Sustainable Growth Through Energy Exports and Implications for Human Capital Investment

Unurjargal Nyambuu

Abstract This paper examines the impact of energy resources financing on investment in human capital through the mechanism of growth dynamics. This is done within a context that includes global financial markets and exports of non-renewable energy. These are frequently related to issues of debt accumulation, which naturally raises questions relating to sustainability and welfare—both present and future. Energy export's contribution to economic growth is emphasized and the distinction between resource-rich and resource-poor countries is highlighted. Major external disturbances for sustained resource-driven development, which can make a country more vulnerable to economic shocks, are discussed. Numerical analysis using Nonlinear Model Predictive Control confirms the empirically observed long-run patterns when non-renewable resources decline monotonically and become depleted. The solutions also confirm typical boom/bust cycle phenomena, where excessive debt may effectively strangle growth. In addition, the implications of investment in human capital for inequality are discussed.

1 Introduction

As the chemist Nathan Lewis once said: "The currency of the world is not the dollar, it's the joule." (Lewis 2007). Thus, modern industrial processes are so valuable precisely because they allow energy, literally, to be concentrated, through machines, into production. Unfortunately, although there are myriad sources of energy which do not release or produce greenhouse gases, e.g., solar, wind, river, the technology of transmission and storage of energy is still in its infancy. Fossil fuels represent one of the most efficient forms of stored (potential) energy. Further, as liquids they can be moved via pipelines and tanker ship. Even coal can be moved via pipes using slurry lines. Thus, they are also a convenient way to transport energy to the places where it is needed, and therefore represent a vast storehouse of the potential

U. Nyambuu
Department of Social Science, New York City College of Technology, The City University of New York, 300 Jay Street, Brooklyn, NY 11201, USA
e-mail: UNyambuu@citytech.cuny.edu

B. Bökemeier, A. Greiner (eds.), *Inequality and Finance in Macrodynamics*, Dynamic Modeling and Econometrics in Economics and Finance 23, DOI 10.1007/978-3-319-54690-2_9 energy we need to run our world. From suburban "car-culture" to manufacturing, fossil fuels are woven throughout the fabric of not only our economy, but into our society as well. As has been discussed in Nyambuu and Semmler (2014, 2017b), the extraction of fossil fuels is an economic problem unto itself. The economic effects and interrelationships of our system of energy production, storage, and transmission impinge themselves on world politics, on everyday life, on financial markets, and on the macro-economy both directly and indirectly, through a vast array of feed-back mechanisms.

In this paper, I concentrate on the problem of the "harvesting" of stored energy, specifically oil, and how this process, through financial markets, impacts economic factors. Bearing in mind the words of Nathan Lewis, we can think of economic growth as the concentration of energy expenditures within national boundaries. Thus, I present a modified model of long-run economic growth which explicitly links in energy harvesting for both closed and open economies. Exhaustible energy, i.e., energy derived from limited resources, e.g., oil, is an essential input into production. This process, through leveraged borrowing, using the resource as collateral, gives rise to external debt accumulation, and affects patterns of development and consumption over time. Despite recent volatility resulting in lower prices, empirical indicators suggest we should expect an increase in the demand for energy, particularly from emerging Asian economies. In addition, empirical evidence found in Nyambuu and Bernard (2015) and Nyambuu and Semmler (2017a) show that a number of energy exporting countries are experiencing excessive debt burdens and increased default risk. Thus, it is important to assess the impact of these constraints on sustainable economic growth and the welfare of society. In order to replicate empirical facts, optimal movements of capital accumulation, extraction rates for non-renewable energy sources, and external debt accumulation are analyzed. Thus, this paper presents research on how resource-rich countries finance their development and growth through the export of their natural resources and how this process affects human capital and, indirectly, inequality.

Different definitions of sustainability are suggested by economists. Pearce et al. (1990, p. 4) emphasize that the "key necessary condition for achieving sustainable development" is "constancy of the natural capital stock." They point out that "natural capital stock should not decrease over time" for development to be sustainable. Brundtland states that "sustainable development is development that meets the needs of the present without compromising the ability of future generations to meet their own needs" (WCED 1987, p. 41). One of the ways a society may "compromise its ability to meet the essential needs of its people in the future—by over-exploiting resources" (p. 42). Economic rapid growth in developing economies based on massive energy production and consumption has come with a significant externality especially in terms of environmental cost (see Bernard and Nyambuu 2015).

Economists have been debating the impact of energy booms on economic growth for many years. On balance, energy-rich countries can benefit significantly from energy discoveries. Wright and Czelusta (2007) highlight the fact that, with new discoveries, investment projects and new technological progress, e.g., in Latin America, particularly in Chile, Peru, and Brazil, is growing quickly (Exploration in South America 2001). Huge discoveries of energy deposits can increase a country's welfare by allowing for a build-up of infrastructure, the development of certain industrial sectors, and by generally improving living standards. Because of rising commodity prices, resource-rich[1] countries were often encouraged to develop their mining sector and to promote its exports. Export revenues of energy are the main source of the fiscal revenue of many energy-rich Latin American and African countries, which are also very vulnerable to volatilities of commodity prices. Openness to foreign markets, together with an increased dependence on global demand for exported goods and external financing and capital flows, also makes a country more vulnerable to external shocks. This relationship is shown in a large body of research including Bekaert et al. (2006), Di Giovanni and Levchenko (2008), Easterly et al. (2001), Kose et al. (2006), Nyambuu and Tapiero (2017), and Rodrik (1997).

Capital inflow can rise due to an increase in consumption (as in Latin American countries), in investment (as in Asian countries), and in debt service. As Harberger et al. (1985) state, Mexico and Venezuela borrowed heavily using their oil as debt collateral to finance consumption; they failed to invest in long-term projects. Stein and Paladino (2001, p. 6) highlight the importance of the "purpose of the foreign borrowing" and argue that current account deficit will create growth in the future if the productive investments are financed by it. Therefore, trade surplus would increase following the rise in competitiveness. Although some energy-exporting countries have succeeded, a number of studies show that, certain energy-rich countries have worse growth performance compared to those classified as energy-poor.[2] This is mainly due to side effects such as Dutch disease and excessive external debt.

Manzano and Rigobon (2001) argue that resources have a negative effect on growth only in cross-sectional data, and the low growth observed might be related to imperfect credit markets and debt overhang. The World Economic Outlook (IMF 2013) summarizes empirical results of recent studies that have shown a debt overhang threshold, which suggests a negative impact for growth due to debt, e.g., as in Caner et al. (2010), Cecchetti et al. (2011), and Ursua and Wilson (2012). IMF (2013, p. 36) notes that these empirical studies show: "above a threshold of about 95% of GDP, a 10 percentage point increase in the ratio of debt to GDP is associated with a decline in annual growth of about 0.15 to 0.20 percentage point a year." Pattillo et al. (2002) find a nonlinear correlation between external debt and per capita GDP for 93 developing countries (1969–1998): debt impact become adverse at debt:GDP ratios above 35–40% and debt:exports ratios above 160–170%.

Various aspects of globalization and its impact on inequality are analyzed in a large number of research papers. For example, Kanbur (2015) presents a connection

[1]According to IMF (2012) and Berg et al. (2012), countries with at least a 20% share of natural resource exports in total exports (average of 2006–2010) are classified as resource-rich countries.

[2]A number of studies including Auty (1990, 1993, 2001), Sachs and Warner (1995, 1999, 2001), and Smith (2004) show a negative relationship between natural resource abundance and economic performance.

between trade and inequality in developing economies and reviews literature on different crises and increased trade. While some early studies, e.g., Levy and Murnane (1992) and Krugman (1994), have argued that trade was not the major contributor to the increase in inequality, other literature including Bhagwati and Kosters (1994), Bloom and Brender (1993), Richardson (1995), Autor (2010), Goldberg and Pavcnik (2007), Irwin (2008), Katz (2008), and Krugman (2008) highlight the role of trade in explaining an increase in inequality. East Asian, as well as Latin American economies experienced rising inequality together with high trade in the 1980s and 1990s when trade was liberalized. Piketty (2014, pp. 326–327) examined income inequality in selected emerging economies and showed that, since the 1980s, inequality has risen rapidly with a very high share of top percentile[3] in overall income: Argentina (17%), South Africa (17%), Indonesia (13%), India (12%) and China (11%) in 2010. Piketty defines fundamental inequality in terms of the rate of return on capital (r) and the rate of economic growth (g), where r>g (see Piketty 2014, p. 25). He argues that "the process by which wealth is accumulated and distributed contains powerful forces pushing toward divergence,[4] or at any rate toward an extremely high level of inequality" Piketty (2014, p. 27). For more detailed analysis of the Latin American case we can refer to Wood (1999), for the Asian case we see Goldberg and Pavcnik (2007), and for a review of empirical evidence for selected developing countries, see Alvaredo and Gasparini (2015).

Assuming exogenous trade and technology, Richardson (1995) presents a basic model and shows that trade might cause a growth in relative wages. Richardson (1995, p. 40) concludes that differences in wages would be higher and inequality would be greater due to increased trade in two cases: "One is an opening of the economy, along a given production possibility frontier" and another one is "exogenous sectoral technical change that is more rapid for investment goods than for other sectors (that is, relatively faster total factor productivity growth in investment goods)". Richardson (1995, p. 46) also argues that "a trade opening will also cause a movement of relative factor prices against pure, less-skilled labor, and in favor of skilled workers, experienced workers, knowledge workers, and physical and technological capital. In fact, there is also a "magnification effect"—the proportional impact on factor prices is a multiple of the product-price change. Thus, one cannot assume that a relatively small decline in relative prices, as a result of trade opening, will guarantee a correspondingly small effect on factor prices."

Specification of production inputs and differentiated commodities, emphasizing the heterogeneity of production process, workers, and firms, was used in studies by Wood (1994), Davis (1996), Kanbur (2015), Helpman et al. (2010), Feenstra and Hanson (1996), Feenstra and Hanson (1997), Grossman (2013), and Costinot (2009). For example, in Feenstra and Hanson (1996, 1997), the production process

[3] This refers to the "top 1 percent highest incomes" in Piketty (2014, p. 273).

[4] Piketty (2014, p. 23) highlights the destabilizing force of divergence "associated with the process of accumulation and concentration of wealth when growth is weak and the return on capital is high".

takes intensities of labor skills, where the developed economy (with abundant skilled labor) requires skilled intensive production, into account. In their model, due to lower trade costs, a higher skill-intensity of production expands the wage gap between workers with different skills. Semmler (2003) examines different mechanisms of inequality, highlighting human skill formation and its financing with credit, as well as spillover knowledge and growth in production, which can lead to an increase in persistent aggregate inequality. Assuming heterogeneous agents, he specifies individual and aggregate human capital and proposes a dynamic model based on utility theory. Semmler (2003, p. 1) stresses that "if those mechanisms are present the models become highly nonlinear and may give rise to thresholds and poverty traps. The capability of the individual to successfully compete is achieved only after some thresholds are passed, since there are often multiple steady state equilibria and there is path dependency as to the outcome of the dynamics." Major factors that also affect inequality are skill-biased technical change and education, as explored in studies by Katz and Murphy (1992), Krugman (1994), Acemoglu (2002), and Aghion (2002). Greiner et al. (2004) examined the interaction between economic growth, skill-biased technical change, and wage inequality; they presented a growth model that incorporates innovation-based technical change and considered different skill groups, emphasizing the elasticity of substitution. In addition to the rate of technological change, they defined other drivers of wage inequality, namely the technological "spill-over effect, the relative supply of skilled and unskilled labor and the elasticity of substitution between high- and low-skilled workers" (Greiner et al. 2004, p. 615). Empirical findings on macroeconomic level in Greiner et al. (2004) show that Europe, as compared to the US, has less inequality in wages.

In Sect. 3, I introduce a theoretical model that incorporates non-renewable energy sources into a basic growth model. Building on the work of Dasgupta and Heal (1974), I employ maximization of discounted utility. The model has constraints of capital stock accumulation and a rate of extraction for non-renewable energy. Furthermore, I extend this closed economy growth model to an open economy model where the current account is financed by external debt. Following Blanchard and Fischer (1989), balance of payment's macroeconomic accounting is used in the derivation of the current account. Most importantly, in the modeling of growth, a distinction between energy-rich and energy-poor countries is taken into account. Next, the open economy model is modified by including a penalty term in the objective function. In Blanchard (1983), this is shown as 'disutility.' It captures the non-monetary cost implications of the external debt for economic growth. Finally, in addition to physical capital, human capital[5] is incorporated in the open economy growth model.

[5]Human capital was introduced in the basic closed economy growth model by early researchers, e.g., Uzawa (1965) and Lucas (1988). This endogenous Uzawa-Lucas growth model was estimated in Greiner et al. (2005) using data from the US and Germany. Their empirical evidence fits the actual data.

Further, I investigate the paths of capital accumulation, stocks of exhaustible energy sources, external debt, and human capital formation when these state variables are characterized by different initial conditions. Additionally, the movements of welfare-improving consumption, utility levels, and the evolution of output are explored. I investigate whether an increase in external debt is correlated with poor economic performance in energy-rich economies. Welfare implications for household consumption is also explored. All the variations of the growth models are numerically solved using Nonlinear Model Predictive Control (NMPC).[6]

The paper is organized as follows: following this introduction, Sect. 2 illustrates historical and statistical facts on commodity exports and indicators of macroeconomic performance, including the current account and external debt. The theoretical model for the basic closed economy and its numerical solution is presented in Sect. 3. In Sects. 4 and 5, the open economy model and its extensions with human capital are presented with numerical results for the dynamics. Finally, Sect. 6 concludes the paper. Solutions of the theoretical models using a current-value Hamiltonian and optimality conditions are sketched in Appendices 1 and 2. NMPC methodology is explained in Appendix 3. Furthermore, NMPC solutions with different parameters, the sensitivity of the results to specific parameter choice is discussed in Appendix 4.

2 Historical and Statistical Facts

Cross-country information on commodity net exports, energy export's share in total exports, current account balance, and external debt stocks are presented in this section. The data are obtained from the U.S. Energy Information Agency (2016) (EIA), World Bank (2016), and International Monetary Fund (IMF 2016). Most energy-rich countries showed an increase in economic growth following an energy boom; in particular, this was due to the contribution of energy production and exports during the period of high commodity prices in the 1970s. However, most of these economies experienced sharp declines during the debt crisis of the early 1980s, late 1990s, early 2000s, and the recent Economic and Financial Crisis of 2007.

Historical trends on net exports of some commodities are illustrated in Fig. 1. Major net oil exporting regions include the Middle East, Eurasia, and Central, South and North America. For the first time since 1949, the United States became a net petroleum product exporter in 2011. During the most recent financial crisis, the amount of oil exports showed decline due to lower external demand. However, following the economic recovery, both oil and natural gas exports started increasing in most countries. The current decline in oil prices is not discussed here as we are interested in long-term trends not connected to political processes, temporary

[6]NMPC methodology is explained in Appendix 3.

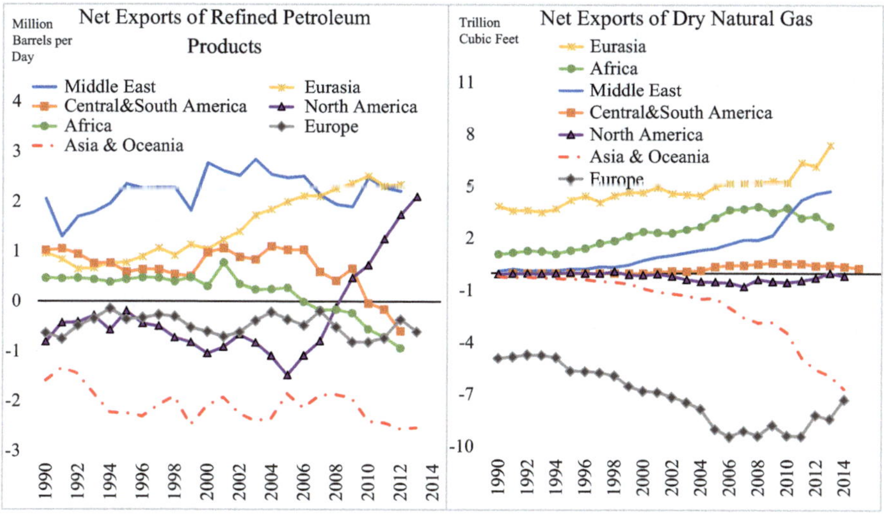

Fig. 1 Net exports of petroleum products and natural gas. Data source: U.S. EIA

technology shocks, e.g., those due to hydraulic fracturing (fracking) and horizontal drilling, and/or short-term market volatility.

The majority of African countries heavily depend on energy exports, e.g., oil exporting in Algeria and Nigeria, and their energy revenues contribute around 80% of total fiscal revenues. According to the World Bank's World Development Indicator database, between 1970 and 2014, as shown in Fig. 2, energy exports of many African as well as Latin American countries dominated the total goods exports. Current account deficits are financed by external debt, thus historical data on current account's[7] share in GDP from IMF's WEO database are shown for chosen countries in Fig. 3 (left-hand side). Countries such as Mexico have current account deficits due to import of consumption goods instead of investment goods. Asian countries have had large current account surpluses; however, in recent years, the surpluses have been declining significantly and even showing deficits.

Historical data on the total external debt,[8] as a share of GDP for selected energy exporting countries, are illustrated in Fig. 3 (right-hand side). Although the external debt:GDP ratio showed a sharp increase during the debt crisis of the 1980s, the ratio declined for most countries, including most developing African, Asian, and Latin American countries. Latin American countries were hit by crises in the early 1980s and in 1997 as well. There were early warning signals of excess debt for Mexico and

[7]Current account data before 2005 are compiled based on Balance of Payments Manual BPM5, but data after 2005 are compiled following the latest manual of BPM6.

[8]According to the World Bank, gross external debt "is debt owed to nonresidents repayable in currency, goods, or services." This includes not only public publicly guaranteed debt, but also private non-guaranteed long-term debt, use of IMF credit, and short-term debt.

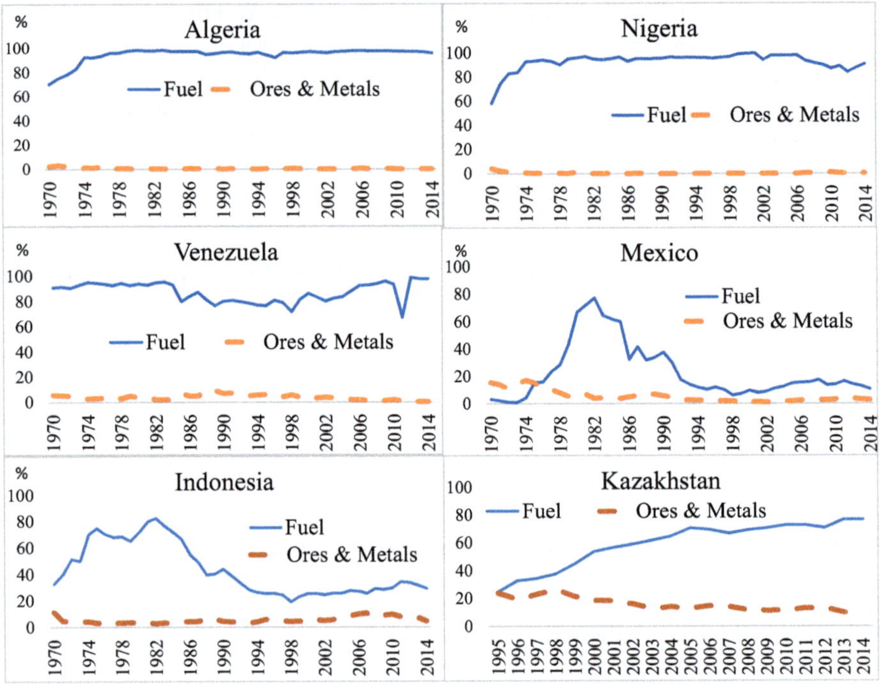

Fig. 2 Fuel and ores and metals export's share in total goods exports. Data source: The World Bank

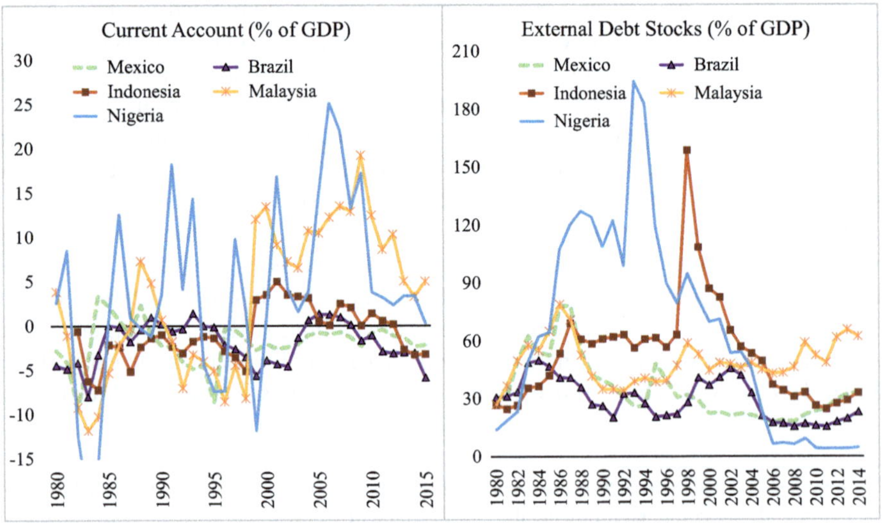

Fig. 3 Current account and external debt shares in GDP. Data source: The World Bank

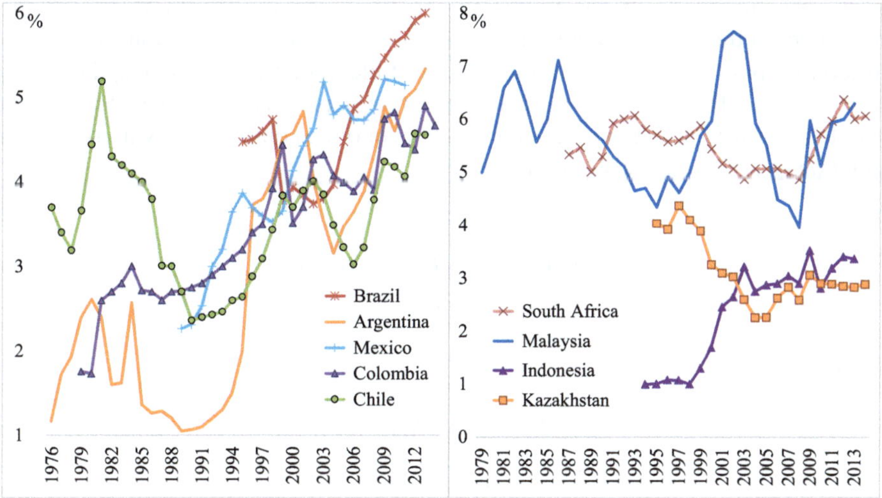

Fig. 4 Government expenditure on education as percentage of GDP. Data source: The World Bank

Venezuela during both of these crises. The promotion of the oil sector and support of domestic spending fueled the Mexican debt crisis. Latin American countries were vulnerable because of external shocks as well as unsustainable policies at home.

While public long term external debt dominates in African countries, in Latin America, both short and long term private debts have been increasing, in particular private long term external debt has exceeded public debt since the mid-1990s. Besides an increasing trend in both public and private long term debt, there is much faster growth in short term debt. In this case, when lenders do not further extend or renew short term loans, these countries can be vulnerable to a crisis. Large capital inflows to developing Asian countries led to a significant increase in their external debt, which were denominated in USD. During the Asian crises from 1997 to 1998, GDP declined significantly and exchange rates depreciated substantially. One consequence here was enormous capital flight, specifically bank loans, which accounted for most of the volatile capital flows. In addition, government expenditures on education as percentage of GDP that have been increasing in recent years are shown in Fig. 4. We should note that the education expenditure has risen first then declined in some countries especially after the opening of the domestic economy to the global market.

3 Basic Model of a Closed Economy and Its Solution

Basic growth models which include the extraction of non-renewable energy sources were introduced in Dasgupta and Heal (1974), in Stiglitz (1974), and in Solow (1973). The present model maximizes a welfare function for households, $U(C)$, with consumption C. It has two state variables—the capital stock, K, and the remainder of the non-renewable energy resource, S,—and two decision variables—the consumption, C, and the flow of the exhaustible resource, X. Following Dasgupta and Heal (1974), and Greiner and Semmler (2008, pp. 141–143), the optimization problem of basic growth model of a closed economy, constrained by a production function $Q(K, X)$, can be shown as:

$$maxV_{[C,X]} = \int_0^N e^{-rt} U(C) dt \tag{1}$$

subject to[9]

$$\dot{K}_t = Q(K_t, X_t) - C_t \tag{2}$$

$$\dot{S}_t = -X_t \tag{3}$$

$$S_t = S_0 - \int_0^t X_i di \tag{4}$$

where $r > 0$ and an initial value of the remainder of the resource S_0 is positive and given. As noted in Greiner and Semmler (2008), since the reserve is finite and being extracted, it is diminishing at an extraction rate of X. Extraction cost is not assumed here as in Greiner and Semmler (2008).[10]

Numerical solutions to this basic model with power utility function, $U(C) = \frac{C^{1-\eta}}{1-\eta}$, and Cobb-Douglas production function, $Q(K, X) = K^{\beta} X^{1-\beta}$, are shown in this section. The NMPC method, developed by Grüne and Pannek (2011), is used for the analysis of the dynamics of the growth models with extraction of non-renewable resources. Dynamic decision of the closed economy model is solved for a shorter and receding time horizon using the NMPC algorithm for different initial values of capital stock $K(0)$ and non-renewable resources $S(0)$. At first, I consider optimal trajectories of the state variables for initial conditions of $K(0) = 1$ and $S(0) = 4$ and

[9]See Appendix 3, on NMPC methodology, for a discussion of the finite upper limit in the integration term.

[10]Solution of the theoretical model using a current-value Hamiltonian and optimality conditions is sketched in Appendix 1. Recent studies show that NMPC is effective in approximating longer-horizon decision problem (Grüne and Pannek 2011; Nyambuu and Semmler 2014). When NMPC is used, infinite decision horizons are truncated; they are replaced with finite horizons. Grüne et al. (2013) show how this method can be used in dynamic decision problems in economics. See Appendix 3 for a discussion of NMPC and, in particular, the finite upper limit in the integration term of Eq. (1).

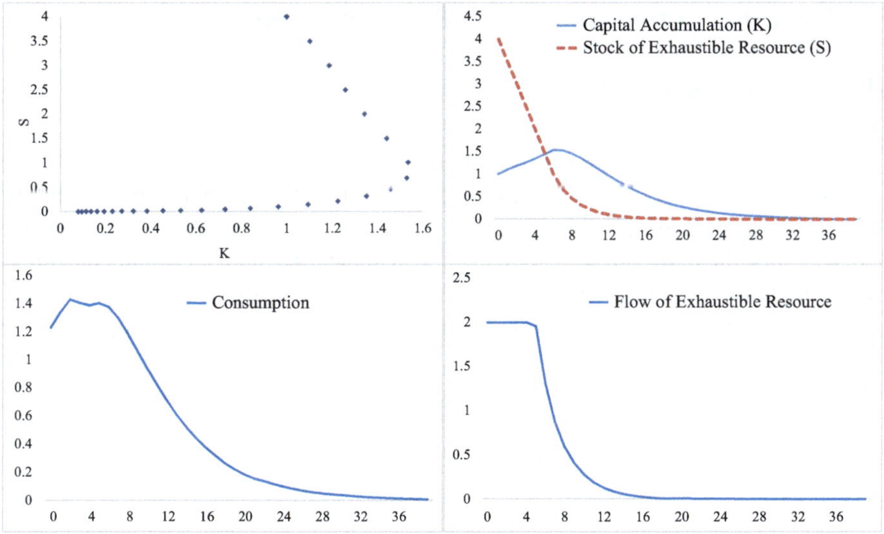

Fig. 5 Optimal trajectories of state variables and controls in a closed economy

parameter values of $\eta = 0.5$, $\beta = 0.3$, and $r = 0.03$.[11] The results of the evolutions of state variables are illustrated in Fig. 5. The parameter value of $(1 - \beta = 0.7)$ in the production function and a high stock of non-renewable resource indicate a resource intensive economy.

We can observe that capital stock increases first with available non-renewable resources but starts falling as the remainder of resources reaches a point near depletion. The stock of non-renewable resources then decreases over time and depletes eventually. Additionally, corresponding optimal paths of consumption level and flows of exhaustible resources are depicted in the second row of Fig. 5. Similar to the evolution of capital stock, these variables illustrate initial increase which is followed by a decrease. Numerical solutions by NMPC indicate an inverted U-shaped capital accumulation movement and monotonically declining stock of non-renewable resources when initial condition of capital stock is not high.[12]

[11]With regard to specific parameter choices, I follow Blanchard (1983), Semmler and Sieveking (2000), and Mittnik and Semmler (2014). Although I have illustrated the ideas using specific parameters, the results presented herein are robust with respect to their variation. In Appendix 4 on NMPC solutions with different parameters, the sensitivity of the results to specific parameter choice is discussed. In addition, results for different initial conditions are shown in Appendix 4.

[12]A basic closed economy growth model is estimated in Greiner and Semmler (2008) using a nonlinear least squares techniques. Their estimated data on capital stock/resources and consumption/resources ratios fit the data well and highlight the periods of oil crises in the 1970s.

4 Open Economy Model with Resources and Its Solutions

4.1 Theoretical Model

In this section, I introduce an extended growth model for an open economy, where the utility function is a function of consumption, $U(C)$. The open economy model introduced has three decision variables—consumption, C, investment, I, and the extraction of the exhaustible resource, X. It is subject to three dynamic constraints which include the changes in capital accumulation, K, the extraction process of the resource imposed by the finiteness of the resource stock, S, and changes in debt accumulation, F, as shown below:

$$maxV_{[C,I,X]} = \int_0^N e^{-rt} U(C)dt \tag{5}$$

subject to

$$\dot{K} = I - \delta K - aX \tag{6}$$

$$\dot{S} = -X \tag{7}$$

$$\dot{F} = \theta F - \left(Y^{+,0} - C - I - \varphi(I, K)\right) \tag{8}$$

where $Y = Q(K, X)$, is the production function,[13] and r stands for the discount rate, X is the extraction rate of the exhaustible resource, δ denotes the capital depreciation, θF is the interest payment on debt, $\varphi(\cdot)$ stands for installation costs of the investment, $a > 0$ derives the cost of extracting one unit of the non-renewable resource, which could be nonlinear.[14]

The country has current account deficits when the sum of spending, $C + I + \varphi(I, K)$, and interest on debt, θF, exceeds the output, Y. The current account balances include not only the trade balance of goods and services but also interest receipts or payments and net transfers. Resource-rich countries with external debt pay the interest rate on their foreign debt, thus it is important to investigate changes in the current account. This derivation of current account is presented in number of papers, e.g., Blanchard (1983), Blanchard and Fischer (1989). The current account deficit is a difference between total spending and total earnings. This can be shown using the macroeconomic accounting relationship of the current account, *CA*, with

[13]For details, see Semmler and Sieveking (2000, p. 1124).

[14]Solution of the open economy model with a current-value Hamiltonian and optimality conditions is sketched in Appendix 2.

private sector ($S - I$) and public sector ($T - G$), where S denotes saving, I is investment, T is tax, and G stands for government expenditure.[15]

For the difference between the resource-rich and resource-poor countries, GDP can be classified as follows: the resource-rich country may have additional sources of revenue from its exports of natural resources. Thus, we can take into account an assumption that while a certain proportion, α, of the produced natural resources are used for domestic production processes, the remainder of the resources, $1 - \alpha$, can be exported abroad. This parameter for the exports can be estimated based on the actual stylized facts that were discussed in an earlier section.

Furthermore, this can be formulated within the GDP of the resource-rich country, which is denoted by Y^{+}. GDP of both resource-rich and resource-poor countries include net exports, NX, as a difference between exports and imports, $EX - IM$, of goods and services. In particular, for the resource-rich country, net exports, NX^{+}, are calculated by taking the difference of exports and imports of manufactured goods as well as raw materials and natural resources. In contrast to other modeling approaches, here, two different types of countries are considered. For the resource-rich countries:

$$Y^{+} = Y^{man} + NX^{+} \tag{9}$$

$$NX^{+} = EX^{total} - IM^{total} = (EX^{nonres} + EX^{res}) - (IM^{nonres} + IM^{res}) \tag{10}$$

$$EX^{res} = (1 - \alpha)X \tag{11}$$

Here, domestic manufacturing output is Y^{man}, and resource exports are, EX^{res}, because the resource-rich country may have more sources for its economic growth. The coefficient α shows how much of the extracted resources are used in domestic production. The higher the α, the larger the share of the resources used in the domestic production. When $\alpha = 1$, there is no export of resources.

The resource-poor country would have the following GDP with net exports of NX^0 :

$$Y^0 = Y^{man} + NX^0 \tag{12}$$

$$NX^0 = EX^{total} - IM^{total} = EX^{nonres} - (IM^{nonres} + IM^{res}) \tag{13}$$

In the numerical solution of the models, the utility function of $U(C) = \frac{C^{1-\eta}}{1-\eta}$, the Cobb-Douglas production function of $Q(K,X) = K^{\beta}X^{1-\beta}$, the installation cost function of $\varphi(I,K) = \frac{1}{2}\left(\frac{I}{K}\right)^2$, and the resource extraction cost function of $\frac{n}{X}$ with a positive values for n are used.

[15]This relationship can be shown from definition of the GDP which is $Y = C + I + G + CA$. From here we derive the current account as $CA = Y - C - I - G$ or $CA = S - I + T - G$ with $S = Y - T - C$.

Following Blanchard (1983), the above model can be extended by introducing a penalty term in the preferences of the objective function that takes into account the non-monetary cost of the external debt for economic growth.[16]

$$max V_{[C,I,X]} = \int_{0}^{N} e^{-rt} \left(U(C) - \kappa(\tau - \tau^{*})^{2}\right) dt \quad (14)$$

subject to

$$\dot{K} = I - \delta K - aX \quad (15)$$

$$\dot{S} = -X \quad (16)$$

$$\dot{F} = \theta F - \left(Y^{+,0} - C - I - \varphi(I, K)\right) \quad (17)$$

where $\tau = F/K$, $\tau^{*} = (F/K)^{*}$ is its steady state value, and κ represents the cost of debt. According to Blanchard (1983), a penalty term can be a function of either debt/efficiency unit, debt/capita, or debt/income.

There should be a distinction between the resource-rich and resource-poor countries under the open economy framework. Thus, the model for resource-rich countries would have three state variables, K, S, and F, and three control variables, C, I, and X. On the other hand, resource-poor countries have two state variables, K and F, and two control variables, C and I.

From the debt dynamics, where the current account deficits occur when spending exceeds earnings, we can analyze whether the country should reduce its deficits through consumption or investment. According to Blanchard (1983), in order to reduce the current account deficit, consumption should be reduced in Brazil. These models are solved for resource-rich and resource-poor countries in the next subsection.

4.2 Numerical Solution of the Open Economy Model

At first, an open economy model is solved for the resource-poor country, where there are no exports of natural resources, using NMPC, with different initial conditions of the state variables of $K(0)$, and $F(0)$. The model has these two state variables and

[16]Potential non-monetary costs of a very high debt can be issues of solvency and political risks as discussed in Blanchard (1983). The penalty term in the objective function, depending on whether the cost of debt is high or low, determines the path of the debt growth. Based on the analysis of marginal cost of debt and implications of different parameters in the penalty term, Blanchard (1983, p. 195) argues that "if a reduction in the growth of debt has to be achieved, it must be done by reducing consumption rather than investment".

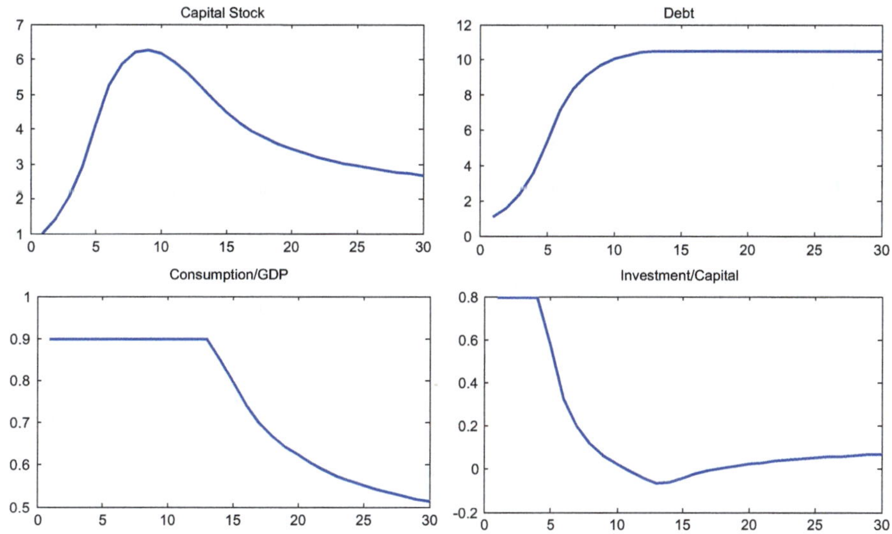

Fig. 6 Optimal trajectories of state variables and controls in a resource-poor country

two control variables, C, and I. In the NMPC code, control variables are expressed as shares, e.g., $c = C/Y$, and $i = I/K$. The results, with initial conditions of the state variables: $K(0) = 0.02, F(0) = 1.1$, and assumed parameter values of $\eta = 0.5$, $\beta = 0.3$, $r = 0.04$, $\delta = 0.1$, $\theta = 0.05$ as well as with a penalty term included in the preference, where the τ^* is assumed to be low, $\tau^* = 0.3$, and $\kappa = 0.0025$, is shown in Fig. 6.[17]

For a resource-poor country, funding from external sources would be more limited, as compared with resource-rich countries. Thus, its external debt increases, but at much slower speed. An initial increase in external debt can be associated with a rise in the capital stock. However, there is a decline in capital stock as time passes and debt further increases; thus, the capital stock has an inverted U-shaped curve. This result is similar to solutions of the closed economy model that was presented earlier. A very large external debt would be associated with a decline in capital stock, as there is less productive investment that would lead to more growth in the future. The excess debt in this case seems to cause consumption to decline after a particular period, when capital stock shows a significant decline and external debt reaches its highest value.

Next, the open economy model for the resource-rich country is solved using NMPC. Since the country is resource abundant, it has a resource constraint in addition to constraints of capital stock and external debt. Thus, there are three control variables, C, I, and X, and three state variables, K, S, and F. A solution to a

[17]Again, with regard to the choice of values for the parameters and the penalty term, I follow Blanchard (1983), Semmler and Sieveking (2000), and Mittnik and Semmler (2014).

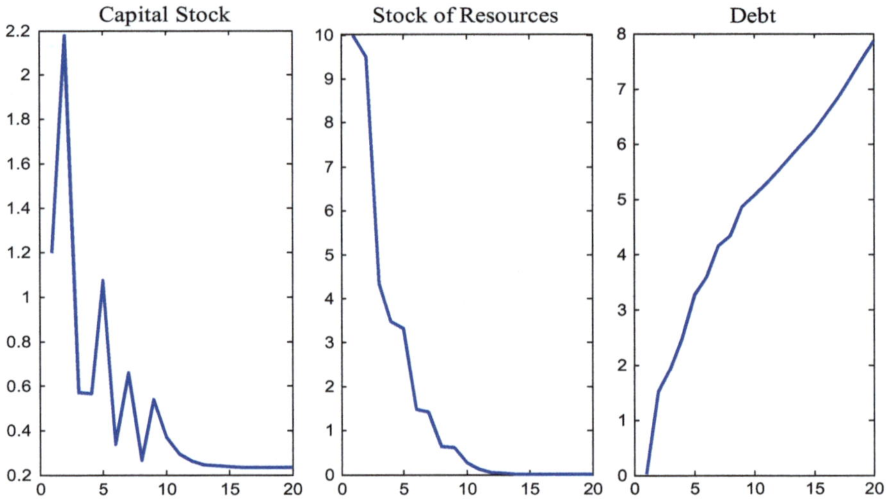

Fig. 7 Optimal trajectories of variables in a resource-rich country

variation of the model with penalty term is shown in Fig. 7, where control variables are expressed as shares, e.g., $c = C/Y$, $i = I/K$, and the flow of resources as share in total stock, $x = X/S$. The initial conditions of the state variables are $K(0) = 1.2$, $S(0) = 10$, and $F(0) = 0.02$ and, as before, the parameter values are $\eta = 0.5$, $\beta = 0.3$, $r = 0.04$, $\delta = 0.1$, $\theta = 0.05$, $\kappa = 0.0025$ and $a = 0.5$; but now with $\alpha = 0.05$, and high τ^*, e.g., $\tau^* = 500$, because of an increasingly large debt value.[18]

As expected, the stock of resources declines until it is completely used up. Debt is increasing, but at much slower growth rate; this is because the high penalty term causes the curve to become more concave. Capital stock initially increases sharply, but it then declines significantly. This cyclical pattern repeats over time until it reaches a stable low value; this is in contrast to the capital stock movement in a resource-rich, but closed economy. Typically, this also occurs in a resource-poor open country, where the capital stock rises first, but later declines monotonically, without any cyclical movements. At the beginning, capital stock increases in the economy where large stocks of a resource is available. However, following a sharp fall in the stock of exhaustible resources, capital stock will decline sharply. Initial borrowing from abroad increases the investment which, in turn, would contribute to an increase in capital stock. Nevertheless, further depletion of the resource, as well as excessive unsustainable debt, will lead to an eventual decline in the capital stock. Thus, the capital stock's path is affected by both available resources and extent

[18] Some of these parameter values are taken from Blanchard (1983), Semmler and Sieveking (2000), and Mittnik and Semmler (2014).

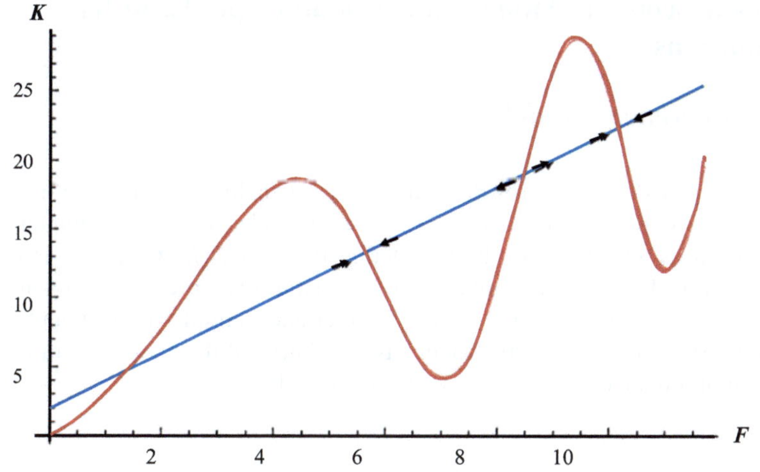

Fig. 8 Boom-bust cycles

of debt. A positive impact of debt on investment and growth turns negative in the presence of excessive debt.

Bernard and Semmler (2012) have shown that many boom-bust cycles are driven by linkages between borrowing and resource prices, yet borrowing could be over-covered or under-covered by resource value in the marketplace. The collateral value is a rather fuzzy concept based on many speculative elements in the marketplace and on the perceived value of the resource producer. Thus, as shown in Fig. 7, a resource boom encourages an overconfidence and overvaluation of assets. This leads over-borrowing and an underestimation of risk. The natural consequence is that too many loans secured by over-valued resources are issued. Finally, the market turns pessimistic and undervaluation ensues. This, in turn, causes the undervaluation of asset prices and a decline in the value of the underlying capital stock.

Technically, there are both absorbing and repelling points in the phase diagram illustration shown in Fig. 8, taken from in Bernard and Semmler (2012). In their paper, Bernard and Semmler show how, in the general case, over-borrowing is undertaken, but unless sufficient return is realized on investment, the capital stock will revert to it's previous value. On the other hand, if sufficient return is realized, a new stability may be achieved. As an example, we can imagine a resource-rich country's borrowing against its exports to improve its transportation system. Initially, this increases the value of its capital stock as there is now an efficient method to move resources to export. However, as the resource depletes and/or becomes more difficult to retrieve, the value of this infrastructure declines. A new discovery or extraction technology may re-ignite the boom, but since the infrastructure is already there, the additional returns to investment are more limited. Obviously the specifics are particular to the situation, but the illustration captures this generality.

5 Open Economy Model with Human Capital and Its Solutions

5.1 Theoretical Model

Here, the open economy growth model with a penalty term and non-renewable resources, as explained in Sect. 4, is further extended to include human capital by adding it as an input into the production function $Q(K, X, h)$; this is done as an addition to the physical capital stock and natural resource. For human capital formation, we follow Semmler (2003), Greiner and Semmler (2008), and Greiner et al. (2005). Thus, the growth model has an additional dynamic constraint on the evolution of human capital formation, h, as shown below:

$$maxV_{[C,I,X]} = \int_0^N e^{-rt}\left(U(C) - \kappa(\tau - \tau^*)^2\right) dt \tag{18}$$

subject to

$$\dot{K} = I - \delta K - aX \tag{19}$$

$$\dot{S} = -X \tag{20}$$

$$\dot{F} = \theta F - (Y - C - I - \varphi(I, K)) \tag{21}$$

$$\dot{h} = \lambda I - \vartheta h \tag{22}$$

where $Y = Q(K, X, h)$, is the production function, and $\lambda > 0$ stands for the "external effect associated with investment"[19] or the contribution of the investment in human capital, and $\vartheta \in (0, 1)$ is the depreciation of human capital.

5.2 Numerical Solutions

The extended model in Sect. 5.1 is solved using NMPC with different initial conditions for four state variables of $K(0)$, $S(0)$, $F(0)$, and $h(0)$. The model has three control variables, C, I, X, which are expressed as shares, e.g., $c = C/Y$, and $i = I/K$ following the previous simulations in the NMPC code. The production

[19]For details see Greiner and Semmler (2008, p. 64) and Greiner et al. (2005, p. 26). Earlier studies by Wan (1970) and Ryder and Heal (1973) highlighted the weighted contribution of the investment in human capital.

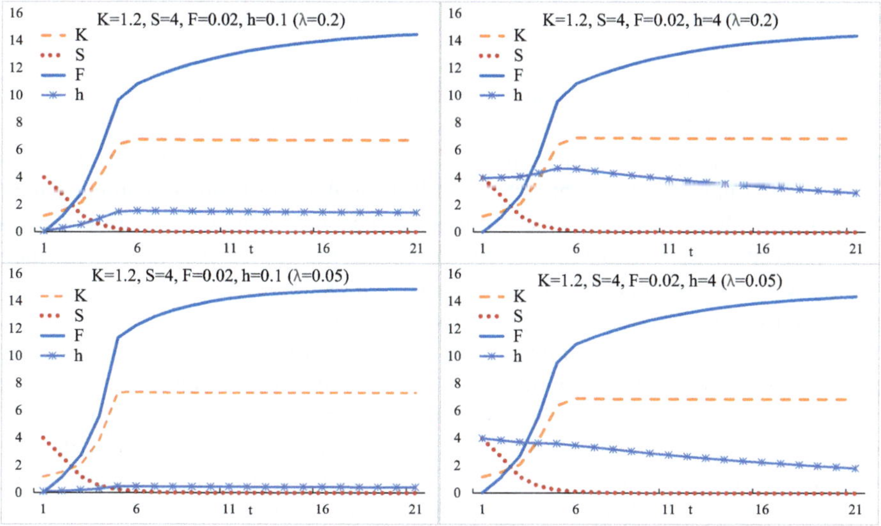

Fig. 9 Optimal trajectories of variables in an open economy model with human capital

function is assumed to be $Q(K, X, h) = K^{\beta}X^{1-\beta-\phi}h^{\phi}$ with $\beta = 0.3$ and $\phi = 0.1$. In addition to the previous parameter values taken from the model in Sect. 4 ($\eta = 0.5$, $\beta = 0.3$, $r = 0.04$, $\delta = 0.1$, $\theta = 0.05$, $\tau^* = 0.3$, $\kappa = 0.0025$), the parameter for human capital depreciation is assumed to be $\vartheta = 0.05$. The numerical results corresponding to initial conditions of $K(0) = 1.2$, $S(0) = 4$, and $F(0) = 0.02$, but different conditions of human capital (low: $h(0) = 0.1$ and high: $h(0) = 4$) as well as investment's contribution $\lambda = 0.2$ and $\lambda = 0.05$ are shown in Fig. 9.

Optimal trajectories of state variables in an open economy with human capital are illustrated in Fig. 9. As expected, resources stocks decline and deplete over time. Similar to previous scenarios, when a country borrows large sums from abroad initially, physical capital stock increases at a fast rate but stops increasing later on. On the other hand, the evolution of human capital depends on the contribution of the investment in human capital. In the first case, if the country spends a higher proportion of its investment on human capital ($\lambda = 0.2$), it increases significantly at first, but then levels off when its initial condition is very low ($h(0) = 0.1$). Alternatively, it immediately starts declining when its initial condition is very high ($h(0) = 4$). The second case, with a very low share of investment in human capital ($\lambda = 0.05$), is not very favorable to the evolution of human capital: human capital stock demonstrates a very slight rise at first and stays constant but at a very low level when its initial condition is very low ($h(0) = 0.1$) or it declines from the beginning at a faster rate when its initial condition is very high ($h(0) = 4$).

6 Conclusion

In this paper, the impact of available non-renewable energy resources, external debt, and human capital on sustainable economic growth has been analyzed. The growth model takes into account constraints on the finite exhaustible resources, external sources for financing development, as well as evolution of human capital. I have extended the basic growth model to apply to an open economy, where countries are involved in trade and borrow from international capital markets, using their energy resources as collateral. In contrast to previous studies, the difference between resource-rich and resource-poor countries is emphasized.

NMPC results show the interactions between economic growth, openness to global markets, and human capital based on different initial conditions of the state variables, as well as different country-specific parameters pertaining to dependence on resources and investment in human capital. Numerical solutions show that the model is consistent with the empirical facts. Solutions for the closed economy show an inverted U-shaped evolution of capital accumulation, together with a monotonically declining stock of non-renewable energy resources. As exhaustible resources deplete, a decline in the capital stock begins to occur. The associated optimal movement of consumption shows an inverted U-shaped trend. Similar paths are found for resource-poor countries in an open economy. In contrast, resource-rich countries with a large penalty term in the preferences have abundant resources as well as external borrowing. In this case, capital stock initially rises; eventually, a decline occurs, but it starts rising again, followed by falling. Thus, this process may exhibit the cyclical movements of typical booms and busts. It reaches a low value when the resources are closer to depletion and debt increases to a high level. Debt may be used for the import of capital goods to increase production domestically, search for more resources, and/or support industrialization. However, in some cases, a debt crisis can occur if the returns on investments fall short of debt obligations, e.g., as in the developing Latin American countries of the 1980s. Excessive borrowing may lead not only to currency and debt crises, but also to increased volatility in the prices of the commodity used to underwrite that same debt (Nyambuu 2016). We also observe a positive correlation between openness and growth, as well as between openness and investment in human capital.

Depletion of non-renewable energy resources can be a warning signal for resource-rich countries that largely depend on export revenues from natural resources. In particular, for some Latin American and African countries, resource revenues account for more than 90% of total fiscal revenue. These countries need to finance productive long-term investments and to develop alternative sectors. Solution of the dynamics, with external debt, indicates a slower growth rate in the long run, which can be related to the increased debt burden. Thus, the solutions of the optimal paths for external debt, capital stock, and consumption, as compared to available natural resources, describe the mechanism by which macroeconomic disturbances may imply the existence of a threat to sustainable growth in the future. Furthermore, from the results shown in the numerical solutions, we suggest that those resource exporting countries that borrow significantly from international capital markets to finance their development should not diminish investment in human capita. Empirical and theoretical findings suggest that inequality is exacerbated by resource/export-based economic growth in resource-rich developing countries.

Appendix 1

The basic growth model is solved with the current-value Hamiltonian with two constraints as follows:

$$H = U(C) + q_1\,(Q(K,X) - C) + q_2(-X) \tag{23}$$

where q_1 and q_2 are co-state variables or shadow prices of capital accumulation and resource constraint respectively. The necessary optimality conditions are obtained as follows:

$$\frac{\partial H}{\partial C} = U'(C) - q_1 = 0 \tag{24}$$

$$\frac{\partial H}{\partial X} = q_1 Q_X - q_2 = 0 \tag{25}$$

$$\dot{q_1} = rq_1 - \frac{\partial H}{\partial K} = q_1(r - Q_K) \tag{26}$$

$$\dot{q_2} = rq_2 - \frac{\partial H}{\partial S} = rq_2 \tag{27}$$

with $Q_X = \frac{\partial Q(K,X)}{\partial X}$, $Q_K = \frac{\partial Q(K,X)}{\partial K}$, $\frac{dq_1}{dt} = \dot{q}_1$ and $\frac{dq_2}{dt} = \dot{q}_2$.

As shown in Dasgupta and Heal (1974, p. 11) and in Greiner and Semmler (2008, pp. 165–166), differentiating Eq. (24) with respect to time and substituting Eq. (26) yields the following consumption rate along an optimal path:

$$\frac{\dot{C}}{C} = \frac{Q_K - r}{\eta(C)} \tag{28}$$

with $\eta(C) = -\frac{CU''(C)}{U'(C)}$ as an elasticity of marginal utility (Dasgupta and Heal 1974, p. 5). Greiner and Semmler (2008) note that from the optimal path of consumption we can observe that higher discount rate is associated with further fall of the rate of consumption over time.

Appendix 2

The current-value Hamiltonian of the open economy growth model with three constraints is as follows:

$$H = U(C) + q_3(I - \delta K - aX) + q_4(-X) + q_5\left(\theta F - (Y - C - I - \varphi(I, K))\right) \tag{29}$$

The optimality conditions are:

$$\frac{\partial H}{\partial C} = U'(C) + q_5 = 0 \tag{30}$$

$$\frac{\partial H}{\partial I} = q_3 + q_5\left(1 + \varphi_I\right) = 0 \tag{31}$$

$$\frac{\partial H}{\partial X} = -q_3 a - q_4 - q_5 Q_X = 0 \tag{32}$$

$$\dot{q}_3 = rq_3 - \frac{\partial H}{\partial K} = (r + \delta)q_3 + (Q - \varphi_K)q_5 \tag{33}$$

$$\dot{q}_4 = rq_4 - \frac{\partial H}{\partial S} = rq_4 \tag{34}$$

$$\dot{q}_5 = rq_5 - \frac{\partial H}{\partial F} = (r - \theta)q_5 \tag{35}$$

with $\varphi_I = \frac{\partial \varphi(I,K)}{\partial I}$, $\varphi_K = \frac{\partial \varphi(I,K)}{\partial K}$, $Q_X = \frac{\partial Q(K,X)}{\partial X}$, $Q_K = \frac{\partial Q(K,X)}{\partial K}$ and $Q_S = \frac{\partial Q(K,X)}{\partial S}$ and q_3, q_4, and q_5 are co-state variables or shadow prices of capital accumulation, resource constraint, and external debt respectively. Here, initial values for all the state variables $K(0)$, $R(0)$ and $F(0)$ are given. The path of consumption is:

$$\frac{\dot{C}}{C} = \frac{r - \theta}{\eta(C)} \tag{36}$$

with $\eta(C) = -\frac{CU''(C)}{U'(C)}$ as an elasticity of marginal utility.

Appendix 3

Nonlinear Model Predictive Control (NMPC) is a methodological paradigm for the analysis of nonlinear feedback control systems. Although it was developed in the early 1960s, the computational complexity of the algorithms created rendered it impractical until relatively recently. However, NMPC algorithms developed by Grüne and Pannek (2011), as I have written previously in Nyambuu and Semmler (2014), can now be implemented on modern computers and used for the analysis of the dynamics of our closed economy model with exhaustible resources. The extension of the model in an open economy with external debt constraint is also solved with NMPC.

NMPC is an optimization-based method of feedback for control of a nonlinear system. For example, consider a controlled and predicted process where at each instant of time a control input, $w(n)$, can be chosen. The control inputs affect the future behavior of the state of the system variable, $v(n)$, which is measured at discrete time instants. The control inputs, $w(n)$, are determined to track the state of the system, $v(n)$, so that it is as close as possible to a given reference, $v^{ref}(n)$ (Grüne and Pannek 2011). When the reference is constant and equal to zero, as shown in Eq. (37), the tracking problem reduces to a stabilization problem of the following type[20]:

$$v^{ref}(n) = v^* = 0 \tag{37}$$

Note, however, the procedure also works—and thus is the novelty of the Grüne-Pannek procedure—if the reference value, for example the steady state as terminal condition, is not known. Greiner et al. (2013) point out that in the case of a very long decision horizon, NMPC can approximate the infinite time horizon solution well. Even with a short decision horizon ($N = 10$), one can still investigate important issues raised in the context of the model.

Following Greiner et al. (2013, pp. 16–18), suppose the optimal decision problem is given as follows:

$$max \int_0^{\infty} e^{-\delta t} g(v(t), w(t))dt \tag{38}$$

with $\dot{v}(t) = f(v(t), w(t))$, $v(0) = v_0$.

An approximate discrete time problem can be written as:

$$max \sum_{i=0}^{\infty} \beta^i g(v_i, w_i) \tag{39}$$

[20]See Grüne and Pannek (2011).

where $v_{i+1} = \Phi(h, v_i, w_i)$ and $\beta = e^{-\delta h}$ with $h > 0$ discretization time step (Greiner et al. 2013, p. 17). Accordingly, instead of the infinite horizon of Eq. (39) we can use a finite horizon functional as shown in Eq. (40).

$$max \sum_{k=0}^{N} \beta^i g(v_{k,i}, w_{k,i}) \tag{40}$$

where the index, i, is the number of the iteration. Here, the decision horizon is "truncated" and is represented by a "finite" horizon, N, where $v_{k+1,i} = \Phi(h, v_{k,i}, w_{k,i})$ after the decision has been made. The converted static nonlinear optimization problem can be solved numerically using the Matlab *fmincon* solver (Greiner et al. 2013, pp. 16–18).

The solution (v_i, w_i) of the problem converges to the accurate solution of Eq. (39) as $N \rightarrow \infty$. This convergence is ensured under the assumptions such that an optimal equilibrium for the infinite horizon problem in Eq. (39) exists (Greiner et al. 2013, p. 18). Thus, in accordance with the recent studies already mentioned, I also assume that the solution converges to the optimal solution even without knowing the equilibrium values (Grüne 2013).

Appendix 4

I solve the optimal control models described in this paper by using NMPC for different initial values of state variables and various parameters. In Sect. 3 of the paper, I showed the optimal trajectories of the variables for initial conditions of $K(0) = 1$ and $S(0) = 4$ and parameter values of $\eta = 0.5$, $\beta = 0.3$, $r = 0.03$ for the resource intensive production function. The results of the evolutions of state variables are illustrated in Fig. 5. Now, I present the optimal trajectories of the state and control variables for capital intensive economy with $\beta = 0.7$ and other parameter values as before: $\eta = 0.5$, $r = 0.03$ and initial conditions of $K(0) = 1$ and $S(0) = 4$. The result is shown in Fig. 10.

Similar to the previous case, we observe an inverted U-shaped movement of capital stock with monotonically decreasing stock of the remainder of exhaustible resources. However, it has taken a longer period for the capital stock to start declining. This indicates that the stock of the capital has begun to fall when the resources had reached a much lower level in comparison to the level in previous result. In addition, corresponding optimal paths of consumption level, flows of exhaustible resources, utility level and output, depicted in Fig. 11, show similar movements as we saw in the first scenario described in Sect. 3. Thus, after reviewing numerous variations and their numerical solution, we generally observe an inverted U-shaped capital accumulation movement and monotonically declining stock of non-renewable resources when initial condition of capital stock is not high. Besides

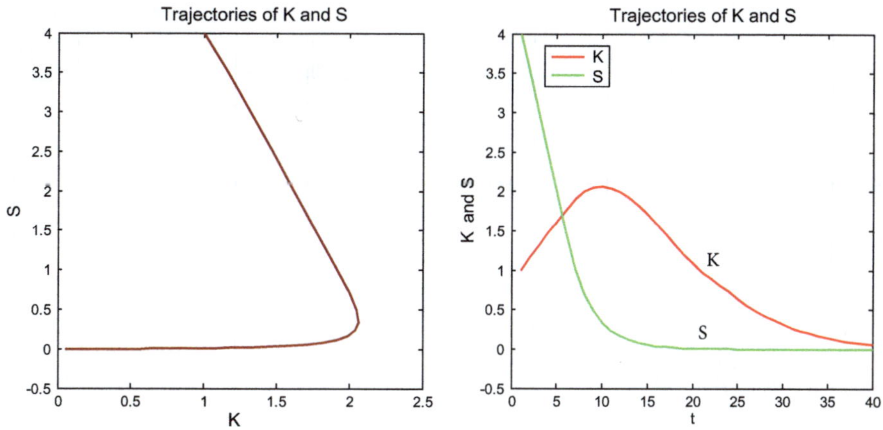

Fig. 10 Optimal trajectories of K and S in the capital intensive economy

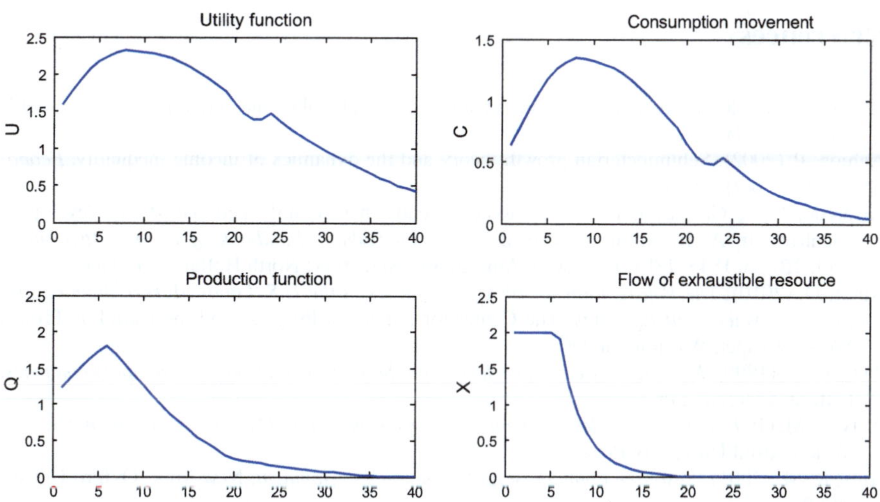

Fig. 11 Evolution of consumption, utility, output and flow of resources in the capital intensive economy

the different values of the parameters, we can be interested in the solutions with the different initial conditions of the state variables. These are shown in Fig. 12.

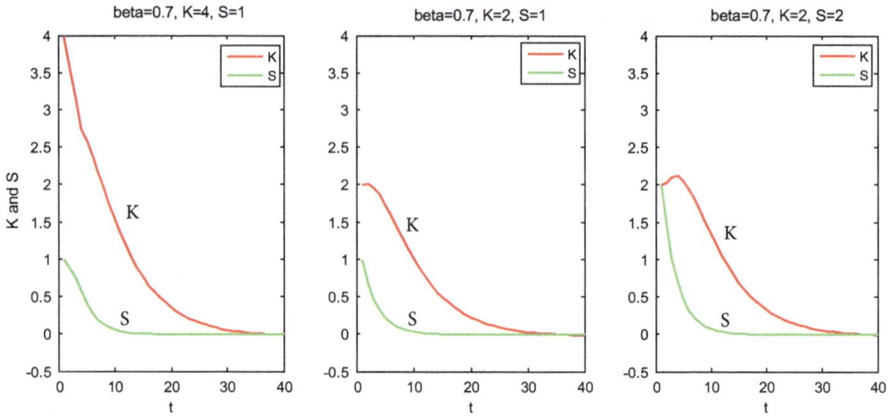

Fig. 12 Optimal trajectories of *K* and *S* with different initial conditions of the state variables

References

Acemoglu, D. (2002). Technical change, inequality, and the labor market. *Journal of Economic Literature, 40*(1), 7–72.

Aghion, P. (2002). Schumpeterian growth theory and the dynamics of income inequality. *Econometrica, 70*(2), 855–882.

Alvaredo, F., & Gasparini, L. (2015). Recent trends in inequality and poverty in developing countries. In A. B. Atkinson & F. Bourguignon (Eds.), *Handbook of income distribution* (Vol. 2B. pp. 1845–1881, 697–805). Amsterdam, New York: North Holland, Elsevier.

Autor, D. (2010). *The Polarization of Job Opportunities in the U.S. Labor Market: Implications for Employment and Earnings*. The Center for American Progress and the Hamilton Project Working Paper, Washington, DC.

Auty, R. M. (1990). *Resource-based industrialization: Sowing the oil in eight developing countries*. Oxford: Clarendon Press.

Auty, R. M. (1993). *Sustaining development in mineral economies: The resource curse thesis*. New York: Oxford University Press.

Auty, R. M. (2001). *Resource abundance and economic development*. New York: Oxford University Press.

Bekaert, G., Harvey, C. R., & Lundblad, C. (2006). Growth volatility and financial liberalization. *Journal of International Money and Finance, 25*, 370–403.

Berg, A., Portillo, R., Yang, S. S., & Zanna, L. F. (2012). *Public Investment in Resource-Abundant Developing Countries*. IMF Working Paper No. 12/274. International Monetary Fund, Washington, DC. https://www.imf.org/external/pubs/cat/longres.aspx?sk=40105.0.

Bernard, L., & Nyambuu, U. (2015). *Global warming and clean energy in Asia. OUPblog. Oxford University Press's Academic Insights for the Thinking World*. http://blog.oup.com/2015/07/global-warming-clean-energy-asia/.

Bernard, L., & Semmler, W. (2012). Boom–bust cycles: Leveraging, complex securities, and asset prices. *Journal of Economic Behavior and Organization, 81*(2), 442–465. ISSN: 0167-2681. doi:10.1016/j.jebo.2011.07.002.

Bhagwati, J., & Kosters, M. H. (1994). *Trade and wages*. Washington, DC: AEI Press.

Blanchard, O. J. (1983). *Debt and the current account deficit in Brazil*. NBER Chapters. National Bureau of Economic Research, Inc. http://www.nber.org/books/arme83-1.

Blanchard, O. J., & Fischer, S. (1989). *Lectures on macroeconomics*. Oxford: MIT Press.

Bloom, D. E., & Brender, A. (1993). Labor and the emerging world economy. *Population Bulletin, 48*(2), 2–39.

Caner, M., Grennes, T., & Koehler-Geib, F. (2010). *Finding the Tipping Point—When Sovereign Debt Turns Bad*. World Bank Policy Research Working Paper No. 5391. World Bank, Washington, DC. doi:10.1596/1813-9450-5391.

Cecchetti, S., Mohanty, M., & Zampolli, F. (2011). *The real effects of debt*. Bank for International Settlements Working Paper No. 352. http://www.bis.org/publ/work352.htm.

Costinot, A. (2009). An elementary theory of comparative advantage. *Econometrica, 77*, 1165–1192.

Dasgupta, P., & Heal, G. (1974). The optimal depletion of exhaustible resources. In *The Review of Economic Studies, Vol. 41, Symposium on the Economics of Exhaustible Resources* (pp. 3–28).http://www.jstor.org/stable/2296369.

Davis, D. R. (1996). *Trade Liberalization and Income Distribution*. NBER Working Paper No. 5693, National Bureau of Economic Research, Cambridge, MA

Di Giovanni, J., & Levchenko, A. (2008). *Trade Openness and Volatility*. IMF Working Paper, WP/08/146. http://www.imf.org/external/pubs/ft/wp/2008/wp08146.pdf.

Easterly, W., Islam, R., & Stiglitz, J. E. (2001). Shaken and stirred: Explaining growth volatility. In B. Pleskovic & N. Stern (Eds.), *Annual World Bank Conference on Development Economics*.

Exploration in South America (2001). *Mining Journal*, April 20.

Feenstra, R., & Hanson, G. (1996). Foreign investment, outsourcing and relative wages. In R. C. Feenstra et al. (Eds.), *Political economy of trade policy: Essays in honor of Jagdish Bhagwati* (pp. 89–127). Cambridge, MA: MIT Press.

Feenstra, R., & Hanson, G. (1997). Foreign direct investment and relative wages: Evidence from Mexico's maquiladoras. *Journal of International Economics, 42*(3–4), 371–393.

Goldberg, P., & Pavcnik, N. (2007). *Distributional Effects of Globalization in developing Countries*. NBER Working Papers, No. 12885.

Greiner, A., Grüne, L., & Semmler, W. (2013). Economic growth and the transition from non-renewable to renewable energy. In *Environment and development economics*. Cambridge: Cambridge University Press. doi:10.1017/S1355770X13000491.

Greiner, A., Rubart, J., & Semmler, W. (2004). Economic growth skilled-biased technical change and wage inequality: A model and estimations for the U.S. and Europe. *Journal of Macroeconomics, 26*(4), 597–621.

Greiner, A., & Semmler, W. (2008). *The global environment, natural resources, and economic growth*. New York: Oxford University Press.

Greiner, A., Semmler, W., & Gong, G. (2005). *The forces of economic growth: A time series perspective*. New Jersey: Princeton University Press.

Grossman, G. (2013). *Heterogeneous Workers and International Trade*. NBER Working Paper No. 18788.

Grüne, L. (2013). Economic receding horizon control without terminal constraints. *Automatica, 49*(3), 725–734.

Grüne, L., & Pannek, J. (2011). *Nonlinear model predictive control: Theory and algorithms*. Berlin: Springer.

Grüne, L., Semmler, W., & Stieler, M. (2013). *Using nonlinear model predictive control for dynamic decision problems in economics*. doi:10.2139/ssrn.2242339.

Harberger, A. C., Smith, G. W., & Cuddington, J. T. (1985). Lessons for debtor-country managers and policymakers. In *International debt and the developing countries*. World Bank. http://www.econbiz.de/Record/lessons-for-debtor-country-managers-and-policy-makers-arberger-arnold/10002651663.

Helpman, E., Itskhoki, O., & Redding, S. (2010). Inequality and unemployment in a global economy. *Econometrica, 78*(4), 1239–1283.

IMF (2012). *Macroeconomic Policy Frameworks for Resource-Rich Developing Countries*. IMF Policy Paper. https://www.imf.org/external/np/pp/eng/2012/082412.pdf.

IMF (2013). *World Economic Outlook*. April 2013. http://www.imf.org/external/pubs/ft/weo/2013/01/pdf/text.pdf.

IMF (2016). *International Financial Statistics*. http://elibrary-data.imf.org/finddatareports.aspx?d=33061&e=169393. Accessed 8 July 2016.

Irwin, D. (2008). Trade and wages, reconsidered: comments and discussion. *Brookings Papers on Economic Activity, Spring, 38*(1), 138–143.

Kanbur, R. (2015). Globalization and inequality. In A. B. Atkinson & F. Bourguignon (Eds.), *Handbook of income distribution* (Vol. 2B, pp. 1845–1881). Amsterdam, New York: North Holland, Elsevier.

Katz, L. (2008). Trade and wages, reconsidered: comments and discussion. *Brookings Papers Economic Activity, 2008*(1), 143–149.

Katz, L. F., & Murphy, K. M. (1992). Changes in relative wages, 1963–1987: Supply and demand factors. *Quarterly Journal of Economics, 107*, 35–78.

Kose, A., Prasad, E., & Terrones, M. (2006). How do trade and financial integration affect the relationship between growth and volatility? *Journal of International Economics, 69*, 176–202.

Krugman, P. (2008). Trade and wages, reconsidered. *Brookings Papers Economic Activity, 39*(1), 103–137.

Krugman, P. (1994). *Past and prospective causes of high unemployment* (Fourth Quarter, pp. 23–43). Federal Reserve Bank of Kansas City: Economic Review.

Levy, F., & Murnane, R. J. (1992). US earnings levels and earnings inequality: A review of recent trends and proposed explanations. *Journal of Economic Literature, 30*, 1333–1381.

Lewis, N. S. (2007). Powering the planet. *Engineering and Science, 70*(2), 12–23.

Lucas, R. E. (1988). On the mechanics of economic development. *Journal of Monetary Economics, 22*, 3–42.

Manzano, O., & Rigobon, R. (2001). *Resource curse or debt overhang*? NBER Working Paper No. 8390. National Bureau of Economic Research, Inc. http://ssrn.com/abstract=277300.

Mittnik, S., & Semmler, W. (2014). *Overleveraging, Financial Fragility and the Banking-Macro Link: Theory and Empirical Evidence*. Discussion paper No. 14–110, ZEW Center for European Economic Research.

Nyambuu, U. (2016). Foreign exchange volatility and its implications for macroeconomic stability: An empirical study of developing economies. In L. Bernard, & U. Nyambuu (Eds.), *Dynamic modeling, empirical macroeconomics, and finance* (pp. 163–182). Springer International Publishing.

Nyambuu, U., & Bernard, L. (2015). A quantitative approach to assessing sovereign default risk in resource-rich emerging economies. *International Journal of Finance and Economics*, Wiley 1512. doi:10.1002/ijfe.1512.

Nyambuu, U., & Semmler, W. (2014). Trends in the extraction of non-renewable resources: The case of fossil energy. *Economic Modelling Elsevier , 37*(C), 271–279. doi:10.1016/j.econmod.2013.11.020.

Nyambuu, U., & Semmler, W. (2017a). Emerging markets' resource booms and busts, borrowing risk and regime change. *Structural Change and Economic Dynamics, 41*, 29–42.

Nyambuu, U., & Semmler, W. (2017b). The challenges in the transition from fossil fuel to renewable energy. In: T. Devezas, J. Leitão, & A. Sarygulov (Eds.), *Industry 4.0 entrepreneurship and structural change in the new digital landscape* (pp. 157–182). Springer International Publishing.

Nyambuu, U., & Tapiero, C. S. (2017). *Globalization, gating, and risk finance*. England: Wiley. Forthcoming.

Pattillo, C., Poirson, H., & Ricci, L. (2002). *External Debt and Growth*. IMF Working Paper No. 02/69. International Monetary Fund, Washington. http://www.imf.org/external/pubs/ft/wp/2002/wp0269.pdf.

Pearce, D. W., Barbier, E., & Markandya, A. (1990). *Sustainable development: Economics and environment in the third world*. Aldershot: Edward Elgar Publishing.

Piketty, T. (2014). *Capital in the twenty-first century*. Cambridge: Harvard University Press.

Richardson, D. J. (1995). Income inequality and trade: How to think, what to conclude. *Journal of Economic Perspectives, 9*, 33–55.

Rodrik, D. (1997). *Has globalization gone too far?* Washington, DC: Institute for International Economics.

Ryder, H. E., & Heal, G. M. (1973). Optimal growth with intertemporally dependent preferences. *Review of Economic Studies, 40*, 1–31.

Sachs, J. D., & Warner, A. M. (1995). *Natural Resource Abundance and Economic Growth.* NBER Working Paper No. 5398. National Bureau of Economic Research, Inc. http://www.nber.org/papers/w5398.

Sachs, J. D., & Warner, A. M. (1999). The big push, natural resource booms and growth. *Journal of Development Economics, 59*(1), 43–76. http://EconPapers.repec.org/RePEc:eee:deveco:v:59:y:1999:i:1:p:43-76.

Sachs, J. D., & Warner, A. M. (2001). The curse of natural resources. *European Economic Review, 45*, 827–38. http://www.earth.columbia.edu/sitefiles/file/about/director/pubs/EuroEconReview2001.pdf.

Semmler, W. (2003). *On the Mechanisms of Inequality.* Center for Empirical Macroeconomics, Bielefeld and New School University. http://www.wiwi.uni-bielefeld.de/forschung/cemm_wpapers/upload/Mechanisms_of_Inequality.pdf.

Semmler, W., & Sieveking, M. (2000). Critical debt and debt dynamics. *Journal of Economic Dynamics and Control, 24*(5–7), 1121–1144.

Smith, B. (2004). Oil wealth and regime survival in the developing world, 1960–1999. *American Journal of Political Science, 48*(2), 232–246. doi:10.1111/j.0092-5853.2004.00067.x.

Solow, R. M. (1973). Is the end of the world at hand?, In A. Weintraub, E. Schwartz, & J. R. Aronson (Eds.), *The economic growth controversy* (pp. 38–61). New York: International Arts & Sciences Press.

Stein. J., & Paladino, G. (2001). *Country Default Risk: An Empirical Assessment.* Center for Economic Studies & Ifo Institute for Economic Research Working Paper No. 469. Munich, Germany.

Stiglitz, J. (1974). Growth with exhaustible natural resources: efficient and optimal growth paths. *The Review of Economic Studies, Vol. 41, Symposium on the Economics of Exhaustible Resources* (pp. 123–137). http://www.jstor.org/stable/2296377.

Ursua, J., & Wilson, D. (2012). Risks to growth from build-ups in public debt. *Global Economics Weekly* (No. 12/10).

U.S. Energy Information Administration (2016). *Statistics.* Energy Information Administration. http://www.eia.gov/. Accessed 5 July 2016.

Uzawa, H. (1965). Optimum technical change in an Aggregative model of economic growth. *International Economic Review, 6*, 18–31.

Wan, H. Y. (1970). Optimal saving programs under intertemporally dependent preferences. *International Economic Review, 11*, 521–547.

Wood, A. (1994). *North-south trade, employment and inequality: Changing fortunes in a skill-driven world.* Oxford: Clarendon Press.

Wood, A. (1999). Openness and wage inequality in developing countries: The Latin American challenge to East Asian conventional wisdom. In R. Baldwin et al. (Eds.), *Market integration, regionalism and the global economy.* Cambridge: Cambridge University Press.

World Bank (2016). *Databank.* http://databank.worldbank.org/data/home.aspx. Accessed 1 July 2016.

World Commission on Environment and Development (1987). *Our Common Future (The Brundtland Report).* New York: Oxford University Press.

Wright, G., & Czelusta, J. (2007). Resource-based growth past and present. In D. Lederman & W. F. Maloney (Eds.), *Natural resources: Neither curse nor destiny* (pp. 183–212). Washington, DC: World Bank.

Macroeconomic Risk, Fiscal Policy Rules and Aggregate Volatility in Asymmetric Currency Unions: A Behavioral Perspective

Christian R. Proaño and Benjamin Lojak

Abstract This paper studies the dynamics of sovereign risk, fiscal policy and the macroeconomy in a two-country monetary union framework under the assumption of a heterogeneous perception of the determinants of sovereign risk by the government and the market participants. The macro-economic volatility resulting from various types of fiscal policy rules aimed at the stabilization of sovereign debt is investigated through numerical simulations. Among other things, these simulations show that an extreme focus on debt stabilization can be counterproductive if the financial markets care more about the country's output gap.

1 Introduction

The assessment of sovereign risk and debt sustainability of euro area countries by financial markets seems to have experienced significant variation since the outbreak of the still unresolved euro area crisis at the end of 2009. During the 2000–2007 period, sovereign yields in the euro area were mainly driven by common international factors related to the global risk perception and not (or to an almost negligible extent) by country-specific factors (see e.g. Codogno et al. 2003; Geyer et al. 2004); the sovereign spreads of most euro area government bonds vis-à-vis German government bonds were also quite close to zero, reflecting the market's perception of the relative default risk of individual countries as almost nil despite quite differentiated underlying macroeconomic fundamentals. Since the default of Lehman Brothers in September 2008, in contrast, country-specific macroeconomic fundamentals do not only play a much more important role in the pricing of government bonds, but their influence itself seems to increase with the level of general risk aversion (see e.g. Manganelli and Wolswijk 2009; Haugh et al. 2009; Attinasi et al. 2009; Barrios et al. 2009; Schuknecht et al. 2009, 2010; Arghyrou and Kontonikas 2011; Borgy et al. 2012; De Grauwe and Ji 2013).

C.R. Proaño (✉) • B. Lojak
Otto-Friedrich-Universität Bamberg, Bamberg, Germany
e-mail: christian.proano@uni-bamberg.de; benjamin.lojak@uni-bamberg.de

B. Bökemeier, A. Greiner (eds.), *Inequality and Finance in Macrodynamics*, Dynamic Modeling and Econometrics in Economics and Finance 23,
DOI 10.1007/978-3-319-54690-2_10

This indirect evidence for a time-varying and differentiated risk perception of similar macroeconomic fundamentals by the financial market participants suggest that behavioral factors such as perception biases and extrapolative expectations may have played an important role in the recent sovereign bond crisis within euro area. For instance, De Grauwe and Ji (2013) argue that the irrationality of market expectations and its self-fulfilling nature have been the driving factors of sovereign risk premia of government debt of euro area countries not only after the 2007–2008 financial crisis and the outbreak of the still ongoing euro area crisis, but in the period before these events as well. As pointed out by Arghyrou and Kontonikas (2011), "This regime-shift not only explains the sudden escalation of the Greek debt crisis but also the difference in spread values observed between Greece and other periphery EMU countries with not too dissimilar macroeconomic outlook: Compared to Ireland, Portugal and Spain, markets perceive a much higher probability of a Greek voluntary exit from the EMU, and/or a Greek default. In short, Greece's problems are as much about trust as they are about economics."

Along these lines of thought, we set up in this paper a two-country monetary union model characterized by various features which make it useful for the analysis of interaction between the perceived sovereign risk by financial markets and its interaction with the macroeconomy in a rigorous way. The risk premium on government bonds and the related perception of sovereign risk is specified in a behavioral, not necessarily model-consistent manner, in stark contrast to the standard general equilibrium asset pricing specification.[1] Through the explicit modeling and variation of the set of variables used for the assessment of sovereign risk by different market participants, as well as their relative importance in such forecasts, it is possible to analyze the macroeconomic consequences of eventual differences between those sets of variables and the governments' policy targets in a straightforward and clear manner (this approach can be related to Farmer's "belief function", see e.g. Farmer 2010). Indeed, as already pointed out, in the real world the link between the pricing of risk and the observable macroeconomic fundamentals seems to be rather loose and cannot account for the changes observable in bond yield- and credit default spreads which are often utilized as measure of sovereign risk. Therefore, there might be other destabilizing sources which can drive the economy into an economic slowdown (see e.g. De Grauwe and Ji 2012; Corsetti et al. 2012).

The analysis for the present paper thus aims to address the following questions: What if governments pursue other goals than what financial market participants consider as relevant for the pricing of sovereign bonds, and what would be the macroeconomic consequences of such a discrepancy in the financial markets perception and actions? What are the macroeconomic consequences of a strict focus on sovereign debt reduction by the fiscal authorities, if markets do not consider this as a major determinant in their assessment of sovereign risk?

[1]In recent times, various studies have investigated the interaction between sovereign risk and economic activity in otherwise rather standard macroeconomic frameworks, see e.g. Adrian et al. (2010).

The remainder of this paper is organized as follows: In Sect. 2 a two-country behavioral macroeconomic model of a monetary union is introduced. In Sect. 3 the following investigations are being conducted: various transmission mechanisms are outlined through the analysis of the dynamic adjustments of the model to an exogenous shock in sovereign risk; the consequences of alternative specifications based on different sets of macroeconomic fundamentals determining the sovereign risk premia on macroeconomic volatility of the individual economies, and the monetary union as a whole. Finally, in Sect. 4 some policy implications and concluding remarks from this study are drawn.

2 The Model

We consider a two-country monetary union which is populated by a continuum of agents on the interval $[0, 1]$, a segment $[0, n]$ residing in a country labeled H(ome), the other segment living in the other country labeled F(oreign). There is no migration between regions. Both countries are assumed to produce tradable consumption goods, which are however considered as imperfect substitutes due to a standard home bias argument, and to feature otherwise the same characteristics concerning the structure of their behavioral equations. The governments in the Home and Foreign regions dictate their fiscal policy in an independent and sovereign manner, financing their expenditures through the raise of taxes and the issuance of bonds (with a maturity of m periods). Furthermore, there is a single monetary policy authority (the monetary union's central bank or MUCB) which sets the riskless short-term interest rate which acts as reference for the pricing of the Home and Foreign government bonds. Moreover, the Foreign country's government will be assumed to be considered as more solvent by the market's participants, becoming in the limit a "safe haven" for financial capital.[2]

While the agents' behavior is modeled in terms of log deviations from the model's steady state, we make use of the level expressions to link the main economic variables as well as to define market equilibria. Accordingly, we have

$$\log(X_t) = \log(\mathbb{X}_t) + x_t,$$

where X_t represents the level of the variable X_t, $\mathbb{X}_t$ its steady state value and x_t the log deviation of X_t from $\mathbb{X}_t$.

[2]The following exercise could be related in the EU case to the analysis of the interaction between Germany and Italy or Spain, for example.

2.1 Households

The preferences of the Home households concerning the consumption of Home- and Foreign-produced goods $C_{h,t}$ and $C_{f,t}$, respectively, is described by the CES composite

$$C_t^H = \left[\gamma_h^{1/a}(C_{h,t})^{(a-1)/a} + (1-\gamma_h^{1/a})(C_{f,t})^{(a-1)/a}\right]^{a/(a-1)} \tag{1}$$

with $a > 1$ denoting the price elasticity of goods demand and $\frac{1}{2} < \gamma < 1$ representing the degree of home bias towards domestic consumption. As discussed e.g. by Walsh (2010, p. 431), households minimize the cost of achieving a given level of C_t under the cost constraint $P_t^h C_{h,t} + P_t^f C_{f,t}$ taking P_t^h and P_t^f as given, by demanding in each period an amount of home- and foreign produced goods determined by

$$C_{h,t}^H = \gamma_h \left(\frac{P_t^h}{P_{c,t}^H}\right)^{-a} C_t^H, \quad \text{and} \quad C_{f,t}^H = (1-\gamma_h)\left(\frac{P_t^f}{P_{c,t}^H}\right)^{-a} C_t^H \tag{2}$$

where $P_{c,t}^H$ is the corresponding aggregate Consumer Price Index in the Home country, i.e.

$$P_{c,t}^H \equiv [\gamma_h (P_t^h)^{1-a} + (1-\gamma_h)(P_t^f)^{1-a}]^{1/(1-a)}. \tag{3}$$

Assuming an analogous behavior of the Foreign households, their optimal consumption of Home- and Foreign-produced goods is given by

$$C_{f,t}^F = \gamma_f \left(\frac{P_t^f}{P_{c,t}^F}\right)^{-a} C_t^F, \quad \text{and} \quad C_{h,t}^F = (1-\gamma_f)\left(\frac{P_t^h}{P_{c,t}^F}\right)^{-a} C_t^F, \tag{4}$$

respectively, with

$$P_{c,t}^F \equiv [\gamma_f (P_t^f)^{1-a} + (1-\gamma_f)(P_t^h)^{1-a}]^{1/(1-a)}. \tag{5}$$

being the aggregate Consumer Price Index in the Foreign country. Concerning the evolution of (the log deviations of) aggregate consumption c_t over time, we assume the following behavioral Euler-type specification

$$c_t^k = \tilde{E}_t[c_{t+1}^k] - \alpha_{yr}(R_{t-1}^k - \pi_{c,t}^k - r^o) + \varepsilon_t^c, \qquad k = \{H, F\} \tag{6}$$

where R_t^k is the nominal interest rate on k-government bonds, r^o the long-term real interest rate, $\pi_{c,t}^k$ is the price inflation (measured in CPI terms), ε_t^c is a stochastic shock (to be defined further below) and $\tilde{E}_t[c_{t+1}^k]$ represents the households' subjective expectation of their future consumption c_{t+1}. For the sake of simplicity, let households expect that consumption of the next period corresponds to a constant fraction α_c of the log deviation of consumption observed in the previous period to the respective level when the economy is at its potential, formally $\tilde{E}_t[c_{t+1}^k] = \alpha_c c_{t-1}^k$.[3] For the evolution of C_t, it holds

$$C_t = \exp(\log(\mathbb{C}_t) + c_t)$$

where $\mathbb{C}_t$ is the steady state consumption level, assumed to be exogenously given.

2.2 *Firms*

Firms production is assumed to be entirely aggregate demand driven, with firms' adjusting their prices according to a standard backward-looking Phillips curve relationship

$$\pi_t^k = \beta_{py} y_t^k + \alpha_\pi \pi_{t-1}^k + \varepsilon_t^{k,p} \tag{7}$$

where β_{py} is the slope of the Phillips curve, $y_t^k = \log(Y_t^k / \mathbb{Y}_t^k)$ the output gap (defined as the log deviation of current output from its steady state level), and $\alpha_\pi \in [0, 1]$ represents the degree of persistence in the inflation dynamics, and ε_t^p is a cost-push shock to be defined further below. We opt for this simple theory of inflation and against the use of a New Keynesian Phillips curve derived from an intertemporal profit maximizing behavior by monopolistic firms, as discussed e.g. Walsh (2010) to keep the key transmission mechanisms as transparent as possible, and also because assuming a firms' behavior based on rational expectations would have been inconsistent with the households' bounded rationality assumption.

2.3 *Monetary Authorities*

Concerning the behavior of the monetary union's central bank (MUCB), the following simple operational monetary instrument rule is assumed[4]

$$i_{\mathrm{T}} = i_o + \phi_i i_{t-1} + (1 - \phi_i)\phi_\pi(\pi_t^{\mathrm{MU}} - \pi^*) + \varepsilon_t^i \tag{8}$$

[3]Given the behavioral approach of the present paper we do not include the expected future consumption as it is done in the standard Euler equations derived from intertemporal utility maximization under rational expectations.

[4]That is, in the words of Svensson (2003, p. 1), a rule which "expresses the central bank's instrument (usually a short interest rate, the *instrument* rate [...]) as an explicit function of information available to the central bank".

where i_o denotes the steady state nominal interest rate, π_t^{MU} is the current average inflation rate in the monetary union (to be defined explicitly below) and π^* the inflation target (which in the following will be assumed to be equal to the steady state inflation rate π_o),ϕ_π the responsiveness of the monetary policy instrument interest rate to deviations of inflation from its target level (with $\phi_\pi > 1$) and ε_t^i is a random shock. Given the implicit focus of this paper on the EMU, this specification seems to be appropriate for this baseline scenario since it describes a *systematic* conduction of monetary policy which literally comprises the mandate of the European Central Bank as determined by Article 105 of the Maastricht Treaty, whereafter "the primary objective of the ESCB [European System of Central Banks] is to maintain price stability."[5]

The MUCB's targeted inflation rate π^{MU}—which in the case of EMU is given by the aggregate Monetary Union Index of Consumer Prices (MUICP)—, is defined here as

$$\pi_t^{MU} = \omega_{\mathrm{H}}\pi_t^H + (1 - \omega_{\mathrm{H}})\pi_t^F \tag{9}$$

where ω_H represents the weighting parameter for the member country H to be discretionarily determined by the MUCB.

2.4 Fiscal Authorities

As in Beetsma and Jensen (2005), national governments are assumed to purchase only goods produced in their own country in order to highlight the stabilizing role of fiscal policy at the national level, and denote by G_t^H and G_t^F the government spendings within the respective Home and Foreign countries. The fiscal policy rule, expressed as the log deviation of government expenditures from its long term counterpart $\mathbb{G}_t$, is assumed to be determined by

$$g_t^k = -\phi_{gy} y_{t-1}^k - \phi_{b,t}^k \left(\frac{B_{t-1}^k}{Y_{t-1}^k} - \psi^k \right) \tag{10}$$

equation (10) consists thus of an anticyclical term meant to summarize the automatic stabilizers in public finances, and a long-run term aimed at the stabilization of the debt-GDP ratio.[6]

---

[5]This quote, however, goes on as following: "Without prejudice of the objective of price stability the ESCB shall support the general economic policies in the Community with a view to contributing to the achievement of the objectives of the Community [. . .]."

[6]Mayer and Stähler (2013, p. 13), using a DSGE framework, analyze also the performance of a balanced budget rule, finding that "due to erratic spending behavior, the balance budget rule tends to destabilize the economy and gives rise to sunspot equilibria. Cyclical fluctuations tend to be

In contrast, government taxes are assumed to be determined by

$$T_t^k = \mathbb{T}^k \exp(y_t^k)^{\tau_y} \tag{11}$$

where $\mathbb{T}_t^k$ is the long-run component of tax revenues.

The country k's government is assumed to finance its expenditures G_t^k and the interest on outstanding debt $R_{t-1}^k B_t^k$ through tax revenues T_t^k as well as through the issuance of new bonds. Under these assumptions, the governments' flow budget constraint (GBC) in real terms is described by

$$G_t^k + (1 + R_{t-1}^k - \pi_t^k) B_{t-1}^k = B_t^k + T_t^k, \tag{12}$$

where government debt is held by foreign as well as domestic households in each country k ($B_{h,t}^k, B_{f,t}^k$).[7]

2.5 *Sovereign Risk*

A range of papers on sovereign risk and contagion in currency unions tend to emphasize the role of government bond yield spreads (or Credit Default Swap (CDS) spreads) as the reference measure of sovereign risk (see e.g. Metiu 2012; Beirne and Fratzscher 2013; De Grauwe and Ji 2013, among others). In order to employ such measure of sovereign risk, we consider two countries, both members of the Monetary Union, where the Home economy should be prone to sovereign risk induced by the agents' perception based on fluctuations in macroeconomic fundamentals, while the Foreign economy enjoys a "safe haven" status. Therefore, government bonds of the Foreign economy should be treated as a safe investment opportunity.

Following Adrian et al. (2010), the perceived sovereign risk at time t is specified as a linear combination of various macroeconomic fundamentals contained in a vector $\mathcal{F}_{t-1}^H \in \Re^{1\times 3}$, i.e.

$$\zeta_t^H = \begin{cases} \boldsymbol{\xi}_{\mathcal{F}} \mathcal{F}_{t-1}^{H\prime} + \varepsilon_t^{\zeta} & \text{if } \geq 0 \\ 0 & \text{otherwise} \end{cases} \tag{13}$$

more pronounced under this regime and cyclical smoothing does not take place. In terms of welfare considerations, this regime also does comparatively poor."

[7]We assume however that due to a home bias argument households hold primarily the bonds of the own government. Accordingly, and in order to clarify the main transmission channels of the model, we assume that only the yield on the respective government bonds enters into the private consumption equation (6).

where $\boldsymbol{\xi}_{\mathcal{F}}$ is a exogenously given coefficient vector which determines the relative importance of a particular variable in $\mathcal{F}_{t-1}^{H}$ in the assessment of ζ_t^H. More specifically, let

$$\mathcal{F}_{t-1}^{H} = \left[\left(\frac{Y_{t-1}^{H}}{\mathbb{Y}^{H}} - 1 \right), \left(\frac{B_{t-1}^{H}}{Y_{t-1}^{H}} - \psi \right), (R_{t-1}^{H} - R_{t-1}^{F}) \right],$$

so that

$$\zeta_t^H = -\xi_y y_{t-1}^H + \xi_b \left(\frac{B_{t-1}^H}{Y_{t-1}^H} - \psi \right) + \xi_r \left(R_{t-1}^H - R_{t-1}^F \right) + \varepsilon_t^{\zeta}. \tag{14}$$

At this point, we assume that the actual values of the respective variables are not known to the financial market participants so that the agents base their perception of sovereign risk on the previous observed magnitudes in period $t-1$. As it can be easily seen, the values of $\boldsymbol{\xi}_{\mathcal{F}} = \{-\xi_y, \xi_b, \xi_r\}$ are central to the assessment of the Home's sovereign risk and allow us to investigate the consequences of a differentiated perception of sovereign risk by the financial markets at the macroeconomic level. According to Eq. (14), the perceived sovereign risk is a negative function of Home economy's output gap in the previous period (which is observable to all agents in the economy), a positive function of the country's indebtedness (relative to GDP) at date $t-1$, and a positive function of the aforementioned bond spreads $(R_{t-1}^H - R_{t-1}^F)$, where ε_t^{ζ} is a stochastic shock.[8] At this point it should be noted that even though the above specification is linear in all its terms, in the analysis of the next section we will let the coefficients vary in order to reflect recent findings e.g. by De Grauwe and Ji (2013) in the context of the euro area crisis.[9]

At this point it should be noted that, as these expressions are formulated, there is no endogenous variation in the link between the sovereign risk and the different fundamentals contained in $\mathcal{F}_{t-1}^{k}$. Obviously, a possible approach to endogenize the relative importance of these variables would be to follow the increasing literature on macroeconomics and behavioral heterogeneous expectations (see e.g. Branch and McGough 2009; Proaño 2013, 2011; De Grauwe 2012; Bask 2012; Lengnick and Wohltmann 2013) and let the perceived sovereign risk be endogenously determined through different behavioral forecasting rules. However, for the sake of simplicity and clarity of exposition these coefficients will not become endogenously determined, but directly varied in an exogenous manner in the simulation analysis below.

[8] An early investigation referring to the driving forces of government bond yield spreads was carried out by Edwards (1984), who figured out that domestic fundamentals, such that public debt, inflation, etc. are important for the determination of the development of government spreads.

[9] Indeed, as previously mentioned, a stylized fact of the recent euro area debt crisis is the non-linear, apparently country-specific and state-dependent link between the sovereign risk premium and the underlying macroeconomic fundamentals of various euro area countries.

Concerning the pricing of bonds, there is no need to impose a no-arbitrage condition between the two financial assets. Instead it is sufficient to specify the corresponding nominal yield to maturity $R_{k,t}^{(m)}$ (with maturity of 1). The gross return on holding bonds should be considered by

$$R_t^k = \begin{cases} i_t + \zeta_t^H & \text{for } k - H \\ i_t & \text{for } k = F \end{cases} \tag{15}$$

where i_t is the short-term interest rate and ζ_t^H the risk premium on the pricing of bonds in the Home economy. For simplicity, we assume that the yield on bonds of the Foreign economy corresponds to the short-term interest rate i_t. Using the specification of the bond returns in Eq. (15), one can easily see that the bond spread component in Eq. (14) corresponds to the previously perceived sovereign risk ζ_{t-1}^H.

2.6 Market Clearing and External Imbalances

The market equilibrium condition for the Home and Foreign country $k = \{H, F\}$ is given by[10]

$$Y_t^k = C_t^k + G_t^k + TB_t^k \tag{16}$$

The respective trade balances for each country are easily obtained by subtracting the amount of domestically produced goods sold abroad and the domestic demand for foreign goods. Thus we get

$$TB_t^H = \underbrace{(1-\gamma_f)\left(\frac{P_t^h}{P_{c,t}^F}\right)^{-a} C_t^F}_{\text{Exports}} - \underbrace{(1-\gamma_h)\left(\frac{P_t^f}{P_{c,t}^H}\right)^{-a} C_t^H}_{\text{Imports}}, \tag{17}$$

and

$$TB_t^F = \underbrace{(1-\gamma_h)\left(\frac{P_t^f}{P_{c,t}^H}\right)^{-a} C_t^H}_{\text{Exports}} - \underbrace{(1-\gamma_f)\left(\frac{P_t^h}{P_{c,t}^F}\right)^{-a} C_t^F}_{\text{Imports}}, \tag{18}$$

[10]Note that we abstract from aggregate investment in the current analysis, as thus from an endogenously determined potential output, given our focus on short-run effects.

3 Model Analysis

Since the present framework deviates in various dimensions from standard DSGE Models, and a proper estimation of the same is beyond the scope of this paper, the choice of the model's parameters is not quite straightforward. However, whenever possible, we take parameters widely accepted in the literature. For instance, we set the long-run trend components of consumption, government expenditures and lump-sum taxes equal to $\mathbb{C} = 80$, $\mathbb{G} = 20$ and $\mathbb{T} = 20$, respectively, which in turn leads to a steady state output level of $\mathbb{Y} = 100$, and thus to a long-run private consumption to GDP ratio of 0.8, a government expenditures to GDP ratio of 0.20 (and thus a ratio of private to government consumption of four), and a balanced government budget in the long-run, as it is standard in the literature, see e.g. Beetsma and Jensen (2005). The degree of home bias is set equal to 0.8, what implies a share of imports to GDP of 0.15, under a long-run zero trade balance. Further, the cyclical elasticity of government expenditures and tax revenues is set equal to $\phi_y = 0.20$ and $t_y = 0.12$, respectively, following Mayer and Stähler (2013), who set a total cyclical elasticity of the structural budget deficit equal to 0.32. For simplicity, we set the elasticity of the government expenditures to the debt-to-GDP as $\phi_b = 1$. This configuration follows the rationale, that the government cuts its expenditures by the same amount of the increase in sovereign debt. The real interest rate elasticity of consumption demand (more precisely, the log-linear approximation around its steady state level) α_{yr} have been estimated several times by a number of authors. Although these estimates refer to the New Keynesian baseline model including rational expectations, we use the value $\alpha_{yr} = 0.164$ obtained by McCallum and Nelson (1999).

Based on the empirical estimates of Goodhart and Hofmann (2005) for the euro area, we set the past inflation rate coefficient to 0.75 where the slope of the Phillips curve is set equal to 0.15. Concerning the MUCB monetary policy rule, as in Proaño (2012) we assume that $\phi_\pi = 2.0$ and $\phi_y = 0.0$. Further, the weighting parameter for the member country H in the interest rate rule is set $\omega_H = 0.5$, assuming an equal weight of both countries in the MUCB's loss function. With respect to the reaction parameters in the market perceptions of Home's sovereign risk, given the lack of empirical estimates, we set it arbitrarily to $\xi_y = 0.2$ and $\xi_r = 0.2$. The value of the investors sensitivity to fluctuations in the debt-GDP ratio corresponds to $\xi_b = 0.0744$ for Home and is taken from the estimates provided by De Grauwe and Ji (2013, Table 1). Table 1 summarizes all these parameter values.

Finally, concerning the stochastic shocks to the system, we assume in a standard manner that all of them follow an AR(1) process and that the corresponding autoregressive coefficients and standard deviations of the innovations are given by the values summarized in Table 2.

Table 1 Parameter values

Parameter	Symbol	Value
Long-run component of consumption	$\mathbb{C}$	80
Elasticity of substitution in consumption	a	1.1
Degree of home bias in consumption	γ_h	0.8
Interest rate elasticity of consumption	α_{yr}	0.164
Elasticity of past consumption on consumption	α_c	0.7
Long-run component of government expenditures	$\mathbb{G}$	20
Output gap elasticity of cyclical government expenditures	ϕ_{gy}	0.2
Debt elasticity of cyclical government expenditures	ϕ_b	0.5
Target Debt-to-GDP ratio	ψ	0.6
Lump-sum Taxes	$\mathbb{T}$	20
Output gap elasticity of cyclical tax revenues	τ_y	0.12
Output gap coefficient in Phillips Curve	β_{py}	0.15
Lagged inflation coefficient in Phillips Curve	α_π	0.0
Inflation gap coefficient in interest rule	ϕ_π	2.0
Output gap coefficient in interest rule	ϕ_y	0.0
Home's weight in the interest rate rule	ω_H	0.5
Country k's sovereign debt coefficient in risk	ξ_b	0.0744
Output gap coefficient in the risk expectation equation	ξ_y	0.2
Return differential coefficient in risk expectations	ξ_r	0.2

Table 2 Standard deviations of stochastic shocks

Variable	Autoregressive term	Std. Dev.
Home and Foreign consumption	0.7	0.20
Home's perception of sovereign risk	0.7	0.24
Home and Foreign price mark-up	0.7	0.14
Monetary policy	0.7	0.24

3.1 Transmission Mechanisms and Dynamic Adjustments

Before we turn to our main analysis where we focus on the macroeconomic consequences of alternative fiscal policy rules aimed towards either debt or aggregate demand stabilization, it is worthwhile to discuss the model's dynamic adjustments to a one-time exogenous increase in the perceived sovereign risk of the Home country's government.[11]

As Fig. 1 clearly illustrates, an increase in the market's perception of Home's sovereign risk reduces economic activity through one main channel, namely through the immediate rise in the nominal yield on bonds and thus in the respective bond

[11]All numerical simulations in this paper were done using Dynare 4.3.3, see Adjemian et al. (2011). The codes are available from the author upon request.

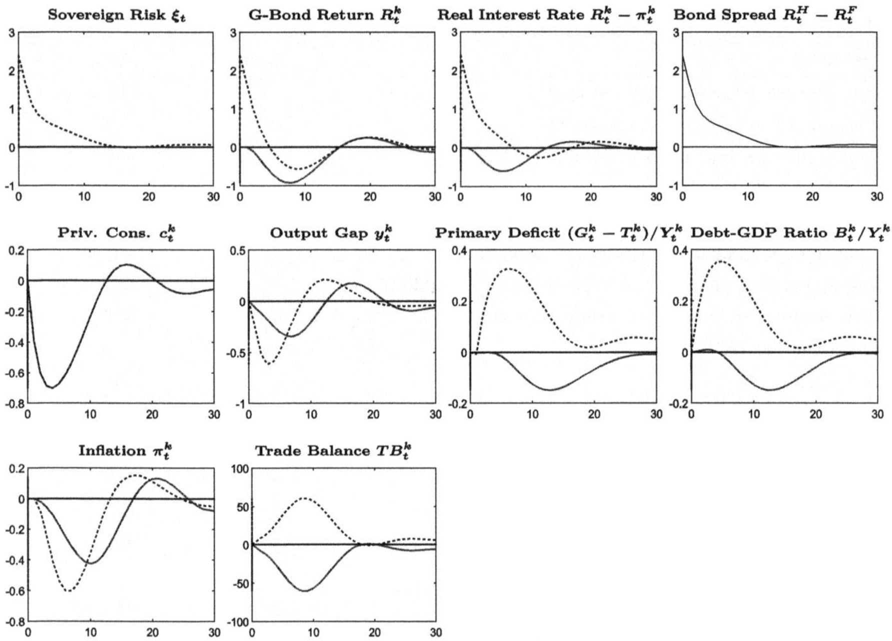

Fig. 1 Dynamic responses in Home (*dashed line*) and Foreign (*dotted line*) to a one-time one std. dev. increase in the market's expectations of Home's sovereign risk

spreads. According to the fiscal policy rules, an increase in sovereign risk which is associated with a rise in return on bonds, leads to an increase in Home's sovereign debt and the subsequent fiscal consolidation efforts, see Eq. (10). Given the international trade interactions between Home and Foreign, an increase in ζ_t^H has a negative impact on economic activity not only of Home, but also of Foreign due to the reduction in Home's aggregate demand for Foreign's consumption goods, what leads to a temporary trade account imbalance between the two countries.[12] The decrease in economic activity in the Home country leads to a downward pressure on inflation and, by extension, an improvement in the relative competitiveness of the Home economy. This—together with the decrease in aggregate consumption and imports—leads to an expansion of Home's trade balance and—by definition—a deterioration of Foreign's trade balance. The boom in net exports leads in turn to an increase in economic activity in Home, which allows aggregate consumption to recover.

Due to the anticyclical terms in the government expenditures and tax collection rules, the downturn in Home's economic activity leads to an increase in the government deficit in the Home country. This effect is magnified by the higher

[12]As previously mentioned, while Home could be related to Spain or Italy in the current euro area crisis, Foreign would represent Germany.

Home government bond yields, which increase the financing costs of the fiscal deficits. Finally, it is worthwhile to highlight the fact that this asymmetric shock in the perceived sovereign risk of Home's government bonds leads to an increase in the Home's government bond spreads w.r.t. the monetary policy interest rate, and to the opposite development in the Foreign economy, which profits from its "safe haven" status.

3.2 Debt Stabilization Policy and Its Effect on the Macroeconomic Stability

In this subsection, we assess the macroeconomic consequences in our two-country model for the case in which a policy of procyclical austerity is implemented in the Home country. For this purpose, we compute the cumulative impulse response functions (IRF's) for increasing values of the sensitivity parameter linking the debt-to-GDP ratio to the fiscal policy rule in a range of $\phi_b \in [0.5, 1]$. For the computation of the adjustment processes we use the values listed in Table 1. In the following we associate the occurrence of large and persistent deviations of the key variables from their corresponding steady state levels with an excess volatility which is likely to infer macroeconomic and, from a wider perspective, welfare costs.

The cumulative IRF's are depicted in Fig. 2. For illustrative purposes, we select the cumulative IRF's of the Home country's output gap (left panel), the agents' perception of sovereign risk (middle panel) and the debt-to-GDP ratio (right panel) over a time span of $t = 30$ quarters. As it can be clearly observed, the cumulated output gaps of the Home country are a positive function of the debt stabilization coefficient ϕ_b^H. It follows that if the government puts too much emphasis on sovereign debt stabilization and enforces it by means of expenditure cuts, both the duration and dimension of the deviation of output from its equilibrium level increase significantly. The economic rationale for this outcome is relatively

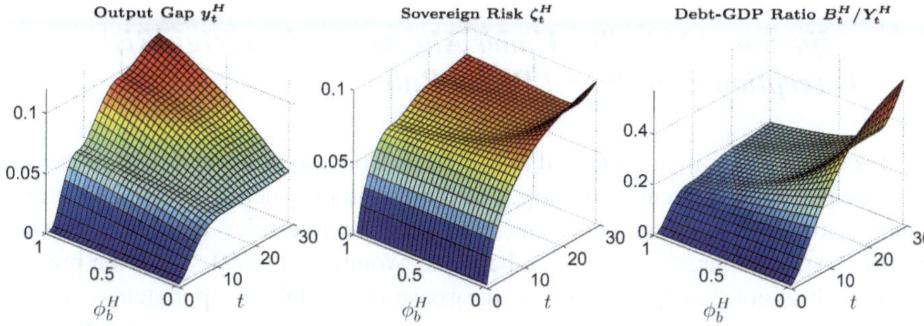

Fig. 2 Cumulative dynamic adjustments in the Home country to a one-time one std. dev. increase in the Home's sovereign risk perception for different values of ϕ_b

straightforward: An initial shock to the agents' perception of sovereign risk immediately raises Home's government yield spreads as it can be clearly observed by Eq. (15). Consequently, aggregate consumption and thus total demand and inflation decrease whereas the associated higher fiscal debt burden cuts the Home country's government expenditures. Its extent critically depends upon the debt stabilization coefficient ϕ_b^H. If the Home's government turns its primary objective on stabilization of the debt-to-GDP ratio (reflected by *ceteris paribus* larger values of ϕ_b^H), the role of output stabilization in the fiscal policy rule becomes increasingly unimportant. Hence, the associated spending cuts succeed more abruptly, what leads to a higher output volatility.

It follows that the increased risk premium—which in turn raises the interest payments on government bonds—makes a stabilization of real economic activity difficult to achieve. The greater the spending cuts in the Home economy in response to exogenous perturbations, the larger is the fall in aggregate demand for Homes's private consumption goods within the economy and abroad. It follows a decline in Foreign's exports as well as imports (primarily due to the change in the relative prices of Home's and Foreign's consumption goods) which in turn eases the output gap adjustment volatility of the Foreign country. A second channel through which the Foreign economy is affected concerns to the interest rate channel. As the MUCB reacts to the price inflation developments on both countries to the same extent and the adjustments of inflation are driven by the fluctuations of the output gap, increased volatility in Home leads to higher short-term interest rate volatility, which in turn affects Foreign's economy through its effect on private consumption and on the pricing of Foreign's government debt.

It is interesting to note that a stronger orientation of fiscal policy towards debt stabilization may reduce sovereign risk only at low ranges of ϕ_b^H, as the middle graph in Fig. 2 clearly shows. Accordingly, all-too excessive austerity endeavours seem to hit a lower bound in the sovereign risk assessment, leading thus only to excessive output fluctuations. This point will be further investigated in the next section.

3.3 Adjustment Volatility Under Alternative Sovereign Risk Perceptions and Fiscal Policy Rules

As it is well-known, in the standard DSGE framework the performance of monetary and fiscal policy is analyzed using a welfare criterion derived from the utility function of the representative agent(s) and the flexible-price equilibrium under rational expectations, see e.g. Rotemberg and Woodford (1997). Since the present framework is not "microfounded" in the sense of the DSGE approach, and does not rely on the rational expectations assumption, but on the notion of behavioral heuristics concerning the agents' perception of sovereign risk eventually determined by specific fundamentals only, such an evaluation strategy is not applicable here.

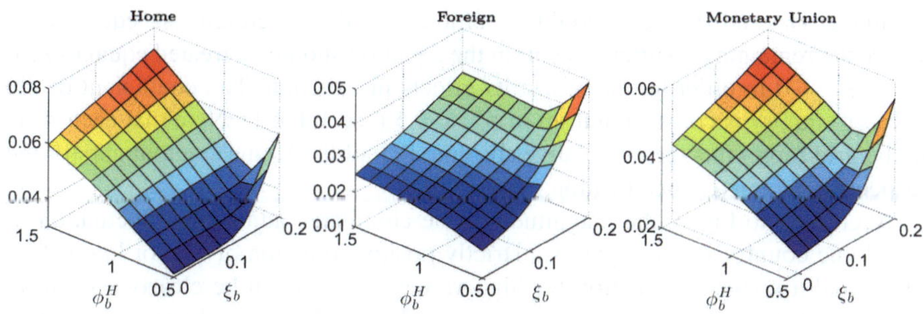

Fig. 3 Scenario 1: Adjustment volatility of Home's, Foreign's and the union wide output gap for varying parameter values of ϕ_b^H and ξ_b after a one-time shock in the agents' perception of sovereign risk

In the latter subsection we illustrated how the cumulated dynamic adjustments of some key variables after a one-time exogenous shock in the perception of sovereign risk are the more long-lasting the more the fiscal authority focuses on debt stabilization. In the following, we compute the same cumulated dynamic adjustments, now however under different fiscal policy rules and perceived sovereign risk specifications. Doing so, we distinguish between four different scenarios. The first scenario considers the case where policy makers as well as market participants both consider sovereign debt as the main determinant of sovereign risk. The second scenario discusses the effect on the cumulated macroeconomic adjustment when the government focuses on fiscal consolidation while the markets consider the output gap as the main variable driving the sovereign risk. In the third scenario, our aim is to analyze the cumulated macroeconomic adjustment in a consensus case where both, fiscal authority and the markets, consider aggregate demand stabilization as the main driver of a sovereign risk reduction. The final scenario emphasis the interaction between a Home government that aims to stabilize the economy by strict debt stabilization programs where the markets focus on the development of the bond yield spreads instead. For this purpose, the cumulative absolute IRF's are employed as evaluation measure for the cumulated macroeconomic adjustment.[13]

3.3.1 Scenario 1: Consensus Between Government and Markets About Sovereign Debt as the Main Determinant of Sovereign Risk

Figure 3 illustrates the cumulative absolute IRF's of the output gap of Home, Foreign and the Monetary Union as a whole for variations of the debt stabilization

[13]Note that Proaño (2013) analyzed the impact of different fiscal policy rules under various sovereign risk perception configurations on aggregate volatility. He computed a loss function over several simulation runs which he employed as evaluation measure of aggregate volatility. He also highlighted the role of the trade balance as determinant of sovereign risk perceptions and its effect on the fluctuations of real economic activity in a monetary union.

parameter ϕ_b^H and the debt-to-GDP coefficient in the sovereign risk equation ξ_b, where the remaining coefficients within the equation should be treated equal to zero $\xi_y = \xi_r = 0$ in order to emphasize its effects in isolation. The coefficient of the Foreign country is constant and set to $\phi_b^F = 0.5$ (see Table 1). Note that each node in the grid represents the cumulative absolute IRF of the Home's output gap for the respective parameter specification.

According to Fig. 3, the magnitude of the cumulated adjustment fluctuation of the Home country's output gap is a strictly positive function of ϕ_b^H for low values of ξ_b, and a nonlinear function for high values of ξ_b as can be clearly seen from the slope of the surface w.r.t. both parameters. Figure 3 delivers thus two important insights. First, when financial markets do not care particularly for debt stabilization, a pronounced austerity policy is harmful for the economy, as it produces excessive fluctuations in the economic activity in the Home and Foreign economies, as thus at the monetary union level. And second, when financial markets do care particularly about debt stabilization—a situation represented here by higher values of ξ_b—, an accordant fiscal policy is beneficial only for small values of ϕ_b^H, but not for larger values, when the counterproductive effects overweight the positive effects of a reduction in the sovereign risk perception due to a fiscal policy oriented towards debt stabilization. As in the previous figure, we find here that an all-to-strong orientation of fiscal policy towards debt stabilization is harmful for the economy, as recent empirical findings e.g. by Blanchard and Leigh (2013) have shown.

3.3.2 Scenario 2: Disagreement About the Main Determinant of Sovereign Risk (Government: Sovereign Debt, Markets: Output Gap)

The second scenario to be considered refers to the case where Home's perceived sovereign risk is driven by the output gap ξ_y. Strictly speaking, the scenario describes what happens if the perceived sovereign risk is indeed driven by the fluctuations in GDP but fiscal policy merely focuses on sovereign debt stabilization. In order to provide an answer to this question, we choose the range $\xi_y \in [0, 0.2]$ and $\phi_b^H \in [0.5, 1.5]$ for the respective coefficients. The counterpart abroad should be held constant at its default value of $\phi_b^F = 0.5$ depicted in Table 1.

The absolute cumulative IRF'S after a one-time one-standard deviation increase in the sovereign risk perception are represented in Fig. 4. The results are similar to the results obtained in the previous scenario, though of a larger dimension. The intuition behind that mechanism is relatively straightforward: The initial shock affects Home's market participants perception of risk in the same manner as mentioned above. The immediate increase in the risk premium and thus the nominal yield on the government bonds of the Home country lead to a massive cut in Home's government expenditures. It follows a stark decline in aggregate demand. This time, the financial markets react with a stronger response regarding the updating of sovereign risk associated with a sharp increase of the risk premium. In this case, where the fiscal authority focuses on debt stabilization whereas the financial

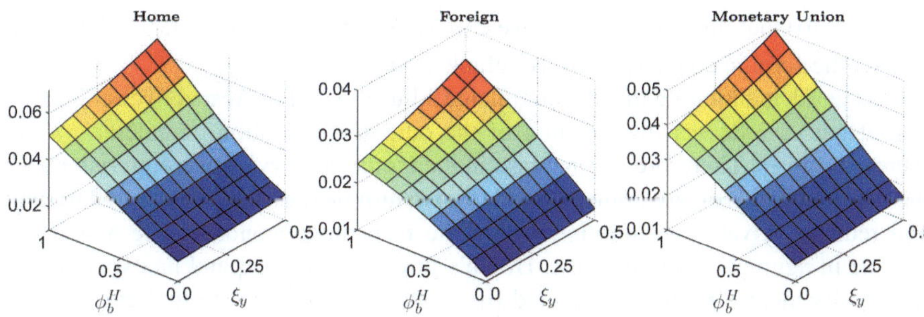

Fig. 4 Scenario 2: Adjustment volatility of Home's, Foreign's and the union wide output gap for varying parameter values of ϕ_b^H and ξ_y after a one-time shock in the agents' perception of sovereign risk

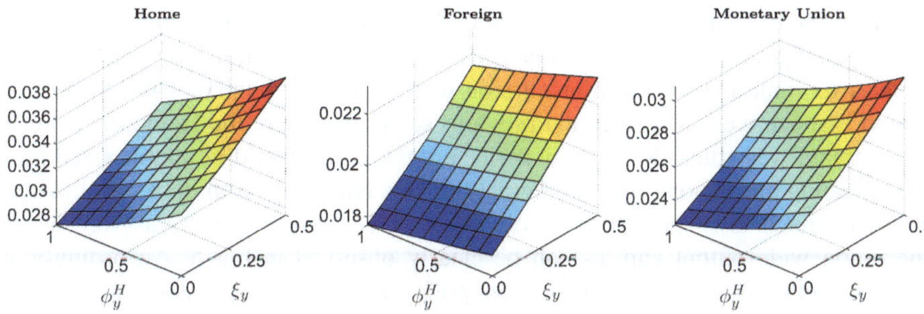

Fig. 5 Scenario 3: Adjustment volatility of Home's, Foreign's and the union wide output gap for varying parameter values of ϕ_y^H and ξ_y after a one-time shock in the agents' perception of sovereign risk

markets assess the sovereign risk as being mainly driven by fluctuations of the Home country's output gap, the result is a more volatile scenario as in the previous consensus case where both put their emphasis on sovereign debt. The decline in Home's aggregate demand for consumption goods shortens not just aggregate output, it also decreases the Home's imports and therefore Foreign's exports. The decline of Home's prices for consumption goods amplifies this process. As the figure illustrates, this configuration increases not just the adjustment volatility in Home, but also in Foreign and in the whole Monetary Union.

3.3.3 Scenario 3: Consensus Between Government and Markets About the Output Gap as the Main Determinant of Sovereign Risk

Figure 5 illustrates the cumulated adjustment of real economic activity in the case that Home's government implements a pronounced countercyclical fiscal policy, and the markets assess sovereign risk as mainly driven by output gap fluctuations. In order to carry out this investigation, we choose ϕ_y^H and ξ_y as parameters to be varied and set the foreign counterpart to its default value $\phi_y^F = 0.2$.

It is worth mentioning that the cumulated adjustment of the respective economy's output gap is a decreasing function of the output-stabilization parameter ϕ_y^H but increasing in ξ_y. Therefore, an increase in the Home country's sovereign risk raises the nominal yields on government bonds that in turn increases the debt-to-GDP ratio and reduces private consumption within the respective economy. The associated decline in aggregate output causes Home's government to expand its expenditures (due to its increasing focus reflected by variations of ϕ_y^H). Note that even in this scenario, we did not drop the assumption of a debt stabilization mechanism ($\phi_b^H = 0.5$) which counteracts the positive stabilizing effects of countercyclical fiscal policy (with a lag of unity). If we had done this, the adjustment volatilities would be even lower than in the current case illustrated in Fig. 5. Since the countercyclical fiscal policy of Home's government neutralizes the destabilizing forces, resulting from spending cuts due to the increase in sovereign risk and thus government debt, at least to some extent, the transmission to the foreign country of the Monetary Union is much weaker than in the previous scenarios where the Home economy's government was stuck down to austerity programs. It should be emphasized that the spending expansions that counteract the cuts emerging from debt stabilization result in much lower adjustment volatilities of Home's, Foreign's as well as of the union wide output gap, as can be clearly observed in Fig. 5. Accordingly, a more pronounced countercyclical fiscal policy is more advantageous in terms of macroeconomic stabilization.

3.3.4 Scenario 4: Disagreement About the Main Determinant of Sovereign Risk (Government: Sovereign Debt, Markets: Country Bond Spread)

In the final scenario we stress the interaction of increasing importance of debt stabilization policy while the markets increasingly focus on the development of the bond spreads as an early indicator for sovereign risk and its contemporaneous effects on output volatility. In this case, the market participants follow extrapolative rules. This case is perhaps the most important since the fluctuations of the returns on bonds, and thus the bond yield spreads, are widely used as representative for sovereign risk in the strand of the literature focusing on the EMU sovereign debt crisis and contagion dynamics within the EMU (see e.g. Metiu 2012; Baldacci and Kumar 2010, among others). For the computation of the absolute cumulative IRF's, we widely used the parameter values depicted in Table 1.

Figure 6 illustrates the consequences for the cumulated macroeconomic adjustment of an increasing importance of debt stabilization for the government of the Home economy (as already carried out in scenario 1 and 2) when the market participants increasingly spend their emphasis on the nominal bond yields. As the figure clearly suggests, increasing values of ϕ_b^H and ξ_r raise Home's, Foreign's as

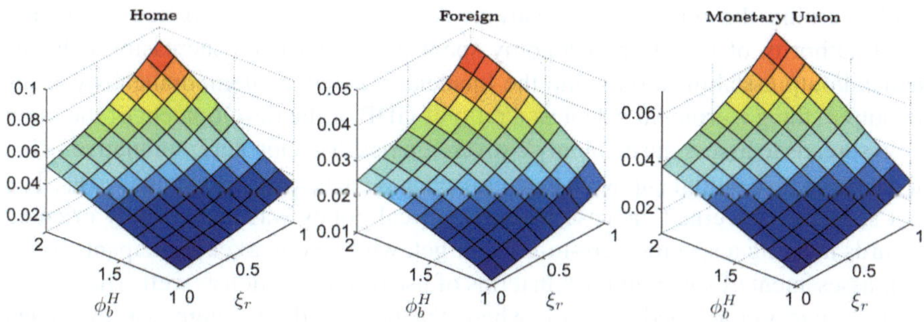

Fig. 6 Scenario 4: Adjustment volatility of Home's, Foreign's and the union wide output gap for varying parameter values of ϕ_b^H and ξ_r after a one-time shock in the agents' perception of sovereign risk

well as the union wide adjustment volatilities after an initial shock in the agents' perception of sovereign risk in Home, which can be observed by the increasing slope of the surfaces in both directions. This time, the magnitude of the respective adjustment paths proceeds more volatile than observed in the previous cases. This finding is not necessarily surprising, as the sovereign risk is assumed to be decisively responsible for the risk premium on bonds, see Eq. (14), and thus for its nominal yields, Eq. (15). The Home government's debt service increases its fiscal solvency concerns and feed immediately back into real economic activity by increasing spending cuts and through a decline in private consumption. The rise in the bond spreads triggers the markets to reassess their subjective valuation of sovereign risk upwards which in turn reinforces the whole process and thus the adjustment volatility of the output gap. The spread to the Foreign economy takes place in the same manner as discussed before.

4 Concluding Remarks

The main motivation of this paper was to assess the interaction between fiscal policy and the financial markets in a monetary union under occasional discrepant perceptions concerning the main determinants of sovereign risk. We aimed to address the following question: What if governments pursue other goals that what financial market participants consider as relevant for the assessment of sovereign risk—and the subsequent pricing of sovereign bonds—and what would be the macroeconomic consequences of such discrepancy?

In order to shed some light on these issues, a theoretical model of a two-country monetary union was developed which featured a variety of innovative modeling aspects existent in the literature so far. A behavioral specification of the risk premium on the prices of government bonds has been used. In this context, the risk premium on government bond returns was determined by the agents' sovereign risk perceptions.

Concerning the methodology, various scenarios were investigated where the fiscal authority of the Home economy spent its emphasis on increasing debt- or demand/output-stabilization while the financial markets contemporaneously focus on output, the government's debt as a ratio of GDP or the development of the bond yield spreads to assess the economy's sovereign risk. The absolute cumulative IRF's of the respective output gap were then utilized as a measure of adjustment costs. The first scenario concerned the case where the fiscal policy was increasingly oriented towards meeting a specific debt-to-GDP target, while the market participants based their assessment of sovereign risk in terms of fiscal deterioration as well. The second main scenario concerned the case where the perceived sovereign risk depended exclusively on the current state of the business cycle. Up to this point, we found that increasing efforts of Home's government towards debt stabilization increased the volatility of the dynamic adjustments of the output gap significantly for the Home-, the Foreign-country as well as for the Monetary Union as a whole. The third scenario, the case where the Home government faced countercyclical fiscal policy represented the most stable regime. From that approach we inferred that the adjustment volatilities were decreasing when the fiscal policy put increasing attention towards output stabilization. This finding could be observed at the union-wide level. The final case, where Home's government was stuck down to austerity programs and the financial markets utilized the government bond yield spreads as an indicator for the overall sovereign risk, provided us with the most volatile dynamic responses.

The numerical computations of these scenarios highlighted in a clear manner the pitfalls of the conduction of economic policy in the real world, where it cannot be taken for granted that markets and governments may share the same goals, targets and expectations, and where a learning mechanism along the lines of Evans and Honkapohja (2001) may not be feasible due to various reasons. Furthermore, in the context of the current euro area crisis, this paper highlighted the dangers of a too restrictive fiscal policy aimed at the stabilization of sovereign debt. Indeed, as acknowledged even by IMF staff (Anderson et al. 2013), a too restrictive fiscal consolidation is quite likely to affect a country's macroeconomic activity, especially if the markets do not share the same views or targets as the governments following such a fiscal austerity path. At this point it should however be pointed out that these results concern the deviations of aggregates from their (exogenously given) steady state level. An incorporation of aggregate investment, and thus of capital accumulation and by extension of an endogenously determined potential output level in the present framework seems to be an interesting venue to pursue to better understand the long-lasting effects of fiscal austerity.

On more general grounds, this paper highlights the importance of the analysis of situations which may not be accurately represented by a rational expectations model, where agents share the same information sets and have consistent beliefs with respect to the future evolution of the economy. Indeed, as the wide empirical evidence on behavioral finance as well as recent studies on euro area sovereign spreads suggest, the pricing of sovereign debt seems to be much more complex than what the rational expectations framework may allow for.

Acknowledgements We gratefully acknowledge useful comments and suggestions by Willi Semmler, Lopamundra Banerjee, Lena Dräger, Philipp Engler, Sabine Stephan, Daniele Tavani, Ramaa Vasudevan and seminar participants at Colorado State University, the 2013 Eastern Economic Association conference, the 2012 FMM conference, the 2012 CFE conference, and the 2015 EEA conference, as well as Mary Borrowman for excellent research assistance. This is a significantly revised version of Proaño (2013). Financial support by the Macroeconomic Policy Institute (IMK) in the Hans-Böckler Foundation is gratefully acknowledged.

References

Adjemian, S., Bastani, H., Juillard, M., Mihoubi, F., Perendia, G., Ratto, M., et al. (2011). Dynare: Reference manual, version 4, Dynare Working Papers 1, CEPREMAP.

Adrian, T., Moench, E., & Shin, H. S. (2010). Macro risk premium and intermediary balance sheet quantities. *IMF Economic Review*, **58**, 179–207.

Anderson, D., Hunt, B., & Snudden, S. (2013). Fiscal consolidation in the euro area: How much pain can structural reforms ease? Technical report, International Monetary Fund.

Arghyrou, M. G., & Kontonikas, A. (2011). The EMU sovereign-debt crisis: Fundamentals, expectations and contagion, Economic Papers 436. European Commission, DG Economic and Financial Affairs.

Attinasi, M.-G., Checherita, C., & Nickel, C. (2009). What explains the surge in euro area sovereign spreads during the financial crisis of 2007–2009? Working Paper 1131, European Central Bank.

Baldacci, E., & Kumar, M. S. (2010). Fiscal deficits, public debt and sovereign bond yields. IMF Working Paper 10/184, IMF.

Barrios, S., Iversen, P., Lewandowska, M., & Setzer, R. (2009). Determinants of intra-euro area government bond spreads during the financial crisis. Economic Papers 388. European Commission, DG Economic and Financial Affairs.

Bask, M. (2012). Asset price misalignments and monetary policy. *International Journal of Finance & Economics*, **17**(3), 221–241.

Beetsma, R. M., & Jensen, H. (2005). Monetary and fiscal policy interactions in a microfounded model of a monetary union. *Journal of International Economics*, **67**, 320–352.

Beirne, J., & Fratzscher, M. (2013). The pricing of sovereign risk and contagion during the European sovereign debt crisis. *Journal of International Money and Finance*, **34**(C), 60–82.

Blanchard, O., & Leigh, D. (2013). Growth forecast errors and fiscal multipliers. Working Paper WP/13/1, International Monetary Fund.

Borgy, V., Laubach, T., Mésonnier, J.-S., & Renne, J.-P. (2012). Fiscal sustainability, default risk and euro area sovereign bond spreads. Document de Travail 350, Banque de France. First version Oct. 2011.

Branch, W. A., & McGough, B. (2009). A New Keynesian model with heterogeneous expectations. *Journal of Economic Dynamics and Control*, **33**(5), 1036–1051.

Codogno, L., Favero, C., & Missale, A. (2003). Yield spreads on EMU government bonds. *Economic Policy*, **18**(37), 503–532.

Corsetti, G., Kuester, K., Meier, A., & Mueller, G. K. (2012). Sovereign risk, fiscal policy and macroeconomic stability. Working Paper 12/33, International Monetary Fund.

De Grauwe, P. (2012). Booms and busts in economic activity: A behavioral explanation. *Journal of Economic Behavior and Organization*, **83**(3), 484–501.

De Grauwe, P., & Ji, J. (2013). Self-fulfilling crises in the Eurozone: An empirical test. *Journal of International Money and Finance*, **34**, 15–36.

De Grauwe, P., & Ji, Y. (2012). Mispricing of sovereign risk and macroeconomic stability in the Eurozone. *Journal of Common Market Studies*, **50**(6), 866–880.

Edwards, S. (1984). LDC foreign borrowing and default risk: An empirical investigation, 1976–1980. *American Economic Review*, **74**(4), 726–734.

Evans, G. W., & Honkapohja, S. (2001). *Learning and expectations in macroeconomics*. Princeton, NJ: Princeton University Press.

Farmer, R. E. A. (2010). How to reduce unemployment: A new policy proposal. *Journal of Monetary Economics*, **57**(5), 557–572; Carnegie Rochester Conference Series on Public Policy.

Geyer, A., Kossmeier, S., & Pichler, S. (2004). Measuring systemic risk in EMU government yield spreads. *Review of Finance*, **8**(2), 171–197.

Goodhart, C., & Hofmann, B. (2005). The Phillips Curve, the IS curve and monetary transmission: Evidence for the US and the Euro area. *CESifo Economic Studies*, **51**(4), 757–775.

Haugh, D., Ollivaud, P., & Turner, D. (2009). What drives sovereign risk premiums? An analysis of recent evidence from the Euro area. Working Paper 718, OECD Economics Department.

Lengnick, M., & Wohltmann, H.-W. (2013). Agent-based financial markets and new keynesian macroeconomics: A synthesis. *Journal of Economic Interaction and Coordination*, **8**, 1–32.

Manganelli, S., & Wolswijk, G. (2009). What drives spreads in the euro area government bond market? *Economic Policy*, **24**(58), 191–240.

Mayer, E., & Stähler, N. (2013). The debt brake: Business cycle and welfare consequences of Germany's new fiscal policy rule. *Empirica*, **40**, 39–74.

McCallum, B. T., & Nelson, E. (1999). Performance of operational policy rules in an estimated semiclassical structural model. In *Monetary policy rules* NBER Chapters, National Bureau of Economic Research, Inc, pp. 15–56.

Metiu, N. (2012). Sovereign risk contagion in the Eurozone. *Economics Letters*, **117**(1), 35–38.

Proaño, C. R. (2011). Exchange rate determination, macroeconomic dynamics and stability under heterogeneous behavioral FX expectations. *Journal of Economic Behavior and Organization*, **77**(2), 177–188.

Proaño, C. R. (2012). Should monetary policy take account of national labor market asymmetries in a currency union? *Economics Bulletin*, **32**(3), 1878–1889.

Proaño, C. R. (2013). Monetary policy rules and macroeconomic stabilization in small open economies under behavioral FX trading: Insights from numerical simulations. *Manchester School*, **81**(6), 992–1011.

Proaño, C. R. (2013), Macroeconomic risk, fiscal policy rules and aggregate volatility in asymmetric currency unions: A behavioral perspective. IMK Working Paper 129–2013, IMK at the Hans Boeckler Foundation, Macroeconomic Policy Institute.

Rotemberg, J. J., & Woodford, M. (1997). An optimization-based econometric framework for the evaluation of monetary policy. In B. S. Bernanke & J. J. Rotemberg (Eds.), *NBER macroeconomics annual 1997* (pp. 297–346). Cambridge, MA: MIT Press.

Schuknecht, L., von Hagen, J., & Wolswijk, G. (2009). Government risk premiums in the bond market: EMU and Canada. *European Journal of Political Economy*, **25**(3), 371–384.

Schuknecht, L., von Hagen, J., & Wolswijk, G. (2010). Government bond risk premiums in the EU revisited. The impact of the financial crisis. Working Paper 1152, European Central Bank.

Svensson, L. E. (2003). What is wrong with Taylor rules? Using judgement in monetary policy through targeting rules. *Journal of Economic Literature*, **41**, 426–477.

Walsh, C. E. (2010). *Monetary theory and policy*, 3rd edn. Cambridge, MA: MIT Press.

Asset Accumulation with Heterogeneous Households: The Rise of Wealth Disparity

Willi Semmler and Damien Parker

Abstract We present a formal model with heterogeneous households examining the dynamics of wealth disparity. We concentrate on financial wealth and study asset accumulation in terms of net wealth. Households borrow to finance investment and consumption. If the asset value is larger then the liability, then new net wealth is accumulated. The question then becomes what variables drive differences in net asset accumulation among households. The returns on assets, saving rates and borrowing capacity are major driving forces behind the differences in asset accumulation among households. The empirical part utilizes US Survey of Consumer Finance (SCF) data and supports the theoretical model. Specifically, the paper finds substantial evidence suggesting that when income groups are subdivided into those that dominantly borrow for consumption and those that dominantly borrow for investment (functioning as consumption smoothers), the former group suffers losses in net wealth while the latter maintains a steady increase in net worth.

1 Introduction

This paper addresses if and why there are shifting wealth shares over time in the US economy. A stochastic dynamic model of (net) wealth accumulation with heterogeneous households is used to examine this possibility.[1] We then study the prediction of the model by using US Survey of Consumer Finance (SCF) data.

[1]Our model resembles the one by Brunnermeier and Sannikov (2014) as well as Stein (2012).

W. Semmler (✉)
Department of Economics, New School for Social Research, New York, NY, USA

University Bielefeld, Bielefeld, Germany
e-mail: semmlerw@newschool.edu

D. Parker
Department of Economics, New School for Social Research, New York, NY, USA
e-mail: parkd855@newschool.edu

B. Bökemeier, A. Greiner (eds.), *Inequality and Finance in Macrodynamics*, Dynamic Modeling and Econometrics in Economics and Finance 23,
DOI 10.1007/978-3-319-54690-2_11

The model as well as the empirics allow for heterogeneous households that borrow for different primary purposes. One type of household dominantly borrows for investment into assets, while the other type of household dominantly borrows for consumption expenditure. Empirically, our time period of reference are the years since the end of the 1980s and early 1990s. Formal presentation of the modeling of the dynamics of net wealth or net worth helps us to understand the empirical trends. We employ the US Survey of Consumer Finance (SCF) data allowing us to study the issue empirically.

Our theoretical model is closely related to Brunnermeier and Sannikov (2014), Stein (2012) and Schleer et al. (2014). However, previous work concentrates either on gross wealth or gross debt when computing wealth or debt dynamics. The model used here utilizes net wealth (or net worth), as the difference between assets and liabilities. This model can be viewed as a more simplified version of the more complex portfolio model presented in Chappe and Semmler (2016). Furthermore, our model stresses aspects not developed in the latter paper and provides empirical validation of the previous work's hypotheses.

As known in the literature, asset price returns and the dynamics of capital gains greatly affect net worth.[2] Moreover, not only asset prices and returns, but also credit expansions affect net worth.[3] For example, the large asset price meltdown in the years 2007–2009 (in particular the meltdown of housing prices) led to a declining share of the 80 or 90% wealth owners, see Wolff (2012). Therefore, both the increasing mortgage and consumption borrowing and boom (before the 2007–2009 crisis) as well as the asset price meltdown (with locked in mortgages) produced adverse effects on the distribution of net wealth among households.

If these results hold over a long time horizon, then the well-known Kaldor (1956) and Pasinetti (1962) debate must be re-examined in a new light. In Pasinetti's response to Kaldor, he assumes, as Kaldor does, that savings from wages are lower than that from capital income. However, Pasinetti also assumes that the asset returns from labor income are the same as from capital income. Although saving rates can indeed be different while the returns on assets are the same, this occurrence keeps the wealth shares constant. If this does not hold empirically, and returns are higher for one group as compared to the other and leveraging is allowed, the wealth shares are likely to shift over time.

This paper is organized into the following sections. Section 2 introduces the primary model with heterogeneous households, borrowing and lending and (net) asset accumulation. In this model, we allow for both high and low net worth households. Section 3, utilizes the Survey of Consumer Finance for U.S. household behavior to provide empirical evidence for our theoretical model.

---

[2]Gordon writes in his new book : "If the stock market continues to advance, we know that the inequality will increase, for capital gains on equity accrues disproportionally to the top income brackets" (Gordon 2016).

[3]See Wolff (2012) and Cynamon and Fazzari (2016).

In Sect. 3.1, we begin by first examining the wage income provided by the Federal Reserve. The purpose is to utilize the fuller sample data of observations starting in the year 1989 according to the same lower 80th percentile and upper 20th percentile split discussed in Sect. 2. Each subsetting and classification of the data set will cordon off observations that are of interest for various reasons. Hence, our analysis will pursue several courses of inquiry in order to increase the empirical scope of this analysis on earnings and net wealth trends in the U.S. For example, Sect. 3.1 will introduce an algorithmic method by which we may separate strict labor or wage earners from other members of society. Although we will begin with analyzing wage income, the purpose here is also to derive the capability of studying various forms of income in relation towards one another. Section 3.2 will expand this method so as to examine the wealth distribution according to this wage bracketing of society.

Section 3.3 will introduce a newer method of income analysis which will be utilized to include a more comprehensive look at societal earnings trends. Once this newer normalized income variable is included as a classifier, it allows for increased analytical utility, but for a shorter time horizon. The variable also allows for analysis of sufficient observations to begin to break the data down into lower, middle and upper percentiles so as to look at the wealth shares within a class system within the U.S. This allows for an examination of the sources of income regarding financial and non-financial income. Additionally, one area of focus in Sect. 3.3 is the distinction between those individuals at the top and those individuals at the bottom of the earnings ladder. Hence, Sect. 3.3 extends the analysis to empirically study those in the middle and examine whether observable trends regarding the lower 80th percent and top 20th percent of earners are also persistent when comparing the more comprehensive classification of middle 40th percentile with the top 10th percentile earners as well.

Section 3.4 then analyzes the wealth disparity via the previously defined income percentiles as well as by age grouping and income sources as developed in the algorithms of Sect. 3.1. For example, we look at income trends by distinguishing between wage earners and those with primarily other sources of income. Wage income through standard employment is distinguished from forms of capital gains earnings as well as business and farm income. Once again, Sect. 3.4 will utilize the normalized income variable as a classifier so as to account for transitory fluctuations involving temporary economic shocks to individuals interviewed for inclusion within the Survey of Consumer Finances.

Lastly, Sect. 3.5 continues to utilize the normalized income variable as a classifier. Furthermore, the purpose of the empirical analysis is to continue to examine the depth of the wealth disparity when subsetting the top earners in society as separate from both the middle and bottom income percentiles as laid out in Sect. 3.4. However, Sect. 3.5 focuses the examination on the greater reliance on capital income in the U.S. through recent history. Our analysis in this section centers on observable trends regarding household capital earnings through the period of the financial crisis of 2008 in the U.S. economy.

Finally, each section contains further detailed explanations of data analysis where necessary and frequently references the Survey of Consumer Finance code book and SAS program provided by the Federal Reserve.[4] Section 4, provides relevant policy remarks. Section 5 concludes the paper.

2 A Model with Heterogeneous Households

Our model of heterogeneous households that consume, invest and borrow builds on a dynamic portfolio approach. We model two types of households. Although both households have wealth, the size of their wealth may be different and therefore their wealth dynamics are different. Additionally, when we refer to wealth, we are considering both financial and non-financial assets.

The first type of households are high net worth households. They have high returns from capital, capital gains, high labor income, and are borrowing to invest into assets. This type of household behaves as consumption smoothers and undertakes optimal consumption-saving decisions. High net worth households can borrow to invest which means, using a dynamic portfolio approach, the household can use short sales of the risk free asset.[5]

The second type of households also have initial wealth, though possibly less than the first type of households. However, they have lower returns on assets, lower capital gains, lower labor income, and obtain income from a risk free rate, for example from bank accounts.[6] Whereas the first type of households represent consumption smoothers, the second household borrows to finance consumption. Consumption decisions for the second group are not necessarily optimal.[7] It is important to note that both types of household can accumulate capital.

This paper presents the dynamics of the appropriate shares of capital for these two distinct households. The capital shares are modeled to evolve as Brunnermeier and Sannikov (2014) predict, see below. It is also likely to evolve as sketched by Pasinetti (1962). However, in this case, the shares even at the steady state can change, due to a difference in—and possibly changing—returns, saving rates and borrowing capacities between the two groups of households. Additionally, we will assume the second type of households are disadvantaged with respect to asset

[4]Codebooks for the SCF are provided at www.federalreserve.gov/econresdata/scf/scfindex.htm.

[5]For a dynamic model of net worth (of financial intermediaries), see Mittnik and Semmler (2013).

[6]For EU countries, for example it is estimated that "The most prevalent financial assets are deposits (sight or saving accounts), which are owned by 96.4% of households; voluntary private pensions/whole life insurance is held by 33.0% of households; all other financial assets are owned by less than 15% of households", see European Central Bank (2013).

[7]There is a large body of literature that starts with Veblen (1899) and his theory of conspicuous consumption, then there is Duesenberry (1949) and recent work, see Drechsel-Grau and Schmid (2014) that shows that consumption of the latter group arises from some social interaction and adaptive behavior of households.

returns, capital gains and labor income. Such households borrow to keep up their standard of living.[8] This second group can be viewed as adapters, in the sense of Veblen (1899), or upward looking households, as in Drechsel-Grau and Schmid (2014). Lastly, the first household may have a positive saving rate while the second household has a very low or even negative saving rate.

To introduce borrowing and net worth dynamics for our stylized two types of households in a finite horizon decision model, we begin with a low-dimensional stochastic model specification. The essential features of the model we employ here may be found in Brunnermeier and Sannikov (2014), and also in Stein (2012).[9] The former specifies the two types of households in their model, while the latter has only one household. Both papers make their specifications stochastic. However, they do not explicitly model further macro feedback loops. We, therefore, abstract from the macro feedback effects as discussed in Mittnik and Semmler (2016), and Gross and Semmler (2017).

In the Brunnermeier and Sannikov (2014) type of model, for both groups of households consumption and leveraging to invest or to consume are decision variables, while the main state variable is net worth, for example $x_{1,t}$, in (2). However, in this paper we will, as mentioned, let the second type of households decide on consumption and thus saving through social interaction mechanisms, as in Veblen (1899).

We solve the stochastic model variant through NMPC (Grune et al. 2015) and add a stochastic shock sequence where necessary, see Eq. (3), representing another state variable. In BS, capital returns are—due to capital gains—stochastic, as is the interest rate. BS initially start with a model where only the capital return is stochastic. They then add a stochastic interest rate by referring to time varying borrowing cost, reflecting the cost of screening and monitoring.[10]

2.1 High Net Worth Households

We first specify our model for the high net worth households which are consumption smoothing households. We here, as Brunnermeier and Sannikov (2014), specify our model for the purpose of presentation, we take the instantaneous payout, $c_{1,t}$, and leveraging, $\alpha_{1,t}$, as decision variables.[11] The model is stated in Eqs. (1)–(3), where preferences for the consumption smoothing households are given by (1), the

[8]For details on roughly zero or negative saving rates of low income households, see Dynan et al. (2004).

[9]For further details of the Stein model and the derivation of this, see Gross and Semmler (2017).

[10]Stein (2012) adds something important to this type of model by providing a static solution for the model in terms of a mean variance approach that allows to measure then excess leveraging, see Schleer et al. (2014).

[11]Note that Brunnermeier and Sannikov (2014) use x_t as a notation for their decision variable leveraging and assume that $x_t > 1$.

dynamics of the capital stock, k_t for the appropriate group is left in the background,[12] but we have in (2), net worth, $x_{1,t}$, and in (3), the stochastic shock process, z_t.

Note that we have written here the stochastic process in discrete time, since we solve the model in discrete time anyway. In our solution method we adopt a discrete–time framework and decision horizon of N periods, executed for each time period, $t = 1 \ldots T$ time periods, and Δ_h time steps and our model is given by

$$V = max_{\{c_{1,t},\alpha_{1,t}\}} \sum_{k=0}^{N} (1+\rho)^{-t} U(c_{1,t} x_{1,t}) \tag{1}$$

such that

$$x_{1,t} = x_{1,t-1}(1+\Delta_h)(\alpha_{1,t}(y_1 + v_1 lnz_t + r_1) \\ + (1-\alpha_{1,t})(i - v_2 lnz_t) - \varphi(x_{1,t}) - c_{1,t}) \tag{2}$$

$$z_t = \exp(\rho \ln z_{t-1} + \epsilon_t). \tag{3}$$

In the above model (1)–(3), c_1 and α_1 are our two decision variables,[13] with the payout $c_1 = C_1/x_1$, and $\alpha_1 = 1 + f_1$, with $f_1 = d_1/x_1$ the leverage ratio, measured as liability over net worth, d, is debt, y_1, capital gains, driven by a stochastic shock, $v_1 lnz_t$. Furthermore, r_1, is the return on capital, i, an interest rate paid on debt, also driven by a stochastic shock, $v_2 lnz_t$,[14] $\varphi(x_{1,t})$ is a convex adjustment cost, ρ, a persistence parameter, with $\rho = 0.9$, and ϵ_t is an i.i.d. random variable with zero mean and a variance, $\sigma = 0.05$. Moreover, note that in the current context of high net worth households, that dominantly borrow for asset purchases, we can have $\alpha_1 = 1 + f_1$, thus $f_1 > 0$. If $\alpha_1 > 1$ there is no short sales constraint of risk free assets and thus there can be active borrowing for investment.[15]

[12]In the solution procedure used here, we neglect the dynamics of the capital stock for this type of households as for the households borrowing for consumption. In Brunnermeier and Sannikov (2014), it represents the aggregate capital (with g the growth rate and δ the resource use for managing the assets) of financial specialists and households. A larger fraction of it will be held by financial specialists, since they can borrow. Those details can be neglected here. Aggregate capital is more specifically considered below, and see also Brunnermeier and Sannikov (2014).

[13]Note that in our model version here, we neglect the dynamics of (2). In Brunnermeier and Sannikov (2014), it represents the aggregate capital of financial specialists and households (with g the growth rate of capital, another decision variable, and δ the resource used for managing the assets). A larger fraction of the assets will be held by financial specialists, since they have a lower discount rate, and with $\alpha_t > 0$, they can borrow. Those details can be neglected here. The dynamics of aggregate capital is more specifically considered in Brunnermeier and Sannikov (2014).

[14]Stein (2012) posits that the interest rate shocks are highly negatively correlated with capital gains' shocks, we have thus a negative sign in (3) for the effect of the shocks on interest rates. We here also assume that the interest rate shocks have smaller variance than the capital gains' shocks.

[15]Note that wage income in such a portfolio model could be treated as in Chiarella et al. (2016).

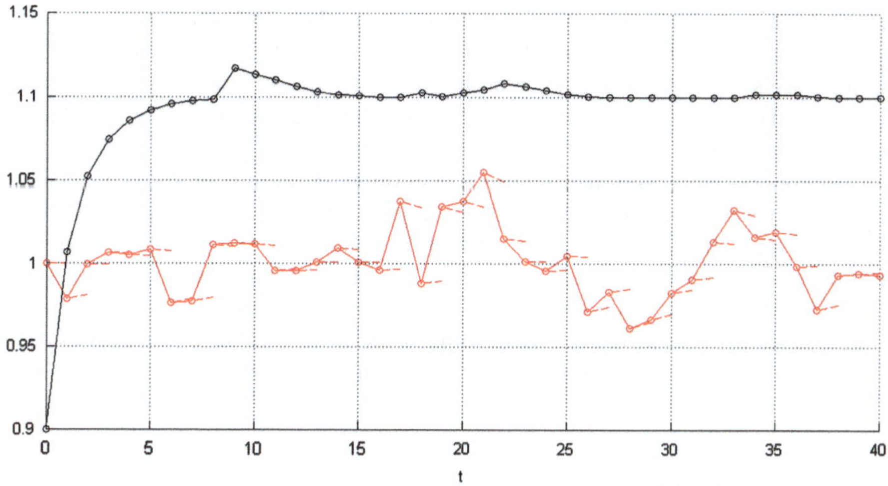

Fig. 1 Borrowing by investors: Path of net worth, *dark line*, and stochastic process, *red line*

We can solve the model variant (1)–(3) through a stochastic version of NMPC, see Grune et al. (2015). Figure 1, the black line presents the path of net worth, the red line the stochastic process, and leveraging $\alpha_{1,t} = 1 + f_{1,t}$, is generated separately, which is always above one. We want to note that we solve here only for optimal leveraging and optimal consumption. Since $\alpha_{1,t}$ is a choice variable, both Brunnermeier and Sannikov (2014) and Stein (2012) assume that debt is redeemed in each period and, without cost, can be obtained on the market without frictions.

The leveraging of this group is not without perils. There could be a macro feedback loop, namely an externality, i.e., endogenous volatility, that is triggered below some steady state, see Brunnermeier and Sannikov (2014), which creates an unstable steady state and unstable downward spiral. In Brunnermeier and Sannikov (2014), such feedback loops can arise from large shocks below some steady state, triggering fire sales of assets, fall of asset prices and fall in net worth, generating a downward spiral.[16] Although model (1)–(3) does not yet directly model instability as a result of an over leveraging of the high net worth households, Fig. 1 depicts the volatility only in the stochastic process, the issue not pursued here further.[17]

Through our numerical computations we can also observe the impact of leveraging on the accumulation of capital, not depicted here, see BS (2014, sect I). As

[16]In Stein (2012) the vulnerabilities and possibly adverse feedback loops are triggered by over leveraging, capital losses and rising borrowing cost.

[17]Excess leveraging in Stein (2012) is driven by actual leverage over and above the optimal leverage, caused by a sequence of persistent shock to increase capital gains and lowering interest rates, both giving rise to excess leveraging, For further amplifying feedback loops, as Mittnik, see Mittnik and Semmler (2016). Stein's model can distinguish between optimal debt, actual debt and excess debt. For empirical measures of those, see Schleer et al. (2014).

Brunnermeier and Sannikov (2014) properly state, through leveraging, the capital share of the active investors in total capital—the share of financial experts in their terms—rises with leveraging, even at the stochastic steady state. This is also what creates in Brunnermeier and Sannikov (2014) the source of the wealth distribution and the endogenous risk.

The increasing wealth distribution is also observed in the empirical analysis in Sect. 3.2 by several notable trends. First, the share of high yielding assets owned by high income earners is substantially larger than that of lower income households. For example, stocks and bonds become an increasing component of the net worth of such households while lower income earners may rely more upon pension funds or simply checking accounts, as discussed next. Further discussion of such observations in SCF data is reserved for Sect. 3 of the paper.

2.2 *Low Net Worth Households*

Next we discuss the second part of the model, which defines the less wealthy households that may have some small amount of assets but mainly borrow for consumption. Thus this second group of households have a lower amount of wealth, lower returns on capital, lower capital gains, lower labor income, and obtain income from a risk free rate. These households are not consumption smoothers in the optimal sense, but borrow to finance consumption.[18] In this case, consumption is not necessarily optimal and may in fact result in larger than optimal consumption, given the household constraints.

$$V = max_{\{c_{2,t},\alpha_{2,t}\}} \sum_{k=0}^{N} (1+\rho)^{-t} U(c_{2,t} x_{2,t}) \tag{4}$$

such that

$$x_{2,t} = x_{2,t-1}(1+\Delta_h)(\alpha_{2,t}(y_2 + v_2 ln z_t + r_2) + (1-\alpha_{2,t})(i - v_2 ln z_t) - \varphi(x_{2,t}) - \tilde{c}_2) \tag{5}$$

$$z_t = \exp(\rho \ln z_{t-1} + \epsilon_t). \tag{6}$$

Here we may normally have $\alpha_2 = (1 + f_2)$, with $\alpha_2 < 1$, which means there is income from the risk free rate as households may deposit savings in a bank with risk free returns. For these types of households we also could have for certain time

[18]Note that recently many financial instruments, such as sub-prime loans, collateralized debt obligations (CDOs), mortgage-backed securities (MBOs) and credit cards, have been developed to induce the low net worth households to borrow and to consume.

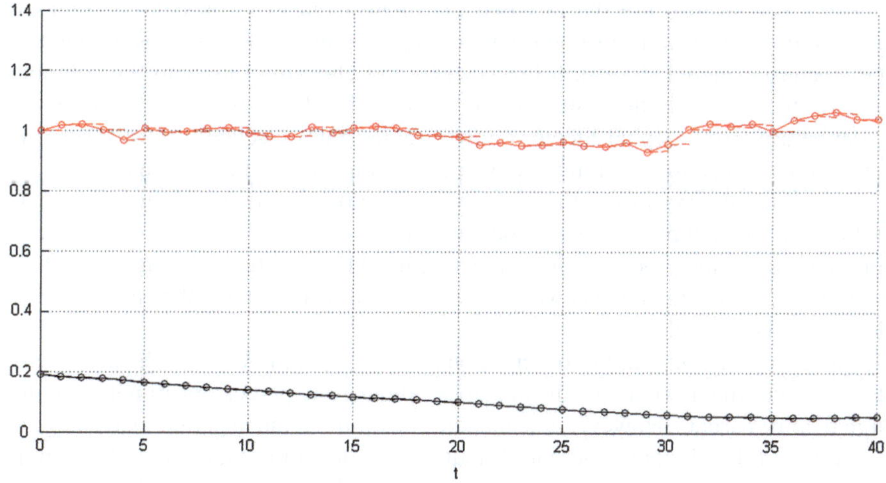

Fig. 2 Borrowing by consumers: Path of net worth, *dark line*, and stochastic process, *red line*

periods $\tilde{c}_2 = const > c_t^{opt}$.[19] We could interpret this as a constant propensity to consume as in some text books of Keynesian economics. Here it is, however, the propensity to consume out of (net) wealth whereby the propensity to consume is taken as constant.[20]

Note that in the context of this section we use NMPC for solving the stochastic difference equations (5) and (6).[21] In other words, we don't use the optimal control problem, this is only simultaneously computed, since we want to know that the condition $\tilde{c}_2 > c_t^{opt}$ holds. Hence, we solve in the system (4)–(6) only Eqs. (5), including (6), a stochastic difference equation. Thus we employ NMPC only as a stochastic difference equation solver. So Eq. (5), where a simulation is run, is solved with non-consumption smoothing behavior or our second type of agents with $\tilde{c}_2 > c_t^{opt}$, which is shown in Fig. 2.

Yet, in order to see whether the condition $\tilde{c} = const > c_t^{opt}$ is fulfilled, whereby c_t^{opt} is the solution of Eqs. (4)–(6), we need also to look at the solution of Eqs. (4)–(6) using NMPC. Then we can check if actually, for some time period, consumption spending for this type of households maybe undertaken with borrowing that does not necessarily fulfill a longer-term sustainability condition of the household debt.

[19]This is for example what some literature has stated on the consumption boom of low income households before 2007/8, see Cynamon and Fazzari (2016).

[20]For this group of households Eq. (4) could be disregarded, since those households are not consumption smoothers. Formally then an additional dynamics, the evolution of debt as in Mittnik and Semmler (2016), may need to be introduced to account for the interest payment on debt. Here too wage income in such a portfolio model could be treated as in Chiarella et al. (2016).

[21]NMPC can also be used to solve stochastic difference equations when instead for the optimal solution $c = c^{opt}$a constant term is used in Eqs. (5).

Here we can also use the fact that the interest rate paid on bank accounts is stochastic, $i - \nu_2 lnz_t$. The same holds for capital gains $y_2 + \nu_2 lnz_t$. Running now the stochastic model (5)–(6), using NMPC, and checking this with the solutions for (4)–(6), we obtain the results as shown in Fig. 2. There we can observe how a lower return (from capital income), lower saving rates—actually mostly negative saving rates—lead to a declining net worth and thus dissipating net wealth.

As can be observed from Fig. 2 the borrowing by low net worth households, to fulfill their consumption needs, produces a declining path of net worth, the dark line. The stochastic process is again the red line in Fig. 2. The declining share of net wealth of low income, low net worth, households is also found in the empirical data demonstrated in Sect. 3.2.

Moreover, the empirical observations found in SCF data in Sect. 3.2 also demonstrates that normal income becomes a decreasing percentage of net worth for high income earners. However, this income constitutes a much greater percentage of net worth for low income households that show a declining share of total wealth in the U.S. Section 3 also demonstrates empirical findings that this trend has become more extreme in recent years.

Although model (5)–(6) does not yet directly model instability, the leveraging and over leveraging of this group of households is also not without macroeconomic perils. This is nicely demonstrated in Cynamon and Fazzari (2016) where a fall of asset prices and fall in net worth, in particular after 2007/8 amplified the downward spiral. Figure 2, upper line, depicts the volatility only in the stochastic process of Eq. (6), the issue not pursued here further.

These observable effects of a persistently downward trend, lower line, may lead to the question of why the households don't rationally expect this trend of dwindling net worth and debt possibly surpassing assets, producing the threat of bankruptcy and default risk. In a model of infinite horizon behavior such a situation would not occur. Through the optimal consumption and the transversality condition this would be prevented.

Since we are working here with a model of finite time horizon, though there is an expectation on some shorter time horizon by households and the lenders are loan pushers for consumption and possibly mortgages for low net worth households, we implicitly assume that there is some near-sightedness allowing for the emergence of over leveraging and the vulnerability of those households.[22] In fact it is this assumption that makes our model more realistic when compared to infinite time horizon models.

[22]For empirical trends like this in the US since the great recession, see Cynamon and Fazzari (2016).

2.3 Wealth Distribution

The next issue to examine is how the share of a country's wealth will evolve through time. Additionally,what effect will our model have for wealth distribution, an imminent issue in the Kaldor-Pasinetti debate. This phenomena can now be studied in the context of the above model variants.[23] In our case we have

$$q_t k_t = x_{1,t} + x_{2,t} \tag{7}$$

In this case the net worth of high net worth households is rising and then becomes constant while the net worth of low net worth households declines. Therefore, the share of high net worth households will rise, which seems to be supported by the empirical evidence.[24] This trend shift is discussed in Sect. 3.2. Note that we have included in the net worth of each type of household the capital gains, resulting from an aggregate, q_t, the asset price of capital, captured also those capital gains by high and low net worth households.[25]

This trend shift in asset holdings would also create a shift of a higher share of capital income in total income of the households holding a higher fraction of assets, a proposition that also seems to be confirmed by our empirical observations in Sects. 3.1 and 3.3.

3 Empirical Trends

Next, we seek to support the above trends generated by the models of (1)–(3) and (4)–(6) by employing data from the Federal Reserve Board's Survey of Consumer Finance.[26] We need to examine the data, using specially created algorithms, the balance sheets of the numerous households' and extract the appropriate data we need for our study.

3.1 Data Sources, Methodology and Measures

The data used herein is from the Federal Reserve Board's Survey of Consumer Finance (2013). The following empirical sections utilize first the data for survey

[23]Brunnermeier and Sannikov (2014) have devoted a considerable part of their study to this issue.

[24]See Sects. 3.2 and 3.3 and also Cynamon and Fazzari (2016).

[25]For a further elaboration on the aggregate q_t, see Brunnermeier and Sannikov (2014, sect. 2).

[26]Data from the 2013 SCF as well as links to the data used herein may be found on the Board's public website at www.federalreserve.gov/econresdata/scf/scfindex.htm. Additional scripts to download all micro data in R programming language, are available at the following url: https://github.com/ajdamico/asdfree/tree/master/Survey%20of%20Consumer%20Finances

observations for the wage income covering the years 1989–2013. A second further breakdown of the data in Sect. 3.3 is utilized for the normalized income variable covering the years 1995–2013. Wage income is analyzed according to the lower 80th percentile and top 20th percentile split discussed in the previous theoretical model.

Additional analysis of empirical trends, in Sects. 3.3, 3.4, and 3.5, utilize the normalized income variable as a classifier which is available for survey years 1995–2013. This is done in conjunction with our categorization of wage earners and capital earners via the determination of a capital difference variable consisting of the difference between wage incomes and capital incomes as a percentage of the sum total of incomes. It should be noted that all appropriate figures are CPI adjusted according to the Federal Reserve's published figures. It was necessary for example, to adjust the component parts of net wealth in such a manner that comparisons would hold across the time period in question while variables such as age, marital status, etc. would remain unchanged.[27]

The capital difference ratio was created by determining the types and sources of income. The wage income variable was slightly limiting in SCF data as it would not account for employment transitions and various other sources of working income that are relevant for additional portions of the population. Hence, the more comprehensive use of normalized income allowed for greater partitioning of observations for empirical analysis and visualization of potential trends in the data sets.

The capital difference ratio utilized the component portions of total income to determine the difference between wage incomes and capital incomes as a percentage of the total sum of these parts. The absolute value was taken since variables in SCF data may turn negative due to outgoing cash flows such as alimony and this analysis is focused on the total impact each income source had on the sum total of individual's incomes. Additionally, the purpose is to determine what gains/losses come from what type of source as a percentage of the whole and then look at the weighted percentiles within society.

$$\mathit{Capital\ Difference\ Ratio} = \frac{ABS(\mathit{Working\ Incomes}) - ABS(\mathit{Capital\ Incomes})}{ABS(\mathit{Working} + \mathit{Capital\ Incomes})}$$

with

$$\mathit{Working\ Incomes} = (\mathit{Wages} + \mathit{Social\ Security\ Retirement\ Income})$$

$$\mathit{Capital\ Incomes} = (\mathit{Business\ Farm\ Income} + \mathit{Interest\ Dividends} + \mathit{Capital\ Gains})$$

[27]All CPI adjustments followed precisely those used by the Federal Reserve in publishing the triennial bulletin "Changes in U.S. Family Finances from 2010 to 2013", associated with the Survey of Consumer Finances and within the provided SAS code used for the publication.

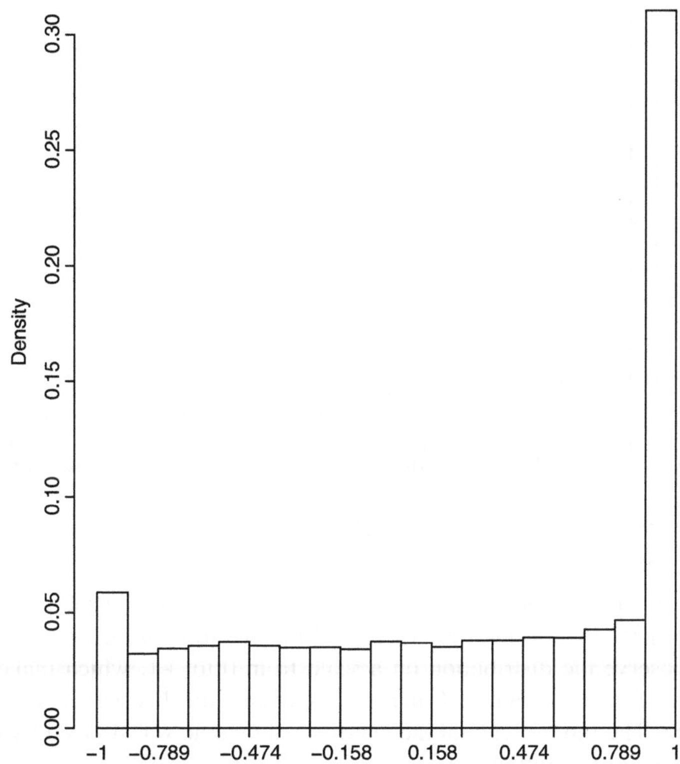

Fig. 3 Histogram of the capital difference ratio

The capital difference variable can be viewed as a histogram in which the number fluctuates from positive one to negative one with most individuals receiving their income from working wages. Therefore, Fig. 3 displays the majority of observations as maintaining a ratio closer to 1 as expected since most individuals work for a living. The few cases of a value of 0, were excluded and found to be extremely uncommon in the sample data. The sample density is demonstrated on the y-axis, while the x-axis displays the capital difference ratio.

Figure 3 can be thought of as showing that when the population is broken down into primarily wage earners as opposed to capital earners the individuals will occupy a position within the range from 0 to 1 on the histogram. However, if individuals primarily earn income from capital sources then they will occupy a position ranging from below 0 to negative 1. All members within society then oscillate within the -1 to $+1$ scale of this variable. Clearly, one can see from the SCF data that most of society will make their incomes from standard wages identified by a value primarily approaching $+1$.

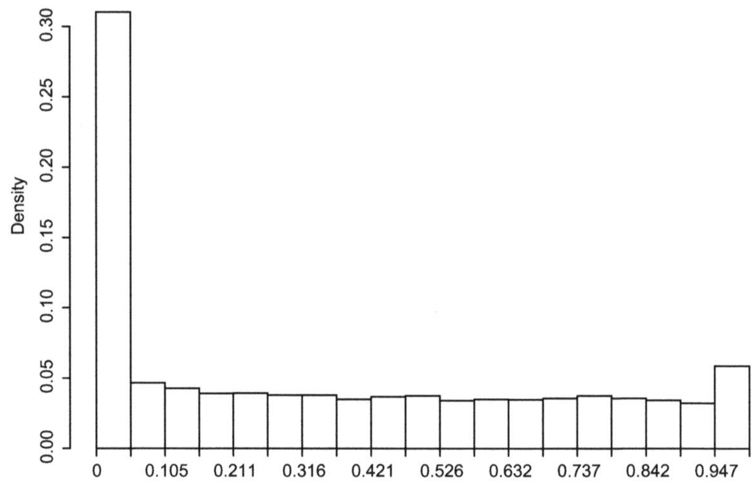

Fig. 4 Histogram of capital income to total income

Another interesting observation utilized for corroboration of the previous findings is to simply look at the ratio of capital income-to-total income. In this ratio, one may observe the distribution on a scale from 0 to +1, which makes greater intuitive sense visually. Figure 4 displays the results for this second societal trend in SCF data. The ratio of capital income-to-total income was derived based on the previous equations as follows.

$$Capital\ to\ Income\ Ratio = \frac{ABS(Capital\ Incomes)}{ABS(Working\ +\ Capital\ Incomes)}$$

As one can see from Fig. 4, the majority of the population rests within a lower placement on the capital-to-total income histogram. The largest distribution is located towards but slightly above 0 on the histogram while substantially fewer individuals exist towards a capital-to-total income ratio of +1. The logic for this finding is similar to that found in the previous figure. In general, the vast majority of people in society work for a living receiving wages and do not receive most of their income from capital sources and investments.

A further breakdown of this variable in society is expanded upon in Sect. 3.5. As we subdivide the data through the years this variable provides a more dramatic visualization of changes in capital income trends within the U.S. population. This helps to provide some enlightenment with regard to such trends as the U.S. approached the housing and financial crises of 2007. As we proceed through the next several empirical sections, we begin to analyze the disparities and greater reliance on capital incomes with regard to changes in individual's net wealth through the last two decades in modern U.S. economic history with a particular emphasis on the heterogeneous households described in Sect. 2.

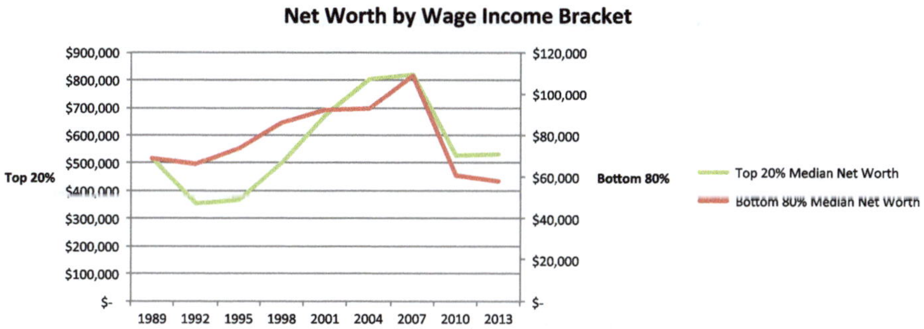

Fig. 5 Level of top 20% net worth households (left scale) and of bottom 80% of net worth households (right scale)

3.2 *Distribution of Net Wealth*

Next we want to evaluate the wealth share of the 20% top net worth households and the bottom 80% net worth households. We are taking here wage income as a variable to define our brackets. The purpose here is to analyze the net wealth dynamics based on wage income from 1989 through the present.

Figure 5 presents the level of net wealth of the top 20% net worth households (left scale) and of the bottom 80% of net worth households (right scale). As one can observe from Fig. 5 the first group owns roughly seven times more net wealth than the second group in 1989, the pre-crisis period before 1990–1991. In the crisis the net worth of the 20% households decreases but with the expansion they catch up and the growth of their net wealth surpasses that of the 80% households. In the great recession, the 80% households net wealth drops faster, mainly due to falling real estate prices.[28] There is a slower drop of net wealth at the 20% households and thus they end up now owning roughly nine times more of net wealth than the 80%.

This first falling but then rising share of the top net worth households, in the period of great moderation, the Clinton boom and then since the great recession 2007–2008, is visible in Fig. 6. Since the general long run upward trend since the recession 1990–1991, with large capital gains, is only interrupted in the boom before the start of the great recession.

Note that our Figs. 1 and 2 ratios are similar. We start there with roughly a ratio of top wealth owners of five and end up with their share in net wealth at roughly ten to eleven times higher. Note however, in the models of Sect. 2 we have assumed constant capital gains, only affected by small shocks, though different for each type of household. Since there is less of a trend in capital gains, this reduces the cyclical movements of net worth.

[28]See Wolff (2012).

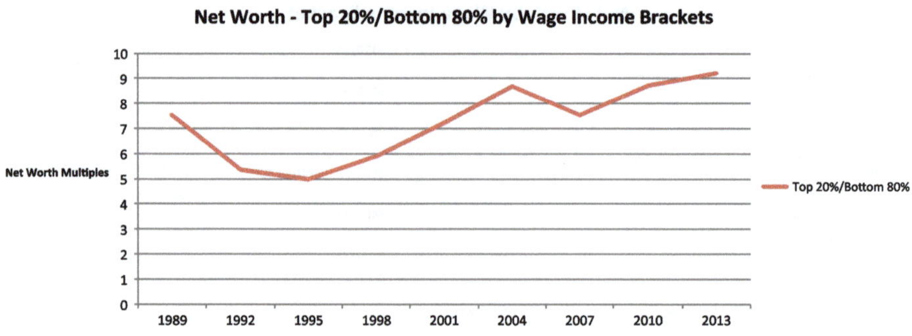

Fig. 6 Ratio of top 20% to the bottom 80% of wealth owners

Similar trends as observed in Figs. 1, 2, 5, and 6 have also been observed by Cynamon and Fazzari (2016). In their data the top 10% wealth owners are computed over the 90% wealth owners, and the downward trend occurs in their computation for financial wealth. Hence, the composition of wealth, the various forms of income as well as the relationships between income and net worth as seen within the data set are of particular interest.

Despite the fact that utilizing the wage income provides some insight with regard to working families of different levels, it also neglects to take into account the full spectrum of U.S. workers. For example, in SCF data, wage income will suffer from both positive and negative transitory shocks. Such deviations reflect temporary income shifts within the data set. Also, business and farm income are neglected when utilizing only wage income as a classifier. Additionally, interest, dividends, capital gains and investment are an increasing portion of the growing disparity between rich and poor in modern U.S. society.[29] Therefore, a more recently derived variable (normalized income) provided by the Federal Reserve serves as a more appropriate classifier for shedding light on the full spectrum of income earners in the U.S.

The Federal Reserve began utilizing the normalized income variable slightly after the release of the 1989 Survey of Consumer Finance.[30] The variable is of particular use in published bulletin releases beginning in 1995 for each year's recorded sample. Although, unlike the wage income classifier, normalized income limits the time horizon under analysis, the normalization process helps prevent bias due to changes across groups when considering both positive and negative economic shocks (Federal Reserve Bulletin 2014).

---

[29] See Gordon (2016).

[30] Technically, the Federal Reserve uses the term usual income, which is calculated after actual income is reported. If respondents reported a temporary deviation or shock to their income in a given survey year, then the classifier was used to smooth away transitory fluctuations in an attempt to better approximate the concept of permanent income. The specific calculations may be found in the 2014 FRB Bulletin as well as the supplemental SAS code.

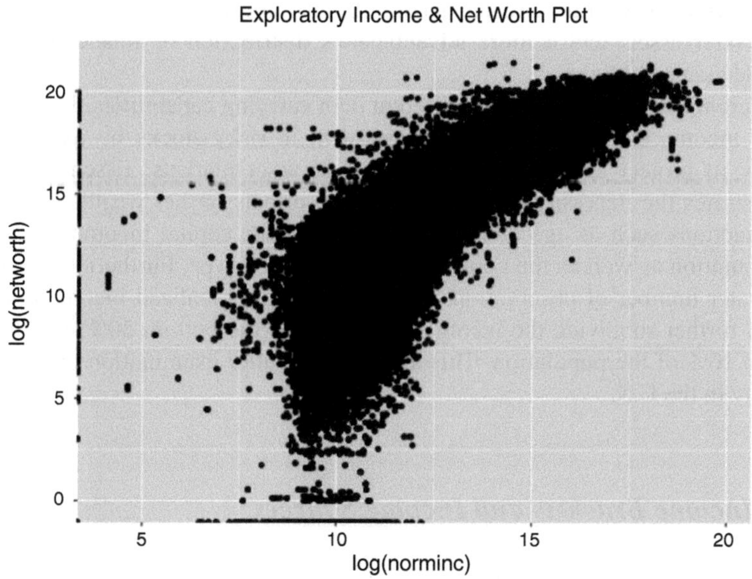

Fig. 7 Normal income and net worth relationship (1995–2013)

In order to analyze how to subdivide society when studying wealth disparities, it is helpful to investigate the nature of the relationship among the variables under consideration. In this case the primary variables of interest are here normal income and net worth. Earnings variables appear to display log normality amongst various subsets of the respondents interviewed throughout the time horizon of the supplied data sample. Thus, looking at changes in these variables and the possibility of a non-linearity in the variables, helps with consideration of the need for a series of partitioning processes with regard to the appropriate data analysis.

As one can see from Fig. 7, the possibility of a non-linearity regarding the relationship between changes in income and net worth is evidenced, requiring one to consider the composition of income, financial, non-financial assets as well as other factors such as respondent age group and point in career. The demonstrated scatter plot displays the log of normalized income (which is set equal to income in the absence of a temporary shock) on the x-axis and the log of net wealth on the y-axis.

Figure 7, demonstrates the importance of examining qualitative aspects of individual's income. For example, one cannot simply look at income brackets and net worth exclusive of important determinants of these variables. Some notable such determinants appear to be respondent's age, accumulated wealth, income sources and types.

The fact that younger income earners would have less accumulated wealth makes it difficult to perform an analysis of changes in net wealth in conjunction with older workers approaching retirement. After all, older workers frequently have had longer periods of higher pay due to greater experience and allowing for greater accumulations of wealth. Greater net wealth also allows for more capital income from riskier assets and a more advantageous distribution of financial and non-financial assets holdings.

Different assets will allow for different debt carrying capabilities. Furthermore, greater income also may allow for investment in risky stocks by using income stability to protect against negative swings in asset markets. Essentially, Fig. 7 demonstrates the necessity for considering the effects on net wealth via a series of distinctions such as age and placement within the capital income holdings of the population as well as the composition of assets by type. Furthermore, utilizing the greater number of observations afforded by the normalized income classifier, we can further subdivide the income percentiles to the bottom 50%, middle 40% and top 10% of the population. This allows for a fuller examination of the wealth disparity in the U.S.

3.3 Income Brackets and Income Sources

The use of an income variable smoothed of the variability of temporary shocks displays the importance of analyzing the full spectrum of income sources. Such a process may then provide insight as to the discrepancies seen in earnings for those at different weighted percentile brackets within the spectrum of earnings agents in the economy.We begin to focus here on the three primary lower, middle and upper classes in the U.S.

Some of the trends that present themselves in the data set regard the utilization of stocks and bonds for determining asset returns. Additionally, a greater incorporation of business assets accompanies increases in the earnings brackets. For the purpose of exposition it helps to differentiate between financial and non-financial asset components of net worth. Lastly, all asset categories follow the definitions laid forth in the survey of consumer finance code book.

It is worth noting that in SCF data sets, bonds do not include bond funds or savings bonds but rather include state and local tax-exempt bonds, mortgage-backed bonds, U.S. government bonds and bills as well as corporate and foreign bonds. Additionally, liquid retirement accounts include IRAs, thrift accounts and future pensions.[31] Other managed assets include trusts, annuities and managed investment with an equity interest.

All abbreviations in Figs. 8 and 9, follow the definitions laid out in the supplemental SAS code found on the Federal Reserves public website.[32] However, for exposition 'retqliq' refers to quasi-liquid retirement accounts such as IRAs, thrift

[31]A full breakdown of the classification of each individual variable, composition of each variable and calculations is included within the SAS code accompanying the SCF 2014 Bulletin release.

[32]Precise definitions and calculations for abbreviations in Figs. 8 and 9 are provided at www.federalreserve.gov/econresdata/scf/scfindex.htm.

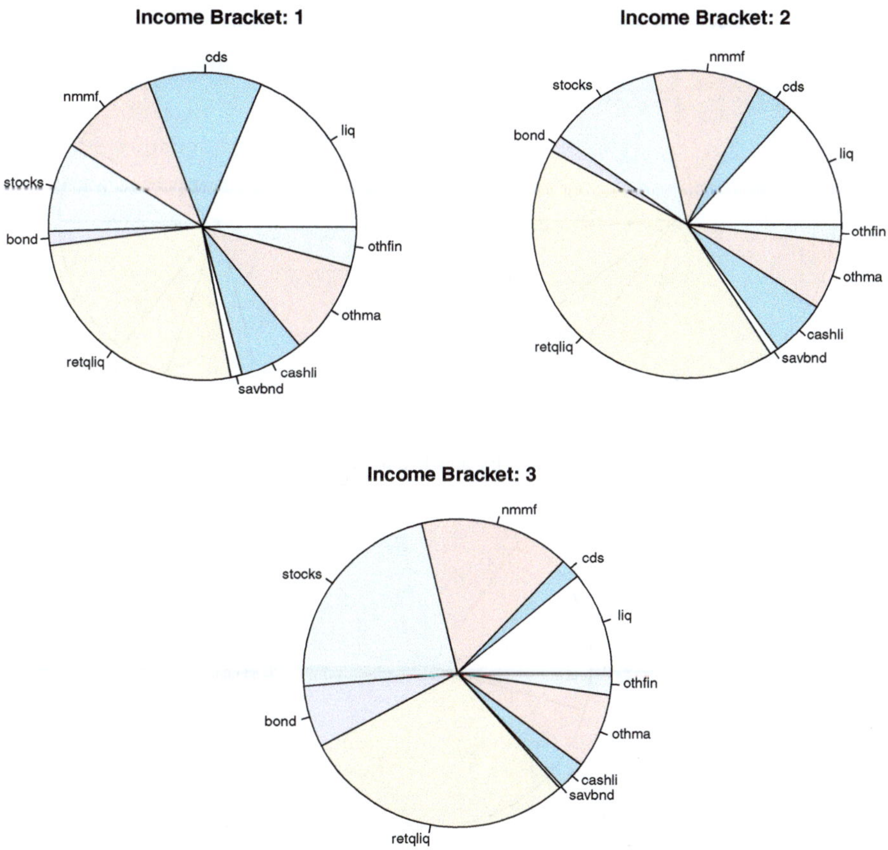

Fig. 8 Financial assets by income percentile

accounts, future pensions and currently received benefits. 'Cashli' refers to the cash value of whole lift insurance and 'othma' refers to other managed assets. 'Liq' refers to liquid assets such as transaction accounts. 'CDs' refers to certificates of deposit; 'othfin' refers to other financial assets. Finally, 'nmmf' refers to directly held mutual funds excluding money market mutual funds.

Figure 8 displays the CPI adjusted financial assets for income brackets based upon percentiles of the lowest 50%, middle 40% and top 10% of income earners for the years 1995–2013. Other than the increases in stocks and bonds for those in the top percentile, it is worth noting that there is at first an increase in retirement assets such as pensions (denoted in yellow) prior to a decrease in the third and highest income percentile.

It would appear that in aggregate there is a decreased ownership of the lower paying assets and an increase in the higher returning assets as one moves up the normalized income ranks. This is particularly noted by the shrinking share of CDs and savings bonds as well.

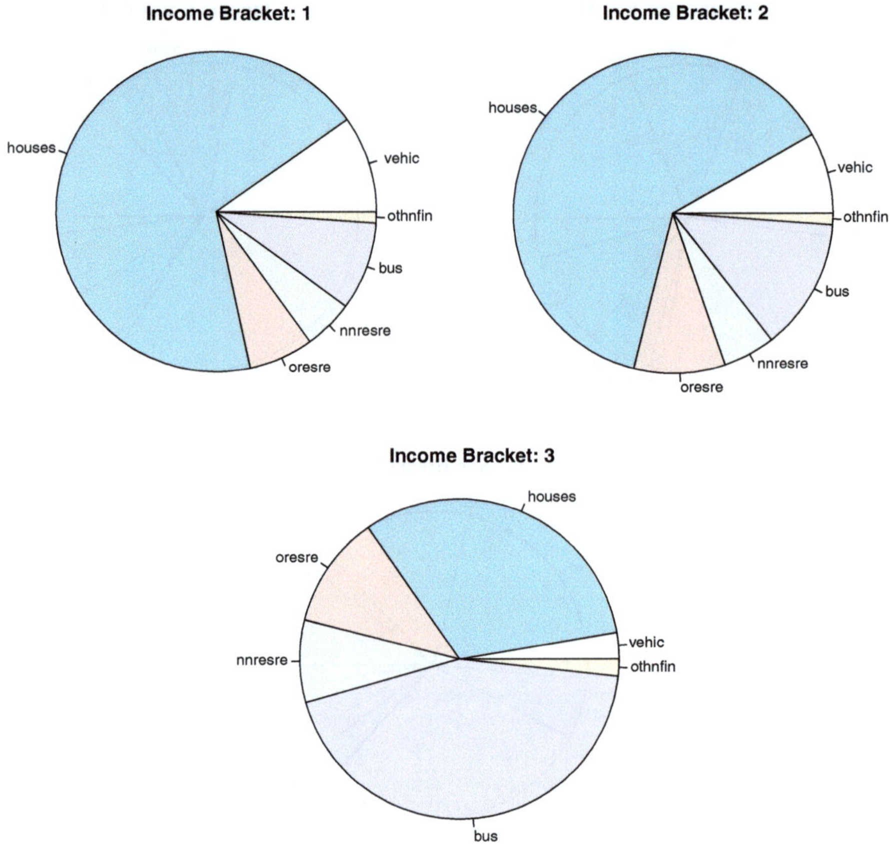

Fig. 9 Non-financial assets by income percentile

It should be noted that these findings are in line with the model simulation in Sect. 2. For example, in Sect. 2 it was suggested that wealthy households engage in consumption smoothing behaviors. The empirical evidence seems to suggest that such families primarily are interested in riskier high yielding instruments to help satisfy consumption when needed. Similar patterns are noticeable when one looks at non-financial assets. Additionally, both Figs. 8 and 9 are based upon average numbers for respective categories.

Figure 9 displays CPI adjusted non-financial assets for the same income brackets for each percentile. One can observe several obvious trends upon simple examination. For example, at the lowest income percentile, vehicle ownership is a larger portion of non-financial assets. As one moves up the income ladder vehicles become a less significant portion of asset holdings.

Once again, the empirical evidence supports the models simulation in Sect. 2. For example, in Sect. 2 it was suggested that lower income households are simply aiming at consuming necessary goods such as vehicles and housing. In most of the United States a car is essential to earning a living. Unfortunately, a car is not typically viewed as an investment while a business affords a substantial income stream to create returns for investors and/or smooth consumption for high net wealth households. However, the most notable trends are those regarding both business and housing ownership.

The portion of business ownership (denoted in purple) is at its largest for the top 10% of income earners. It would appear that successful business owners are able to garner a larger portion of income. Additionally, there is a general increase in real estate holdings that are separate from individual housing. For example, at the lower end of the income spectrum, housing constitutes a dramatically large percentage of non-financial asset holdings, yet this tends to decrease with increases in income percentiles.

Once again, for exposition, 'houses' refers to primary residence. 'Vehic' refers to autos, motor homes, RVs, airplances and boats. 'Oresre' refers to other residential realestate such as properties other than principal. 'Nnresre' refers to net equity in nonresidential realestate. 'Othnfin' refers to other non-financial assets. Finally, 'bus' refers to business interests. Once again, as in Fig. 8, all abbreviations follow the definitions laid out in the SCF codebook.

It should be noted that other non residential real estate included land contracts/notes, properties other than primary occupied residences, time shares and vacation homes. Interestingly, this variable (denoted by "oresre" in the pie charts) tends to increase with income brackets. It makes intuitive sense that as families earn more income they look towards having secondary home options. Clearly this variable is smallest for the lowest 50% of the income earning population in the U.S.

The findings in Figs. 8 and 9 are for aggregate sample data. It is worth noting that the same findings persisted within the data for individual years. For example, the same trends were present in each individual survey year from 1995 to 2013. It appears that this trend is very persistent in the United States for the income brackets under consideration.

3.4 Life Cycle Aspects of Income and Wealth

The topic of age is worth mentioning since, as individuals age and gain experience, income grows while asset holdings change as does the type and size of debt holdings. For example, later in life many individuals will have paid off most mortgages related to housing purchases necessitating a change in net worth. Additionally, some individuals will have acquired more financial assets allowing for greater returns to capital than others for whom the primary source of savings is their income.

Figures 10 and 11 display median normalized income (as defined in Sect. 3.2) over net worth on the y-axis for different income percentiles of the population (noted in the right side legend) based on age for two distinct population groups through time. The first group is the 50th percentile for whom wage income (as defined by

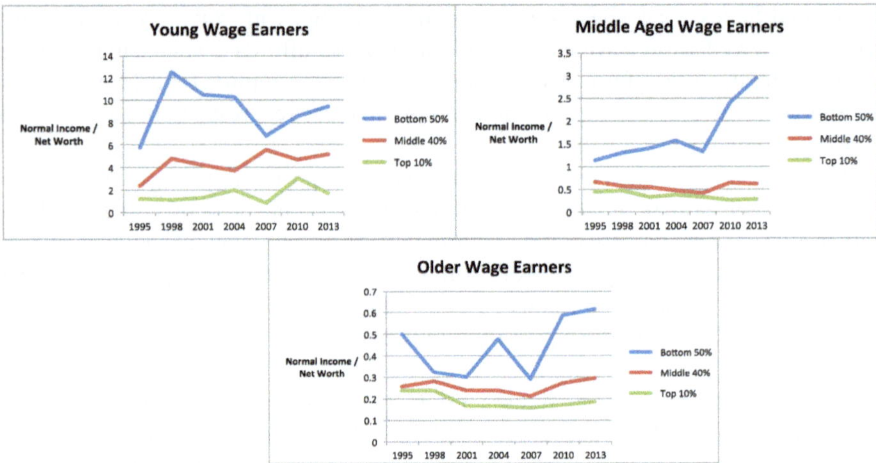

Fig. 10 Lower 50th percentile working-capital ratio

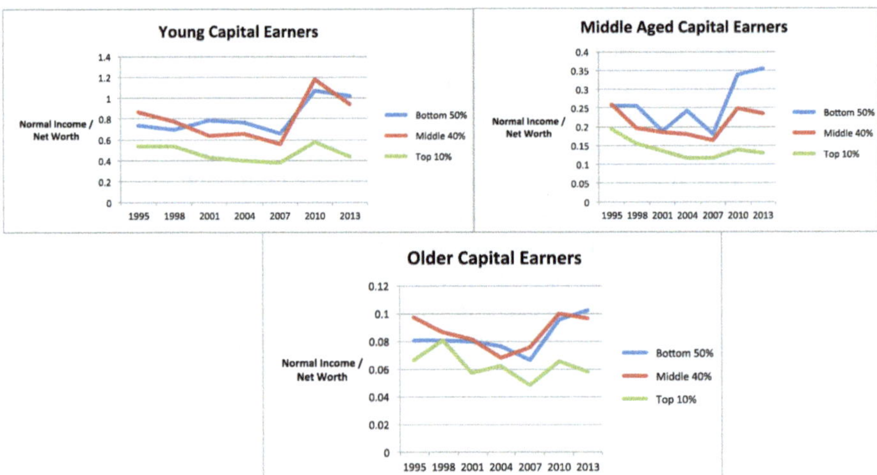

Fig. 11 Upper 50th percentile working-capital ratio

the capital difference ratio in Sect. 3.1) constitutes a primary percentage of total income and the second group is that part of the population for whom capital income becomes an increasing percentage of their total income.

Median numbers are used to omit the effects of larger outlying values within the population that do not fully reflect the performance of the vast majority of individuals within the respective income bracket. Additionally, normalized income was used in the numerator so as to smooth the effects of temporary shocks in respondent's annual earnings. The right hand side of each graph lists the income brackets of the lowest 50%, middle 40% and top 10% of income earners and the horizontal x-axis reflects the years 1995 through the present.

The biggest discrepancies within each graph are those between the lowest 50th percentile (denoted in blue) and the top 10% (denoted in green). For example, when the blue line grows higher and the green line remains even or decreases. This reflects the normalized income of the lowest 50th percent of the population of primarily wage earners becoming a larger percentage of their net worth while the opposite may be said of the top 10%. The reason is primarily that during economic shocks, the lower percentiles appear to take on more debt for consumption, while easy-money policies raise the asset values held by those in the top 10% of income earners. Essentially, the spread between these two lines greatly reflects an inequality in the composition and determinants of net worth for these two groups as well as distinctly different spending habits.

One observation in the data is that debt did not dramatically increase during these same times for those in the top 10% of the population while composite risky asset holdings increase with the income grouping relative to other percentiles. A similar finding is postulated in Semmler (2011)[33] when discussing dynamic portfolio theory and the pension fund problem that, to the extent that income may shield against volatility in asset prices, some may then choose to allocate greater percentages of wealth to risky assets.

It should also be worth noting that as individuals age, they begin to spend more of their net worth, especially during significant economic shocks. This can be seen in the rising normal income-to-net worth line (denoted in green) for the top 10% of older wage earners. In terms of the age designations, the younger workers were aged 20–39, while the middle aged workers were aged 40–59 and the older workers were over the age of 60. Additionally, the capital-difference ratio, outlined in Sect. 3.1, was used to determine whether more income was being earned through wages as opposed to capital sources.

The data demonstrates that for the youngest percentiles of earners, accumulated savings in assets is far from its future realized potential since many of the individuals are still in their twenties. However, as one can see, the trends are not quite as pronounced for capital earners of each age group as they are for individuals dependent upon the wage for contributions to the agent's net worth.

Additionally, it is interesting that, with the exception of the younger wage earners, for all other age groups of both wage and capital earners, the only group to see a CPI adjusted increase in net worth since the 1990s was the top 10% of income earners. Once again, most of this change is due to increases in composite risky asset holdings. Therefore, even if normalized income decreased for the top 10% as well

[33]The discussion, in section 18.5, specifically involves a dynamic portfolio decision with risky assets and labor income for consideration of the pension fund problem currently facing many retirees. In such a model the low frequency components of L_t may be estimated using harmonic estimations.

as the bottom 90% the effect on net worth was not as detrimental to financial well-being for those individuals lucky enough to be in the top 10th percentile of income earners.

Finally, it is important to note the changing scale between the percentages reflected in Figs. 10 and 11. For example, the incomes tended to rise slightly between wage earners and capital earners based upon the capital-difference ratio used and depicted in Sect. 3.1. However, the primary difference between these two groups was their accumulated net wealth. Hence, if normal income increased slightly between equivalent ages between the two groups, net wealth increased far more dramatically changing the scale of the depicted ratios accordingly. Hence, for example, among capital earners normal income appears to be a far smaller percentage of net wealth than for wage earners, which makes intuitive sense.

3.5 Capital Income Trends

Next we discuss income trends that have changed through recent history. In the U.S. the trend has been towards taxing capital gains and business income and wages at different rates. Changes in tax structure may change individual's investment strategies and alter the distribution of wealth in society. As demonstrated in the previous sections, as well as in Sect. 2 of this paper, those dependent upon the wage rate as their primary form of income are not able to garner sufficient returns to grow net wealth.

One way to analyze the disparities in society is to take capital income as a percentage of total income. In Sect. 3.1, we displayed a histogram (Fig. 4) of the distribution of the ratio of capital income as a percentage of total income for the full SCF sample data covering the years 1995–2013. The calculation of capital income was displayed in Sect. 3.1 to consist of the sum of business farm income, interest dividends and capital gains. Total income was the sum of this capital income and wages and social security retirement incomes. We now will analyze this capital-to-total income ratio for various income percentiles and ages in the U.S.

Figure 12 displays normalized income percentiles consisting of the lowest 50%, middle 40% and top 10% as noted on the right hand title for each bar graph. The percentage of capital-to-total income is denoted on the far left of each graph. The age group of each segment of the population is noted by each graph's individual title heading. Finally, the horizontal axis depicts the years 1995 through the present, since we are once again using here normal income for our income brackets.

The most notable trend depicted in Fig. 12 is that the ratio of capital income-to-total income has risen in the U.S. over the last approximately 20 years for all three age groupings of the population. However, the most interesting observation of these figures is the fact that for all three age groups as well as each individual normalized income percentile, the peak year of capital income-to-total income ratios clearly peaked in 2007 just prior to the financial crisis of 2008. Furthermore, the top 10% of

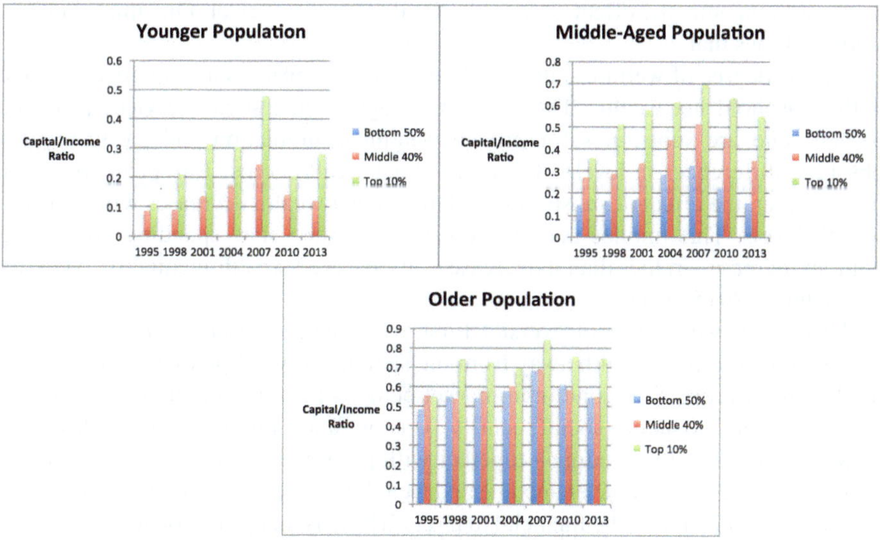

Fig. 12 Capital income-to-total income ratios

the population appears to have experienced the greatest growth in this ratio building up to the financial collapse in the U.S. in 2008.

If one examines the lowest 50% of the population, then it appears that the growth in this ratio has been very slight since 1995 if not almost stagnant. Additionally, the bottom 50% of young workers had extremely low capital income-to-total income ratios of approximately 0.00% for all years. Such trends could be particularly troubling for such individuals and families should they experience a negative financial shock during such periods of time.

Other trends are also noteworthy, though substantially less troubling. For example, the older population has rather high capital income-to-total income ratios. However, this follows the logic that as individuals age their wage incomes begin to drop off and as they approach retirement they are heavily dependent upon investment savings to fund their consumption streams. Although middle-aged individuals also have high capital income-to-total income ratios, they are less dependent upon these cash flows to survive and their incomes may help fund consumption streams for significantly longer than for the aging population for which incomes not only are not rising and may actually decrease to 0 in the short run.

4 Private Wealth and Public Policies

It is challenging to add a note on the US forces and mechanisms on wealth formation and the EU. As compared to the US, though public policies—and the welfare state in the EU—appear to be mitigating income inequality, in particular inequality in disposable income, the effect on wealth distribution is not clear. One important EU study[34] shows that

"...the degree of welfare state spending across countries is negatively correlated with household net wealth. These findings suggest that social services provided by the state are substitutes for private wealth accumulation and partly explain observed differences in levels of household net wealth across European countries. In particular, the effect of substitution relative to net wealth decreases with growing wealth levels. This implies that an increase in welfare state spending goes along with an increase—rather than a decrease—of observed wealth inequality." (Fessler and Schuerz 2015, p. 1).

This may however be controversial. First, what one can observe in recent decades seems to be a shift of wealth away from the state and toward private wealth in both the US and Europe.[35] Second, non-market benefits from the public sector such as unemployment payments, social security, pension funds, public sector employment, as well as health care spending—together with progressive tax rates—seem to be a corrective force for the uneven income and wealth distribution generated through the market mechanisms and the financial markets. Thus, as many empirical studies show the welfare state in the EU seems to have in the end also significant redistributional effects.

5 Conclusions

As the above results show over a longer time period it seems to hold that the net wealth shares are considerably shifting. This was first demonstrated, in a stylized way, in an asset accumulation model based on heterogeneous consumers in Sect. 2 and then confirmed in our empirical study in Sect. 3. More specifically, our empirical results are as follows.

Empirical evidence in Sect. 3.1 demonstrate that the vast majority of individuals in the U.S. are dependent upon wages for savings. Yet the data in Sects. 3.2, 3.3, 3.4, and 3.5 displays the distribution of net wealth as accruing towards the wealthiest portions of society since the 1980s. Additionally, the relationship between income and net wealth demonstrate the need to understand the various attributes of wage and capital earners in society through time.

Through determining the asset compositions of various income percentiles in U.S. society we find that the wealthier portions of society are able to accrue returns from holding assets that are unavailable to the lower-class wage earners. This helps shed light on the aforementioned Kaldor-Pasinetti debate. Clearly Sect. 3.4 demonstrates that this wealth disparity has been growing among the various age groupings within society over the last two decades. This disparity occurs based on

[34]See Fessler and Schuerz (2015).

[35]See Piketty (2013).

income brackets both within the primarily wage earning population as well as that portion of the population that garners most of their income from capital.

Finally, Sect. 3.5 demonstrates the growing reliance upon capital income for each age grouping of the U.S. population and income brackets peaking in the year 2007 prior to the financial crisis of 2008. This capital-to-total income ratio is still at levels much higher than those seen in the early 1990s and is almost even with levels seen prior to the technology bubble of 2001. If the growing wealth disparity continues along recent trajectories, the U.S. may suffer increased instability as a result of socio-economic inequality.

In terms of economic theory our results are also relevant for the Kaldor-Pasinetti debate. In Pasinetti's response to Kaldor he assumes, as Kaldor does, that savings from wages are lower than that from capital income, but then Pasinetti also assumes that the interest income from labor savings are the same as from capital income. This would keep wealth shares constant. Yet our dynamic model with heterogeneous households and our empirics show that this does not seem to hold empirically—instead the wealth shares are changing over time.

Our results have also ramifications for macroeconomics. The rise of debt for the different types of households, in conjunction with the movement in asset prices and returns, the over leveraging of particular households—or in reverse their declining net wealth—is relevant for macroeconomic dynamics. Households with low income and low wealth, and higher debt to income ratios, arising from consumption spending, are more vulnerable in the business cycle and thus are likely to amplify the down turns. They are more threatened by default since they might not be able to meet their debt obligations, which may amplify the economic down turns. They will also be more income and credit constrained households which in turn generate prolonged recessions.[36]

References

Brunnermeier, M., & Sannikov, Y. (2014). A macroeconomic model with a financial sector. *American Economic Review, 104*(2), 379–421.

Chappe, R., & Semmler, W. (2016). *Disparity in wealth accumulation.* SSRN https://ssrn.com/abstract=2727819 or http://dx.doi.org/10.2139/ssrn.2727819.

Chiarella, C., Semmler, W., & Mateane, H. L. (2016). *Sustainable asset accumulation and dynamic portfolio decisions.* Berlin: Springer.

Cynamon, B. Z., & Fazzari, S. M. (2016). Inequality, the great recession and slow recovery. *Cambridge Journal of Economics, 4*(2), 373–399.

Drechsel-Grau, M., & Schmid, K. D. (2014). Consumption savings decisions under upward-looking comparisons. *Journal of Economic Behavior & Organization, 106*(C), 254–268.

Duesenberry, J. (1949). *Income, saving, and the theory of consumer behavior.* Cambridge: Harvard University Press.

[36]See Cynamon and Fazzari (2016) and European Central Bank (2013).

Dynan, K., Skinner, J., & Zeldes, S. (2004). Do the rich save more? *Journal of Political Economy, 112*(2), 397–444.

European Central Bank. (2013). Statistics Paper Series No. 2.

Federal Reserve Bulletin. (2014). *Changes in US Family Finances from 2010 to 2013: Evidence from the Survey of Consumer Finances* (Vol. 100, No 4, pp. 1–41).

Fessler, P., & Schuerz, M. (2015). *Private wealth across European countries: The role of income, inheritance and the welfare state* ECB Working Paper No. 1847. http://ssrn.com/abstract=2664150

Gordon, R. (2016). *The rise and fall of American growth: The U.S. standard of living since the civil war* Princeton: Princeton University Press.

Gross, M., & Semmler, W. (2017, January). *Mind the output gap: The disconnect of growth and inflation during recessions and convex Phillips curves in the euro area.* Task Force on Low Inflation (LIFT). ECB Working Paper, No 2004.

Grune, L., Semmler, W., & Stieler, M. (2015). Using nonlinear model predictive control for dynamic decision problems in economics. *Journal of Economic Dynamics and Control, 60*, 112–133.

Kaldor, N. (1956). Alternative theories of distribution. *Review of Economic Studies, 23*, 83–100.

Mittnik, S., & Semmler, W. (2013). The real consequences of financial stress. *Journal of Economic Dynamics and Control, 37*(8), 1479–1499.

Mittnik, S., & Semmler, W. (2016). *Overleveraging, financial fragility and the banking-macro link.* ZEW - Centre for European Economic Research Discussion Paper No. 14-110.

Pasinetti, L. (1962). Rate of profit and income distribution in relation to the rate of economic growth. *Review of Economic Studies, 29*(4), 267–279.

Piketty, T. (2013) *Le capital au XXIe siècle*. Paris: Seuil.

Schleer, F., Semmler, W., & Illner, J. (2014). *Overleveraging in the banking sector*. ZEW - Centre for European Economic Research Discussion Paper No. 14–066.

Semmler, W. (2011). *Asset prices, booms and recessions, financial economics from a dynamic perspective*. Berlin: Springer.

Stein, J. L. (2012). *Stochastic optimal control and the U.S. Financial Debt Crisis*. New York: Springer.

Veblen, T. (1899). *The theory of the leisure class*. New York: B.W. Huebsch.

Wolff, E. N. (2012). *The asset price meltdown and the wealth of the middle class*. Cambridge: National Bureau of Economic Research.

MIX
Papier aus verantwortungsvollen Quellen
Paper from responsible sources
FSC® C105338

If you have any concerns about our products,
you can contact us on
ProductSafety@springernature.com

In case Publisher is established outside the EU,
the EU authorized representative is:
Springer Nature Customer Service Center GmbH
Europaplatz 3, 69115 Heidelberg, Germany

Printed by Libri Plureos GmbH
in Hamburg, Germany